# FIRST-PRINCIPLES CALCULATIONS FOR FERROELECTRICS

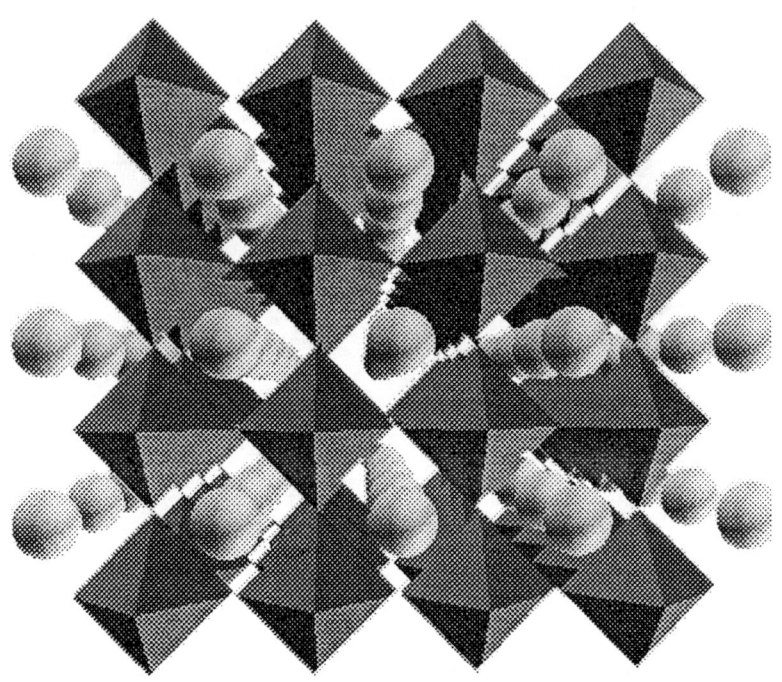

Editor: Ronald E. Cohen

Organizing Committee:
Ronald E. Cohen
Henry Krakauer
Karin Rabe

# Sponsored by
# The Office of Naval Research and
# The Carnegie Institution of Washington

# FIRST-PRINCIPLES CALCULATIONS FOR FERROELECTRICS

Fifth Williamsburg Workshop

Williamsburg, VA   February 1998

EDITOR
Ronald E. Cohen
*Carnegie Institution of Washington*
*Washington, DC*

**American Institute of Physics**

AIP CONFERENCE
PROCEEDINGS 436

Woodbury, New York

**Editor:**

Ronald E. Cohen
Geophysical Laboratory
Carnegie Institution of Washington
5251 Broad Branch Road, NW
Washington, DC 20015-1305

Email: cohen@gl.ciw.edu

Authorization to photocopy items for internal or personal use, beyond the free copying permitted under the 1978 U.S. Copyright Law (see statement below), is granted by the American Institute of Physics for users registered with the Copyright Clearance Center (CCC) Transactional Reporting Service, provided that the base fee of $15.00 per copy is paid directly to CCC, 222 Rosewood Drive, Danvers, MA 01923. For those organizations that have been granted a photocopy license by CCC, a separate system of payment has been arranged. The fee code for users of the Transactional Reporting Service is: 1-56396-730-8/ 98 /$15.00.

© 1998 American Institute of Physics

Individual readers of this volume and nonprofit libraries, acting for them, are permitted to make fair use of the material in it, such as copying an article for use in teaching or research. Permission is granted to quote from this volume in scientific work with the customary acknowledgment of the source. To reprint a figure, table, or other excerpt requires the consent of one of the original authors and notification to AIP. Republication or systematic or multiple reproduction of any material in this volume is permitted only under license from AIP. Address inquiries to Office of Rights and Permissions, 500 Sunnyside Boulevard, Woodbury, NY 11797-2999; phone: 516-576-2268; fax: 516-576-2499; e-mail: rights@aip.org.

L.C. Catalog Card No. 98-71901
ISBN 1-56396-730-8
ISSN 0094-243X
DOE CONF- 980235

Printed in the United States of America

## Contents

**Preface** .................................................................................. vii

**Local Structure and Polarization in Pb Containing Ferroelectric Oxides** ...... 1
    T. Egami, W. Dmowski, M. Akbas, and P. K. Davies

**Ferroelectric Effects in PZT** ............................................................ 11
    L. Bellaiche, J. Padilla, and D. Vanderbilt

**Modeling Cation Ordering in $A(B'_{1/3}B''_{2/3})O_3$ Perovskites** .................... 20
    B. P. Burton, R. P. McCormack, G. Ceder, R. L. B. Selinger, G. Kresse, and J. Hafner

**Effective Hamiltonian for the Ferroelectric Phase Transitions in $KNbO_3$** ...... 32
    U. V. Waghmare, K. M. Rabe, H. Krakauer, R. Yu, and C.-Z. Wang

**First-Principles Study of Piezoelectricity in Tetragonal $PbTiO_3$ and $PbZr_{1/2}Ti_{1/2}O_3$** ................................................... 43
    G. Saghi-Szabo, R. E. Cohen, and H. Krakauer

**Temperature-Dependent Dielectric Response of $BaTiO_3$ from First Principles** .............................................................. 53
    A. Garcia and D. Vanderbilt

**Temperature-Dependent Dielectric and Piezoelectric Response of Ferroelectrics from First Principles** ........................................ 61
    K. M. Rabe and E. Cockayne

**Configuration Dependence of Physical Properties of a Ferroelectric Solid Solution** ............................................................... 71
    E. Cockayne and K. M. Rabe

**Local Lattice and Precursor Effects in $SrTiO_3$ and $PbTiO_3$** ................... 81
    A. Bussmann-Holder

**Self-Ordered Second-Component Clusters in Solid Solutions on the Basis of Ferroelectric Perovskites: Nb-clusters in $KTaO_3$** ....................... 87
    R. I. Eglitis, V. S. Vikhnin, P. A. Markovin, and G. Borstel

**DAFS Study of Local Structure of Ordered Nanodomains in $PbMg_{1/3}Nb_{2/3}O_3$** ................................................................ 96
    A. I. Frenkel, D. M. Fanning, I. K. Robinson, D. L. Adler, and J. O. Cross

**Hartree−Fock Studies of the Ferroelectric Perovskites** ..................... 107
    L. Fu, E. Yaschenko, L. Resca, and R. Resta

**Effect of Nanopolar Regions on Electrostrictive Coefficients of a Relaxor Ferroelectric** ...................................................... 118
    A. E. Glazounov, J. Zhao, and Q. M. Zhang

**Ferroelectricity and Protonic Conductivity in $Cs_{1-x}(NH_4)_xH_2PO_4$** .......... 129
    S. Lanceros-Mendez, S. Meschia, and V. H. Schmidt

**First-Principles Study of $SrTiO_3$ in Cubic and Tetragonal Phases** .......... 139
    C. LaSota, C.-Z. Wang, R. Yu, and H. Krakauer

**Maximally-Localized Wannier Functions in Perovskites: Cubic $BaTiO_3$** ...... 146
    N. Marzari and D. Vanderbilt

**Stress-Induced Phase Transition in $Pb(Zr_{1/2}Ti_{1/2})O_3$** ........................ 156
    N. J. Ramer, S. P. Lewis, E. J. Mele, and A. M. Rappe

**First-Principles Studies of Local Order in Relaxor Ferroelectrics** ........... 165
    M. Wensell and H. Krakauer

The Quantum-Mechanical Position Operator and the
Polarization Problem .................................................. 174
    R. Resta
Quantum Effects in Ferroelectrics ....................................... 184
    S. A. Ktitorov and L. Jastrabik
Dynamical Model for Phase Coexistence in Proton Glass ................... 192
    V. H. Schmidt
Proton Tunneling and Nonlinear Polarizability Effects in
Hydrogen-Bonded Ferroelectrics ......................................... 202
    A. Bussmann-Holder and K.-H. Michel
First-Principles and Semi-Empirical Hartree–Fock Calculations
for $F$ Centers in $KNbO_3$ and Li Impurities in $KTaO_3$ ..................... 207
    R. I. Eglitis, E. A. Kotomin, A. V. Postnikov, N. E. Christensen,
    and G. Borstel
LAPW vs. LMTO Full-Potential Simulations and
Anharmonic Dynamics of $KNbO_3$ ........................................ 217
    A. V. Postnikov and G. Borstel
Self-Consistent Atomic Deformation Calculations for Strontium Titanate ..... 227
    L. L. Boyer, H. T. Stokes, and M. J. Mehl
Pressure-Induced Changes in the Local Structure of $KNbO_3$ ................ 238
    A. I. Frenkel, E. A. Stern, and Y. Yacoby
Weighted Density Functionals for Ferroelectric Materials ................... 251
    I. I. Mazin and D. J. Singh
Structural and Vibronic Properties of Perovskites Studied by Using
the Perdew–Berke-Ernzerhof GGA ....................................... 265
    C. O. Rodriguez, D. L. Novikov, M. G. Stachiotti, and N. E. Christensen
Modeling and Microscopic Dynamics of $KNbO_3$ from First-Principles ....... 274
    M. G. Stachiotti, M. Spliarsky, R. L. Migoni, and C. O. Rodriguez
Mechanisms and Kinetics of the Initial Stages of
Ferroelectrics Island Film Growth under Non-Uniform Conditions .......... 284
    D. A. Grigoriev and S. A. Kukushkin
Prospects for Gigabit Ferroelectric Nonvolatile Memories Using
Strontium Bismuth Tantalate Thin Films ................................. 294
    J. F. Scott

Author Index .......................................................... 301

# Preface

This volume contains papers presented at the Fifth Williamsburg Workshop on First-principles Calculations for Ferroelectrics, which met February 1-4, 1998. This series of meetings, which began in 1990 has been very successful in fostering a growing field, that of understanding the fundamental physics of ferroelectrics, and developing techniques for computations of ferroelectric properties from first-principles. The papers in the present volume indicate the broad and deep progress in the field. Eight years ago, it was just realized that theory could provide insights into ferroelectrics, and now deep understanding is developing that even transcends the field of ferroelectrics, and practical electromechanical properties are being computed for the first-time. This is (at least) the second generation of ferroelectrics research, and it is fascinating to see the problems that plagued earlier studies slowly be solved by the use of new techniques, modern computers, and new methods.

This meeting also contains experimental contributions to prod theorists into understanding the complexities of real ferroelectric materials. Experimental contributions themselves are varied and exciting, and indicate the fertilization that occurs when theorists and experimentalists work together. Experimental contributions range from diffraction studies of complex ferroelectrics, to x-ray absorption, film growth, and applied studies for applications as ferroelectric memories.

Advances in theory range from the first first-principles studies of ferroelectric solid solutions, the first computations of piezoelectric constants in ferroelectrics, development and testing of new, more accurate, density functionals, finite temperature simulations, and developments of new methods.

Ferroelectrics research impinges also on fields of dielectrics, insulators and semiconductors, oxides, and perovskites, which have their own interest groups. The rich activity in ferroelectrics is starting to spill over into these other fields as well, and spark renewed interest.

Progress continues. The discovery of new ultra-large coupling piezoelectrics by Park and Shrout (J. Appl. Phys., 82, 1804, 1997) has introduced even more excitement into the field, and further challenges theory to reach new levels of understanding. The future promises to be fruitful and exciting to anyone with an interest in materials.

Ronald E. Cohen
April, 1998

# Local Structure and Polarization in Pb Containing Ferroelectric Oxides

T. Egami, W. Dmowski, M. Akbas and P. K. Davies

*Department of Materials Science and Engineering,*
*University of Pennsylvania, Philadelphia, PA 19104*

**Abstract.** While the Pb containing ferroelectric and antiferroelectric oxides show a large variety in crystal structure, the pulsed neutron atomic pair-distribution function (PDF) studies indicate that their local atomic structures are surprisingly similar to each other. In particular the environment of Pb is nearly independent of composition, with Pb being strongly off-centered in the $PbO_{12}$ cluster, resulting in a large local polarization. A new model is proposed to describe the interplay between the Pb polarization and the random local structural fluctuation. This model explains the relaxor ferroelectricity from a general point of view, while other models such as nano-domain space charge model and the random field model were introduced specifically for hetero-valent systems such as $Pb(Mg_{1/3}Nb_{2/3})O_3$ (PMN).

## INTRODUCTION

Pb containing transition metal oxides such as $(Pb_{1-x}La_x)(Zr_{1-y}Ti_y)O_3$ (PLZT) and $Pb(Mg_{1/3}Nb_{2/3})O_3$ (PMN) exhibit large varieties of structural and dielectric properties. For instance $PbZrO_3$ (PZ), an end member of the PLZT system ($x = y = 0$), is antiferroelectric and has an orthogonal crystal structure with a large unit cell containing 40 atoms. With $x = 0$ and $y > 0.05$, $Pb(Zr_{1-y}Ti_y)O_3$ (PZT) is ferroelectric with a rhombohedral structure. With a small replacement of Pb by La ($x \sim 0.1$) it shows a relaxor ferroelectric behavior. When $y > 0.5$ the structure changes to tetragonal. PMN, on the other hand, is nearly cubic with a very small rhombohedral distortion. Such wide varieties in the crystal structure and properties imply very complex interaction, and it appears impossible to find a unified explanation for all these disparate behaviors. In fact the relaxor ferroelectricity of PMN is currently explained by the nano-domain space charge model [1] or the random-field model [2], while these models do not seem to apply to PLZT so well. However, the recent results by pulsed neutron scattering atomic pair-distribution function (PDF) analysis demonstrate that *the local atomic structures of these solids are remarkably similar to each other* despite the large differences in the long range structure. This suggests that it may be possible to describe the structure and properties of these solids from a more unified point of view. In this paper we qualitatively discuss such an approach.

## PULSED NEUTRON PDF ANALYSIS

The crystal structure determined by crystallographic analysis of diffraction data represents the long range periodic structure. In these analyses only the Bragg peaks are considered for the structural analysis since the structure is presumed to be perfectly periodic. In reality, however, the atomic positions in crystals are not perfectly periodic, either because of lattice defects or because of alloying effects. For instance in PZT the B-site of the perovskite structure is occupied by either Zr or Ti. Since the atomic radii of Zr and Ti are different (0.72 and 0.61 Å respectively [3]), at an atomic level the structure is not homogeneous and locally varies from site to site. Such local variations apparently have no effect on certain properties. Alloy semiconductors are widely used for optical applications, and metallic alloys have long been used for mechanical applications. However for other properties they can have serious consequences. We believe that relaxor ferroelectricity is one of them.

The local structural variations are observed in the crystallographic structure only through the unusually large Debye-Waller factor. For more direct observation a local probes are necessary. X-ray absorption fine structure (XAFS) analysis is one of such techniques, but the resolution of the XAFS method is limited to very short distances. In contrast the atomic pair-distribution function (PDF) analysis is capable of determining the local structure over a much wider range of distances, covering the short (< 5 Å) to intermediate (5 to 20 Å) ranges. The PDF describes the distribution of distances between atoms in the sample averaged over all angles. In particular the PDF method using pulsed neutron scattering proved to be a most powerful technique in determining the local structure of liquids, amorphous materials and crystalline solids including oxides. In this method the powder diffraction data obtained using a pulsed neutron source is Fourier-transformed to obtain the PDF. The use of pulsed neutron source is particularly beneficial for this method as it provides a high flux of epithermal short wavelength neutrons and enables the structure function to be determined up to large momentum transfers. The high accuracy of this method was demonstrated in a number of studies [4,5].

The pulsed neutron PDF of PMN is shown in Fig. 1 [6]. It is compared with the PDF calculated for the cubic perovskite structure. The peaks in the calculated PDF are broadened to represent thermal and quantum atomic vibrations. In the calculated PDF the first peak represents the (Mg, Nb)-O distances, the second peak the Pb-O and O-O distances, the third the (Mg, Nb)-Pb distances, etc. In spite of the peak broadening the peaks of the calculated PDF are much sharper than those in the experimental PDF,

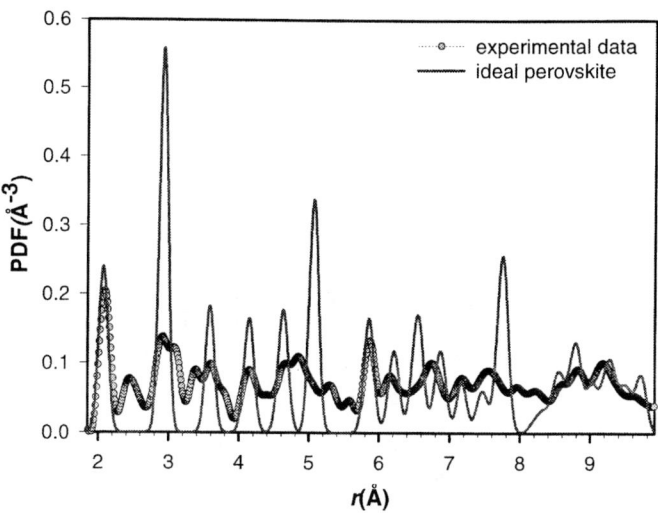

Fig. 1. The PDF of PMN (circles) compared to the PDF calculated for the ideal perovskite structure (solid line) [6].

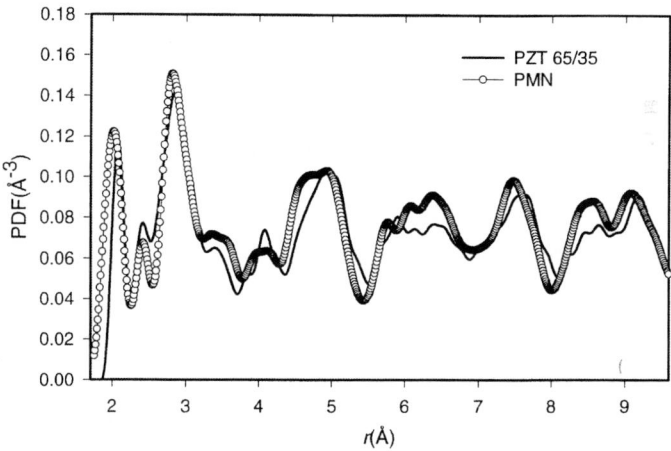

Fig. 2. The PDF of PMN (circles) compared to the PDF of PZT 65/35 (solid line) [7].

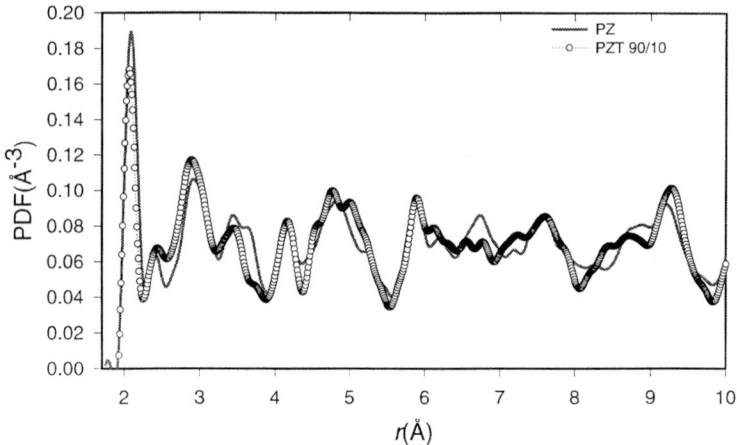

Fig. 3. PDF's of PZ (solid line) and PZT 90/10 (circles). They are similar in spite of their difference in the long range structure [7].

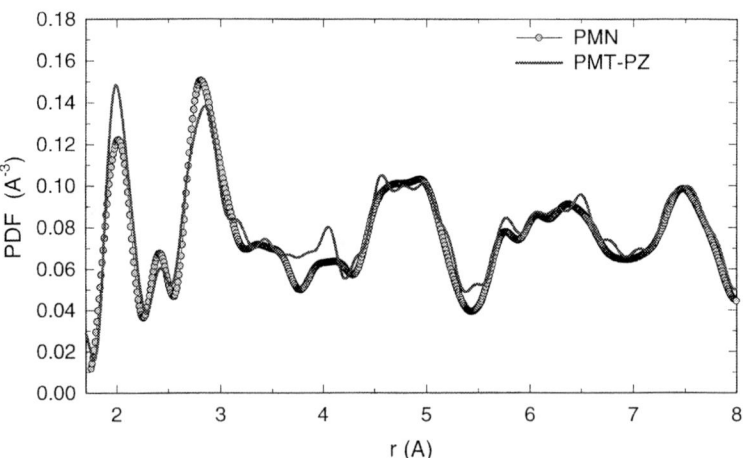

Fig. 4. PDF's of PMN (circles) and PMT-PZ (solid line).

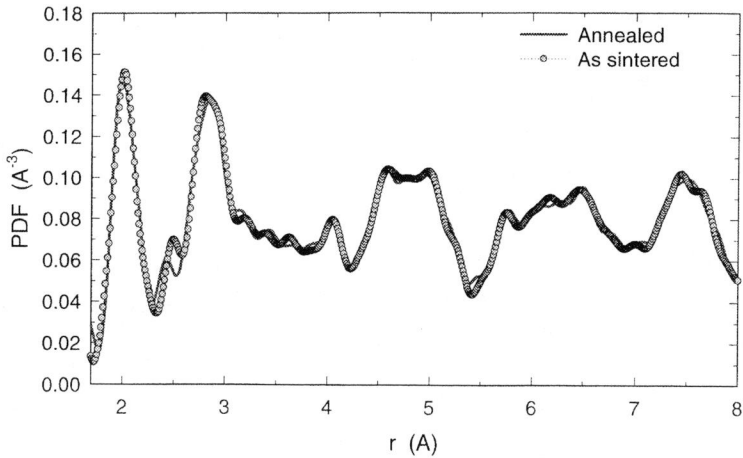

Fig. 5. PDF's of PMT-PZ as sintered and after annealing at 1325°C for 48 hrs.

suggesting that there are significant local deviations from the ideal perovskite structure. In particular the data PDF has a well defined peak at 2.4 Å, while the calculated PDF has no peak at all in that neighborhood.

As it turns out the PDF of PMN is very similar to the PDF of PZT, as shown in Fig. 2 for the comparison with $Pb(Zr_{0.65}Ti_{0.35})O_3$ (PZT-65/35) [7]. This is rather surprising, since the B-site of PMN is hetero-valent ($Mg^{2+}$ and $Nb^{5+}$) while that of PZT is homo-valent ($Zr^{4+}$ and $Ti^{4+}$). Apparently the difference in valence has little effect on the structure. Furthermore the PDF of PZT is very similar to that of PZ as shown in Fig. 3, in spite of the fact that PZT is ferroelectric and PZ is antiferroelectric [7,8]. The PDF of PMN is very similar to that of $Pb(Mg_{0.3}Ta_{0.6}Zr_{0.1})O_3$ (PMT-PZ) as well, as shown in Fig. 4. PMT-PZ as sintered has short range chemical order on B-site, while the chemical order develops into long range order after annealing. Even after long range chemical ordering develops, however, PMT-PZ retains the relaxor behavior [9]. This observation casts a serious doubt in the nano-domain space charge model of the relaxor behavior of PMN as we will discuss below. The PDF of PMT-PZ before and after annealing at 1325°C for 48 hrs. shown in Fig. 5 are again similar except for a slight shift in the position of the 2.4 Å peak, indicating that the development of order has little effect on the local structure.

## LOCAL STRUCTURE AROUND Pb

We have to caution the reader that the agreement in Fig. 2 is rather exaggerated, since the PDF of PZT depends upon the Zr/Ti ratio because the neutron scattering length of Ti is negative. However, the main features remain surprisingly independent of composition, including the peak at 2.4 Å. Since the structure of PZ can be unambiguously determined by crystallographic analysis the nature of this peak is easily understood [10]. This peak originates from the three short Pb-O bonds. The average distance between Pb and O is about 2.8 ~ 3 Å, but Pb is significantly off-centered, and form a few short covalent bonds with O. From this inference we can safely conclude that all the 2.4 Å peaks seen in Figs. 1-5 are due to short Pb-O bonds. This conclusion agrees with the earlier result by XAFS on PMN related perovskites [11]. This means that the Pb environment is always distorted and Pb is off-centered. In these solids Pb has 12 O neighbors, but Pb is strongly off-centered in the $PbO_{12}$ cluster, forming a large electric dipolar moment. While the position and intensity of the 2.4 Å peak is slightly dependent upon composition and ordering, compared to the Pb-O distance in the ideal perovskite (2.8 ~ 3 Å), the variation is rather minor. Therefore we may conclude the magnitude of local polarization of the $PbO_{12}$ cluster is largely independent of composition and ordering.

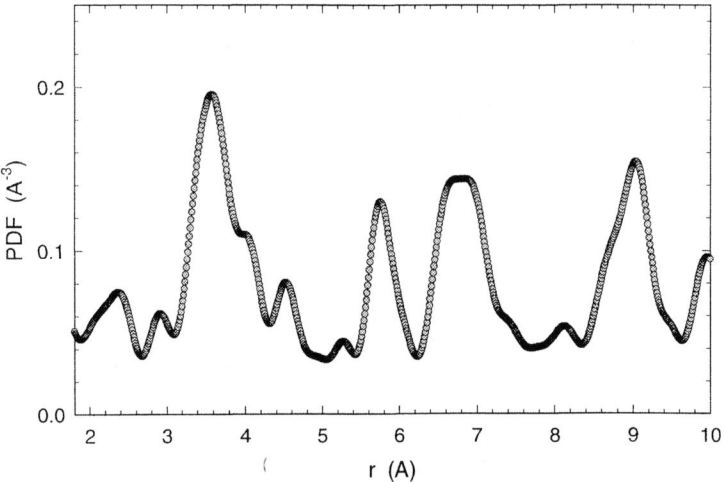

Fig. 6. X-ray PDF of PMT-PZ. The peak at 3.5 Å describes the Pb-Ta distance.

It is interesting to note, however, that the Pb is not much off-centered against the B cations. Shown in Fig. 6 is the x-ray PDF of PMT-PZ obtained at the X-7A synchrotron beamline of the NSLS of the Brookhaven National Lab. Since x-rays scattering amplitude is proportional to the atomic number the x-ray PDF is dominated by the cation-cation correlation, in particular the Pb-Ta correlation. It is noted that the nearest Pb-Ta peak is rather well defined, and is at 3.59 Å, very close to the average A-B distance of 3.60 Å. This result differs slightly from the earlier result on the x-ray PDF of PMN [6]. However, this measurement is difficult due to strong absorption of x-rays by Pb, and the present result appears to be more reliable. Since the ionic sizes of $Ta^{5+}$ (0.64 Å [3]) and $Mg^{2+}$ (0.72 Å [3]) are not much different, if Pb is shifted toward a particular B cation because of the valence (in this case toward $Mg^{2+}$) or size (in this case toward $Ta^{5+}$) the Pb-Ta distance should deviate from the average A-B distance. The absence of deviation indicates that Pb is not much shifted from the average A-site. Therefore the Pb polarization is produced mainly by the displacement of the $O_{12}$ cage. Then the Pb polarization must be closely related to the size and orientation of the $BO_6$ octahedra (B = B site transition metal ions).

## MODEL OF INTERACTING Pb DIPOLES

The magnitude of the Pb polarization can be estimated by examining the structure of PZ. In PZ the center of gravity of the $O_{12}$ cage is shifted from the position of Pb by 0.5 Å. If we assume the nominal Pb charge to be 2+, the $PbO_{12}$ cluster has the local dipolar moment of $p = 1.6 \times 10^{-19}$ Coulomb Å $= 1.6 \times 10^{-29}$ Cm. This corresponds to the volume dipolar moment of $P = 0.22$ C/m$^2$, which is a very significant polarization. The value of p will be even larger if we use the correct Born effective charge. The local moments p interacts with each other via the electric dipolar field, elastic field, and hybridization field through B-site elements if the B-site elements are ferroelectrically active. We may describe this interaction by an effective Hamiltonian,

$$H_c = \sum_{i,j} J(\mathbf{r}_{ij}) \sum_{\ell,m} Q_\ell^m(\mathbf{P}_i) Q_\ell^{-m}(\mathbf{P}_j) \qquad (1)$$

where $\mathbf{P}_i$ is the local polarization of the i-th Pb ion, $Q_\ell^m(\mathbf{P}_i)$ are the spherical harmonics equivalents of $\mathbf{P}_i$, for instance $Q_2^0(\mathbf{P}) = \frac{1}{2}[3P_z^2 - P^2]$, etc. The $\ell = 1$ term describes the pseudo-dipolar interaction, while the $\ell = 2$ term corresponds to quadrupolar interaction. The local dipolar moments, however, interact also with the lattice, since the rotation of the local moments require displacements of oxygen ions which are coupled to the B-site cations. This dipole-lattice interaction may be described by the local anisotropy Hamiltonian,

$$H_a = \sum_i B_\ell^m(\mathbf{r}_{ij}) Q_\ell^m(\mathbf{P}_i) \qquad (2)$$

The $\ell = 1$ term corresponds to the local field, while the $\ell = 2$ term describes the steric or elastic field that tends to confine **P** along one direction which may be called the easy axis. The dielectric properties of the interacting Pb dipoles should be described by the total Hamiltonian which is the sum of (1) and (2).

## ORIGIN OF THE RELAXOR FERROELECTRICITY

The chemical ordering in PMN as evidenced by the superlattice peaks is short range and extends only over about 40 Å [1]. If this ordering is due to Mg/Nb 1:1 ordering, it leaves the ordered domain charge imbalanced, since the composition inside the ordered domain is $Pb(Mg_{1/2}Nb_{1/2})O_3$. This led to the speculation that the growth of the ordered domains may be inhibited by the space charge effect, and the fact domains remain nano-scale results in the relaxor behavior [1]. However, there are negative evidences of compositional imbalance associated with the nano-domain [12,13]. In addition it was found recently that in $Pb(Mg_{0.3}Ta_{0.6}Zr_{0.1})O_3$ (PMT-PZ) the relaxor behavior is retained even after long range chemical ordering develops as we mentioned earlier [9]. These observations conclusively show the absence of composition and charge imbalance in the ordered domains, and the space charge model cannot explain the relaxor behavior. Another model is the so-called random-field model [2]. It is based upon the presence of random local electric fields produced by the randomly distributed heterogeneous ions such as $Mg^{2+}$ and $Nb^{5+}$ in PMN. This model is perhaps more sophisticated than the space charge model.

Both of these models, however, were proposed for the hetero-valence systems such as PMN, and would not explain the relaxor behavior of PLZT which is basically homo-valent. The random fields produced by the mixture of $Pb^{2+}$ and a small amount (only ~ 10%) of $La^{3+}$ in PLZT is much smaller than those due to $Mg^{2+}$ and $Nb^{5+}$ in PMN. Thus the behavior of PLZT has to be explained by a different mechanism. However, the recent PDF studies of these systems as reviewed in this article revealed the strong similarity of the local structure, thus the similarity of the interaction involved, of the Pb containing ferroelectric oxides. It is more reasonable, therefore, to assume that a universal mechanism exists that explains both PMN and PLZT. We propose that the Hamiltonian (1) + (2), with emphasis on the $\ell = 2$ term in (2), would provide a more unified description of the phenomena. In a randomly mixed oxides such as PMN and PLZT the local environments of mixed ions randomly vary from site to site because of the difference in the sizes of the B-site ions. For instance in PMN the $MgO_6$ octahedra must be strongly rotated because of the large size of the $Mg^{2+}$ ion, while $NbO_6$ octahedra would not be. In PLZT the major source of randomness is not the mixing of Pb and La, but the mixing of Zr and Ti. Because of the ionic size the $ZrO_6$ octahedra are always rotated, while the $TiO_6$ octahedra are not. Consequently the local environments of PMN and PZT are severely randomized, and the PDF of PZT strongly resembles that of PMN as shown in Fig. 2. Because of the local random rotation of the BO$_6$ octahedra the direction of the easy axis of the Pb polarization must be random as well. This results in the random anisotorpy for the Pb polarization. By rotating the local coordinates to the local easy axis direction and retaining only the $\ell = 2$, m = 0 term, (2) becomes

$$H_a = \sum_i B_2^0(\mathbf{r}_i) Q_2^0(\mathbf{P}_i') \qquad (3)$$

where $\mathbf{P}_i'$ is $\mathbf{P}_i$ in the rotated coordinates. The total Hamiltonian (1) + (3) is very similar to the Harris-Plischke-Zuckermann (HPZ) Hamiltonian for the random anisotropy ferromagnet (in this case only the $\ell = 1$ term, Heisenberg exchange, is retained in (1)) [14]. The HPZ Hamiltonian was studied in great detail, and the behavior of the system described by this Hamiltonian is now well understood. The critical parameter is the ratio between the average values of the interaction parameter $J_{ij}$ and the local parameter $B(r_i)$, B/J. If B/J is large at low temperatures the local moments are trapped in the local easy axis, and the system shows a spin-glass behavior in the case of a random magnet and relaxor ferroelectricity for a random dipolar system. As the value of J is increased, or the B/J ratio is decreased, the interaction term starts to influence the system and the local dipolar moments become spatially correlated. The correlation length $\xi$ strongly depends upon the B/J ratio, and at a critical value (B/J)$_{crit}$ $\xi$ becomes larger than the sample size, so that the system behaves ferromagnetic (ferroelectric). The consensus is that in three-dimensions strictly speaking the random anisotropy ferromagnet is *always* a relaxor in the presence of the random anisotropy [15,16], but when the value of $\xi$ is large the system is effectively ferromagnetic (ferroelectric).

The value of J can be estimated from the Curie temperature, T$_C$, while the value of B may be deduced from the freezing temperature T$_f$ for relaxor systems. The value of B for PLZT should be as large as that for PMN. In PZT, however, apparently the value of B/J is slightly subcritical, and the system is ferroelectric. By substituting La for Pb apparently the value of J is locally reduced, driving the system eventually to a phase transition. Near the phase boundary the randomness dominates, resulting in the relaxor behavior.

It is instructive to consider how the piezoelectric response depends upon the B/J ratio. High piezoelectric response is derived from the polarization rotation. At a small B/J ratio the system is ferroelectric and domain wall motion dominates the response, so that the piezoelectric response is small. On the other hand at a large B/J ratio polarization is strongly trapped in the local easy direction, and the response is also small. The piezoelectric response becomes maximum at a B/J ratio near (B/J)$_{crit}$, when the polarization locally rotates with ease in the relaxor system. In order to design piezoelectric device with a relaxor ferroelectrics it is therefore advisable to choose a material close to the relaxor-to-ferroelectric phase boundary.

## CONCLUSION

The results of the pulsed neutron PDF studies of Pb containing ferroelectric perovskite type oxides indicate that the environment of Pb is nearly invariant of composition. The $O_{12}$ cage that surrounds Pb is shifted by about 0.5 Å with respect to Pb producing a local dipolar polarization. The system can then be described in terms of the local Pb dipolar moments which are interacting with each other and with the lattice via the dipolar anisotropy energy. In mixed ion systems such as PMN the local easy axis of the dipolar moment randomly varies from site to site. The total Hamiltonian describing the system resembles that of a random anisotropy magnet, and from the knowledge of the behavior of a random anisotropy magnet it is possible to outline the behavior of the random anisotropy ferroelectrics. This model appears to explain the relaxor behavior of both hetero-valent systems such as PMN and more homo-valent systems such as PLZT, while other models apply only to hetero-valent systems.

## ACKNOWLEDGMENT

The authors are grateful to I-W. Chen, A. Rappe, and E. Mele for useful discussions. This work was supported by the National Science Foundation through the MRSEC program with the LRSM, DMR96-32598.

## REFERENCES:

1. 1. E. Husson, M. Chubb, and A. Morell, *Mat. Res. Bull.*, **23**, 357 (1988).
2. V. Westphal, W. Kleemann and M. Glinchuk, *Phys. Rev. Lett.* **68**, 847 (1992).
3. R. Shannon, *Acta Cryst.* **A32**, 751 (1976).
4. B. H. Toby and T. Egami, *Acta Cryst.* A **48**, 336 (1992).
5. T. Egami and S.J.L. Billinge, *Progr. Mater. Sci.,* **38**, 359 (1994).
6. T. Egami, H.D. Rosenfeld, B.H. Toby and A. Bhalla, *Ferroelectrics*, **120**, 11 (1991).
7. T. Egami, S. Teslic, W. Dmowski, D. Viehland and S. Vakhrushev, *Ferroelectrics*, **199**, 103 (1997).
8. S. Teslic, T. Egami and D. Viehland, *J. Phys. Chem. Solids*, **57**, 1537 (1996).
9. M. Akbas and P. K. Davies, *Commun. Amer. Ceram. Soc.*, **80**, 2933 (1997).
10. S. Teslic and T. Egami, *Acta Cryst.* B, in press.
11. I-Wei Chen, P. Li and Y. Wang, *J. Phys. Chem. Solids*, **57**, 1525 (1996).
12. K. Park, L. Salamanca-Riba, M. Wuttig, and D. Viehland, *J. Mater. Sci.* **29**, 1284 (1994).
13. T. Egami, W. Dmowski, S. Teslic, P. K. Davies, I.-W. Chen and H. Chen, *Ferroelectrics*, in press.
14. R. Harris, M. Plischke and M. I. Zuckermann, *Phys. Rev. Lett.* **31**, 160 (1973).
15. R. A. Pelcovitz, E. Pytte and J. Rudnick, *Phys. Rev. Lett.* **40**, 476 (1978).
16. M. C. Chi and T. Egami, *J. Appl. Phys.* **50**, 1651 (1979).

# Ferroelectric effects in PZT

L. Bellaiche, J. Padilla and David Vanderbilt

*Department of Physics and Astronomy, Rutgers University, Piscataway, New Jersey 08855-0849*

**Abstract.** First-principles calculations are performed to investigate alloying and ferroelectric effects in lead zirconate titanate (PZT) with high Ti composition. We find that the main effect of alloying in the paraelectric phase of PZT is the existence of two sets of B–O bonds, i.e., shorter Ti–O bonds *vs.* longer Zr–O bonds. On the other hand, ferroelectricity leads to the formation of very short covalent Ti–O bonds and to the formation of covalent chains of Pb–O bonds. The covalency in the ferroelectric phase is mainly induced by an enhancement of hybridization between Ti $3d$ and O $2p$, and between Pb $6s$ and O $2p$. These hybridizations induce a striking decrease of the effective charges when going from the paraelectric to the ferroelectric phase of PZT.

## INTRODUCTION

The technologically important lead zirconate titanate alloy (e.g., $PbZr_{1-x}Ti_xO_3$ usually denoted as PZT) has an interesting phase diagram [1]. Increasing the Ti $x$ composition yields progressively the following *ground state* phases: an antiferroelectric orthorhombic phase for $x \lesssim 0.1$, a ferroelectric rhombohedral $FE_2$ phase for $0.1 \lesssim x \lesssim 0.4$, another ferroelectric rhombohedral $FE_1$ phase for $0.4 \lesssim x \lesssim 0.5$, and finally a tetragonal ferroelectric phase for $x$ larger than 50%. The high-temperature phase of this alloy at all compositions is the cubic perovskite structure.

Previous theoretical studies [2] focused on the long-range B-site ordering effects in PZT for a Ti composition equal to 0.5, i.e., close to the morphotropic phase boundary between rhombohedral and tetragonal ferroelectric phases. The subject of the present theoretical study is rather different: we will investigate alloying and ferroelectric effects on structural, chemical and dielectric properties in the tetragonal phase of the PZT alloy. In other words, we would like to know what are the effects of alloying and ferroelectricity on bond lengths, chemical bonding and Born effective charges in PZT with high Ti content.

## METHOD

We focus on the ordered structure shown in Figure 1 and exhibiting a Ti composition equal to 2/3. The B-site ordering of this supercell consists of one Zr plane alternating with two Ti planes along the [001] direction. We perform local-density approximation (LDA) calculations on this supercell using the Vanderbilt ultrasoft-pseudopotential scheme [3], and including the semicore shells for *all* the metals considered. Specifically, the Pb $5d$, $6s$ and $6p$, the Zr $4s$, $4p$, $4d$ and $5s$, the Ti $3s$, $3p$, $3d$ and $4s$, and the O $2s$ and $2p$ electrons are treated as valence electrons. We choose the plane-wave cutoff to be 25 Ry and use the Ceperley-Alder exchange and correlation [4] as parameterized by Perdew and Zunger [5]. The first-principles calculations throughout this work are performed using a (6,6,2) Monkhorst-Pack mesh [6]. Further technical details of the procedure used in the present study can be found in Ref. [7].

In fact, we perform two different calculations corresponding to two different symmetries of the structure shown in Figure 1: (1) using a centrosymmetric cell (i.e., exhibiting an inversion symmetry about the central Pb atom); and (2) using a ferroelectric cell (i.e., relaxing the inversion symmetry constraint). Results of calculation (1) identify the alloying effects on various physical properties, while comparison of (1) with (2) allows us to isolate the ferroelectric effects on those properties.

The lattice parameter, the axial ratio c/a and the atomic positions along the [001] (compositional) direction are optimized in calculation (1) by minimizing the total energy and the Hellmann-Feynman forces, the latter being converged to within 0.02 eV/Å.

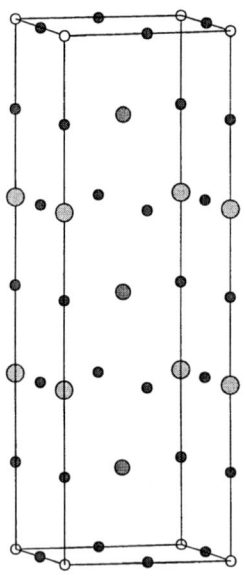

**FIGURE 1.** The [001]-ordered Pb(Zr$_{1/3}$Ti$_{2/3}$)O$_3$ supercell. The central (Pb,O) plane is chosen to have an [001] component equal to zero in the ideal structure. Tables I and Table II give the corresponding atomic positions in the paraelectric and ferroelectric phases.

The ferroelectric experimental ground state of Pb(Zr$_{1/3}$Ti$_{2/3}$)O$_3$ has the tetragonal P4mm point group and does not present any evidence of (long-range) B-site ordering. The material is thus composed of a succession of equivalent planes of composition Zr$_{1/3}$Ti$_{2/3}$ stacked along the tetragonal direction. To mimic this situation, we have chosen the ferroelectric direction in the supercell of Figure 1 to lie along the [100] direction, rather than along the compositionally-modulated [001] direction. The length of the lattice vector along the perpendicular [010] direction is fixed to be equal to the lattice parameter $a$. In our ferroelectric supercell, two axial ratios thus exist. These are the "ferroelectric-related" $a_1/a$, and the "ordered-related" $c/a$, where $a_1$, $a$, and $c$ are the lengths of the supercell lattice vectors along the [100], [010], and [001] directions, respectively. We are thus dealing with a P2mm orthorhombic ferroelectric supercell rather than with a P4mm tetragonal ferroelectric supercell. However, in order to be as close as possible to the experimental situation, we will keep the $c/a$ ratio as equal to the ideal value of 3.0. In this case, we shall refer to our ferroelectric cell as "quasi-tetragonal" along [100] (i.e., tetragonal as regards the axial ratios, although true tetragonal symmetry is broken down to orthorhombic by the B-plane ordering in the [001] direction). The lattice parameter $a$, the axial ratio $a_1/a$, and the atomic positions along the [100] and [001] directions are then optimized in calculation (2) by minimizing the total energy and the Hellmann-Feynman forces (again to within a tolerance of 0.02 eV/Å for that latter).

The determination of the electronic ground state in calculations (1) and (2) is used to investigate the ferroelectric effects on the bond length distribution and on the chemical bonding in PZT with high Ti content. The effective charges in each case (i.e., in both non-centrosymmetric and centrosymmetric cells) will then be calculated from the polarization differences between the ground state and slightly distorted structures, by following the procedure introduced in Ref. [8] and intensively used in Ref. [9].

# RESULTS

## Centrosymmetric case

Optimizing each degree of freedom in the centrosymmetric supercell leads to the lattice vectors $\vec{a}_{1,c} = a_0[1,0,0]$, $\vec{a}_{2,c} = a_0[0,1,0]$, and $\vec{a}_{3,c} = a_0[0,0,2.99]$, where $a_0=7.498$ a.u. is the lattice parameter. The renormalized $c/a$ ratio defined as the actual ratio (i.e., 2.99) divided by the ideal one (i.e., 3.00) is equal to 0.997 and is thus very close to unity. For this reason, our centrosymmetric supercell can be referred as "quasi-cubic", which is consistent with the fact that the experimental paraelectric phase of PZT is cubic.

The atomic relaxed positions and the effective charges in this non-polar structure are given in Table I. It can be seen from Table I that (i) the Pb and O atoms lying between the Zr and Ti planes (i.e., Pb1, O1, Pb3, and O7) move significantly

**TABLE 1.** Structural relaxations and effective charges for the [001] centrosymmetric supercell. The $\Delta z$ are the [001] atomic displacements of the non-polar structure with respect to the ideal ordered structure associated with $\vec{a}_{1,c}$, $\vec{a}_{2,c}$ and $\vec{a}_{3,c}$. $Z_{xx}$ and $Z_{zz}$ are the effective charges along the [100] and [001] direction, respectively.

| Atoms | Relaxed positions (a.u.) | | | Displacements | Effective charges | |
|---|---|---|---|---|---|---|
|  | x | y | z | $\Delta z$ (a.u.) | $Z_{xx}$ | $Z_{zz}$ |
| Pb1 | 3.749 | 3.749 | -7.194 | 0.279 | 3.90 | 4.04 |
| Pb2 | 3.749 | 3.749 | 0.000 | 0.000 | 3.88 | 3.53 |
| Pb3 | 3.749 | 3.749 | 7.194 | -0.279 | 3.90 | 4.04 |
| Ti1 | 0.000 | 0.000 | -3.643 | 0.093 | 6.77 | 6.65 |
| Ti2 | 0.000 | 0.000 | 3.643 | -0.093 | 6.77 | 6.65 |
| Zr1 | 0.000 | 0.000 | 11.210 | 0.000 | 6.33 | 6.69 |
| O1 | 0.000 | 0.000 | -7.222 | 0.251 | -2.58 | -5.39 |
| O2 | 3.749 | 0.000 | -3.638 | 0.099 | -5.58 | -2.34 |
| O3 | 0.000 | 3.749 | -3.638 | 0.099 | -2.72 | -2.34 |
| O4 | 0.000 | 0.000 | 0.000 | 0.000 | -2.53 | -5.57 |
| O5 | 3.749 | 0.000 | 3.638 | -0.099 | -5.58 | -2.34 |
| O6 | 0.000 | 3.749 | 3.638 | -0.099 | -2.72 | -2.34 |
| O7 | 0.000 | 0.000 | 7.222 | -0.251 | -2.58 | -5.39 |
| O8 | 3.749 | 0.000 | 11.210 | 0.000 | -5.17 | -2.94 |
| O9 | 0.000 | 3.749 | 11.210 | 0.000 | -2.33 | -2.94 |

towards the Ti planes; and (ii) the Ti and O atoms belonging to the Ti planes (i.e., Ti1, O2, O3, Ti3, O5, and O6) move very slightly towards the central mirror (PbO) plane. These atomic motions lead to shortened Ti–O bonds and lengthened Zr–O bonds. For example, the Ti–O1 bond length shrinks to 1.89 Å, while Zr1–O7 enlarges to 2.11 Å, to be compared with the unrelaxed B–O bond length of 1.98 Å in the ideal structure. As a matter of fact, the appearance of several different bond lengths associated with the mixed sublattice seems to be a general feature of alloying, and has also been observed and predicted in zinc-blende, wurtzite and rocksalt alloys [10–15].

Alloying effects in the present ordered [001] structure also leads to a change in the lengths of the Pb–O bonds. For example, the Pb3–O bonds can be decomposed into three different groups: shorter Pb3–O bonds (e.g., Pb3–O5 equal to 2.73 Å), roughly unrelaxed Pb3–O bonds (e.g., Pb3–O7 equal to 2.80 Å), and long Pb3–O bonds (e.g., Pb3–O8 equal to 2.91 Å). The three groups are populated in the ratio 4:4:4. Thus, alloying has some significant effects on the B–O bonds ($\sim$4.5% change in bond lengths), and to a smaller extent, on the Pb–O bonds ($sim$2.5% change).

The Born effective charges for our PZT supercell are detailed in Table I. They exhibit the same trends as in cubic bulk $PbTiO_3$ and $PbZrO_3$ compounds [9]: large values of about +4.0 for Pb atoms; large values around +6.5 for the B atoms; and two sets of values for the oxygen atoms, either close to -5.5 for oxygen atoms moving parallel to the B–O–B chain, or or close to -2.5 for oxygen atoms moving

**TABLE 2.** Structural relaxations and effective charges for the non-centrosymmetric supercell.

| Atoms | Relaxed positions (a.u.) | | | Displacements (a.u.) | | | Effective charges | |
|---|---|---|---|---|---|---|---|---|
| | x | y | z | $\Delta x$ | $\Delta y$ | $\Delta z$ | $Z_{xx}$ | $Z_{zz}$ |
| Pb1 | 3.547 | 3.731 | -7.250 | -0.334 | 0.000 | 0.211 | 3.17 | 3.92 |
| Pb2 | 3.551 | 3.731 | 0.000 | -0.331 | 0.000 | 0.000 | 3.37 | 3.56 |
| Pb3 | 3.547 | 3.731 | 7.250 | -0.334 | 0.000 | -0.211 | 3.17 | 3.92 |
| Ti1 | 0.015 | 0.000 | -3.630 | 0.015 | 0.000 | 0.101 | 5.38 | 5.81 |
| Ti2 | 0.019 | 0.000 | 3.630 | 0.015 | 0.000 | -0.101 | 5.38 | 5.81 |
| Zr1 | 0.118 | 0.000 | 11.192 | 0.118 | 0.000 | 0.000 | 6.06 | 6.06 |
| O1 | 0.628 | 0.000 | -7.223 | 0.628 | 0.000 | 0.238 | -2.15 | -4.80 |
| O2 | 4.432 | 0.000 | -3.637 | 0.551 | 0.000 | 0.094 | -4.56 | -1.95 |
| O3 | 0.605 | 3.731 | -3.634 | 0.605 | 0.000 | 0.097 | -2.16 | -2.59 |
| O4 | 0.656 | 0.000 | 0.000 | 0.656 | 0.000 | 0.000 | -2.10 | -4.92 |
| O5 | 4.432 | 0.000 | 3.637 | 0.551 | 0.000 | -0.094 | -4.56 | -1.95 |
| O6 | 0.605 | 3.731 | 3.634 | 0.605 | 0.000 | 0.097 | -2.16 | -2.59 |
| O7 | 0.628 | 0.000 | 7.223 | 0.628 | 0.000 | -0.238 | -2.15 | -4.80 |
| O8 | 4.201 | 0.000 | 11.192 | 0.320 | 0.000 | 0.000 | -4.62 | -2.63 |
| O9 | 0.779 | 3.731 | 11.192 | 0.779 | 0.000 | 0.000 | -2.07 | -2.90 |

perpendicular to these chains. The large values of the effective charges for B and O atoms are due to a (weak) hybridization between the B $d$ and O $2p$ orbitals [9,16]. It is interesting to note that along the [001] axis, the effective charge of Ti is very similar to that of Zr, while the difference between these two effective charges in the bulk parent compounds is larger than 1.0 (i.e., 7.06 *vs.* 5.85 for Ti and Zr respectively according to Ref. [9]). One can also point out that the effective charge along [001] for atom O7 sitting between the Ti and Zr atoms is -5.39, i.e., very close to the average value -5.32 of the corresponding oxygen effective charges in the bulk parents (-5.83 and -4.81 for $PbTiO_3$ and $PbZrO_3$ respectively according to Ref. [9]).

## Ferroelectric effects

We now turn to a consideration of the *ferroelectric* effects on bond-length distributions and effective charges. Optimizing each degree of freedom in the non-centrosymmetric cell previously described yields the following lattice vectors in atomic units: $\vec{a}_{1,nc} = a'_0[1.0404, 0, 0]$, $\vec{a}_{2,nc} = a'_0[0, 1, 0]$, and $\vec{a}_{3,nc} = a'_0[0, 0, 3.00]$, where the lattice parameter is $a'_0$=7.461 a.u., i.e., 0.5% smaller than for the centrosymmetric cell. A similar decrease of the lattice constant of around 0.7% has also been theoretically predicted when going from the paraelectric cubic phase to the tetragonal ferroelectric phase of the bulk $PbTiO_3$ compound [7,17]. By looking at $\vec{a}_{1,nc}$, we also notice that our calculation predicts a "ferroelectric-related" axial ratio $a_1/a$ of 1.040. This prediction must be very close to the true value in $Pb(Zr_{1/3}Ti_{2/3})$, since recent measurements performed on $Pb(Zr_{1-x}Ti_x)$ films [18] for $x=0.6$ found a value of 1.035 for this ratio, to be compared with values of 1.064 and 1.02 for $x=1$ and $x \simeq 0.5$ respectively [1].

Interestingly, the optimized non-centrosymmetric cell has an energy that is 0.15 eV/5-atom-cell lower than that of the optimized centrosymmetric cell. This is consistent with the fact that the experimental ground state of Pb($Zr_{1/3}Ti_{2/3}$) is ferroelectric and tetragonal rather than the paraelectric and cubic. The relaxed atomic positions and the effective charges in the non-centrosymmetric structure are shown in Table II. The quantities $\Delta x$, $\Delta y$, and $\Delta z$ are the [100], [010] and [001] atomic displacements of the ferroelectric structure with respect to the ideal ordered structure associated with $\vec{a}_{1,nc}$, $\vec{a}_{2,nc}$ and $\vec{a}_{3,nc}$. $Z_{xx}$ and $Z_{zz}$ are the effective charges along the [100] and [001] directions, respectively.

A comparison of Tables I and II leads to the following observations. (i) The atomic displacements along the [001] direction are quite comparable between the centrosymmetric and the non-centrosymmetric phases. (ii) The ferroelectricity in tetragonal PZT is mainly characterized by the very large displacement of oxygen atoms along the tetragonal [100] direction (as in tetragonal PbTiO$_3$ bulk), the large displacement of Pb atoms along the [-100] direction, and by the slight displacement of Zr atoms along the [100] direction.

This ferroelectric atomic relaxation yields to two different Ti–O bond lengths along the [100] direction: a very long bond of length 2.33 Å, which is even longer than the longest Zr–O bond (2.16 Å); and a very short bond of length 1.55 Å. This very short Ti–O bond is much shorter than the shortest Zr–O bond of 1.95 Å, and is even much shorter than the shortest Ti–O bond of 1.87 Å occurring in tetragonal ferroelectric PbTiO$_3$.

Ferroelectricity also leads to a drastic change in the Pb–O bonds. There are now some very short Pb–O bonds with an average length of 2.51 Å, "normal" Pb–O bonds with an average length of 2.84 Å, and very long Pb–O bonds with an average length of 3.26 Å. As in the centrosymmetric cell, the population ratio between these three groups is again 4:4:4. However, the oxygens exhibiting the shortest Pb–O bonds now share a common (100) plane, while they share a common (001) plane in the non-polar structure. Pair-distribution function analysis of recent pulsed neutron powder diffraction measurements on ferroelectric PZT alloys clearly confirms the existence of these three different groups [19]. The experimental average value of the three different Pb–O bond lengths is ~2.5 Å, ~2.9 Å, and ~3.4 Å, i.e., in excellent agreement with our predictions.

Comparing Table I and Table II also indicates that ferroelectricity leads to a striking decrease of the Born effective charges. The most spectacular decrease occurs for the atoms exhibiting a large change in their bond lengths. As a matter of fact, the effective charges along the [100] direction for atoms Ti1, Pb3 and O6 all decrease by the ~20% with respect to the centrosymmetric case. As consequence, the effective charges of the Ti atoms and Pb atoms are reduced by 1.4 and 0.7, respectively, relative to their non-polar values. Previous theory has shown that a change of the effective charge by more than one unit of 'e' is indeed not unusual in going from the cubic to tetragonal ferroelectric phase in perovskite compounds

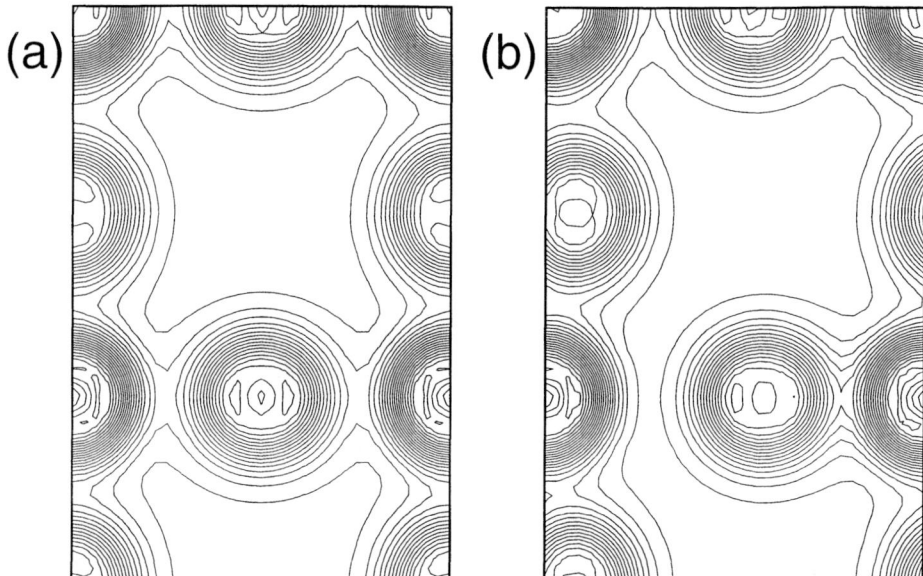

**FIGURE 2.** Electronic charge density plotted in the (B,O) plane for (a) paraelectric, and (b) ferroelectric, Pb(Zr$_{1/3}$Ti$_{2/3}$)O$_3$ supercells. Only the upper half of the supercell is shown; the horizontal and vertical axes lie along [100] and [001] respectively. The sequence of atoms appearing at the left and right edges is Zr (top), O, Ti, O (bottom); and the remaining middle atoms are oxygens.

[20,21]. The most extreme example appears to be for Nb in KNbO3, where the effective charges changes from 9.67 to 7.05 in the direction parallel to the tetragonal axis.

To further understand the ferroelectric effects in PZT, Fig. 2 compares the electronic charge density in the (B,O) planes for the centrosymmetric and non-centrosymmetric cases. Figure 3 shows a similar comparison but in the (Pb,O) planes. Figure 2 indicates that ferroelectricity in PZT leads (i) to a chemical breaking of some Ti–O bonds which generates the long Ti–O bonds of 2.33 Å, and (ii) to the formation of strong covalency between Ti and O, which is the cause of the very short Ti–O bonds of 1.55 Å. In fact, we also found similar behavior for the electronic charge density in the (Ti,O) plane for cubic and tetragonal lead titanate. Thus, as in bulk PT [22], the formation of ferroelectricity in tetragonal PZT leads to an enhancement of hybridization between Ti $3d$ and O $2p$ orbitals. Interestingly, we don't observe in Figure 2 any breaking of Zr–O bonds nor the formation of strong covalent Zr–O bonds. The different chemical behavior between Zr and Ti may perhaps be the cause of the difference in ground states exhibited

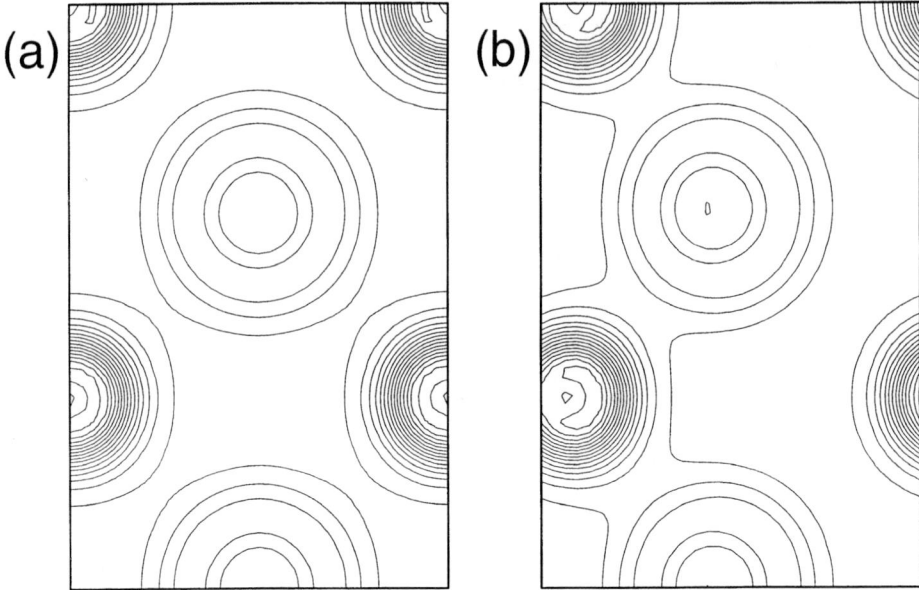

**FIGURE 3.** Electronic charge density plotted in the (Pb,O) plane for (a) paraelectric, and (b) ferroelectric, Pb(Zr$_{1/3}$Ti$_{2/3}$)O$_3$ supercells. Only the upper half of the supercell is shown; the horizontal and vertical axes lie along [100] and [001] respectively. Atoms appearing at the left and right edges are O; atoms in the middle are Pb.

by the corresponding bulk parents (antiferroelectric and orthorhombic for PbZrO$_3$ vs. ferroelectric and tetragonal for PbTiO$_3$). The striking feature of Fig. 3 is the formation of covalent chains between Pb and O atoms, which is the cause of the very short Pb-O bonds of 2.5 Å. We also found similar trends in the (Pb,O) planes of paraelectric and ferroelectric PT. Thus, as in bulk PT [22], the hybridization between Pb 6$s$ and O 2$p$ orbitals plays an important role in the ferroelectric behavior of tetragonal PZT.

## CONCLUSIONS

Using 15-atoms supercells and Vanderbilt ultrasoft pseudopotentials within the local-density approximation, we investigated alloying and ferroelectric effects on the bond lengths, chemical bonding and effective charges in lead zirconate titanate alloys (PZT).

Our principal findings are as follows.:

(i) The centrosymmetric PZT alloy is mainly characterized by two sets of B–O bonds (shorter Ti–O bonds vs. longer Zr–O bonds), while the Pb–O bonds differ only slightly (by ~2.5%) from the ideal structure.

(ii) Allowing ferroelectricity in PZT alloys has two striking chemical effects: enhancement of hybridization between Ti $3d$ and O $2p$ orbitals, and hybridization between Pb $6s$ and O $2p$ orbitals.

(iii) These chemical and ferroelectric effects lead to the formation of very short covalent Ti–O bonds while breaking other Ti–O bonds, and give rise to the formation of covalent chain of very short Pb–O bonds.

(iv) The atoms engaged in covalent bonding exhibit a striking decrease of their effective charges by ~20% relative to the paraelectric phase.

## ACKNOWLEDGMENTS

This work is supported by the Office of Naval Research grant N00014-97-1-0048. We thank Professor T. Egami for helpful discussions and for communicating his results with us.

## REFERENCES

1. M.E. Lines and A.M. Glass, *Principles and Applications of Ferroelectrics and Related Materials* (Clarendon Press, Oxford, 1977).
2. G. Saghi-Szabo and R.E. Cohen, *Ferroelectrics* **194**, 287 (1997).
3. D. Vanderbilt, *Phys. Rev. B* **41**, 7892 (1990).
4. D.M. Ceperley and B.J. Alder *Phys. Rev. Lett.* **45**, 566 (1980).
5. J. Perdew and A. Zunger *Phys. Rev. B* **23**, 5048 (1981).
6. H.J. Monkhorst and J.D. Pack, *Phys. Rev. B* **13**, 5188 (1976).
7. R.D. King-Smith and D. Vanderbilt, *Phys. Rev. B* **49**, 5828 (1994).
8. R.D. King-Smith and D. Vanderbilt, *Phys. Rev. B* **47**, 1651 (1993).
9. W. Zhong, R.D. King-Smith and D. Vanderbilt, *Phys. Rev. Lett.* **72**, 3618 (1994).
10. J.C. Mikkelsen and J.B. Boyce, *Phys. Rev. Lett.* **49**, 1412 (1982).
11. Z. Wang and B.A. Bunker, *Phys. Rev. B* **46**, 11277 (1992).
12. J.L. Martins and A. Zunger, *Phys. Rev. B* **30**, 6217 (1984).
13. N. Marzari, S. de Gironcoli and S. Baroni, *Phys. Rev. Lett.* **72**, 25 (1994).
14. C.K. Shih, W.E. Spicer, W.A. Harrison, and A. Sher, *Phys. Rev. B* **31**, 1139 (1985).
15. L. Bellaiche, S.H. Wei and A. Zunger *Phys. Rev. B* **56**, 13872 (1997).
16. M. Posternak, R. Resta and A. Baldereschi, *Phys. Rev. B* **50**, 8911 (1994).
17. A. Garcia and D. Vanderbilt, *Phys. Rev. B* **54**, 3817 (1996).
18. W. Zhu, Z.Q. Liu, W. Lu, M.S. Tse, H.S. Tan and X. Yao, *J. Appl. Phys.* **79**, 4283 (1996).
19. T. Egami, S. Teslic, W. Domowski, P.K. Davies and I.-W. Chen, *submitted* (1997).
20. Ph. Ghosez, X. Gonze, Ph. Lambin and J.-P. Michenaud *Phys. Rev. B* **51**, 6765 (1995).
21. C.-Z. Wang, R. Yu and H. Krakauer, *Phys. Rev. B* **54**, 11161 (1996).
22. R.E. Cohen, *Nature* **358**, 136 (1992).

# MODELING CATION ORDERING IN $A(B'_{1/3}B''_{2/3})O_3$ PEROVSKITES

B.P. Burton and R.P. McCormack

*Materials Science and Engineering Laboratory, Ceramics Division National Institute of Standards and Technology, Gaithersburg, MD 20899*

G. Ceder

*Department of Materials Science and Engineering, Massachusetts Institute of Technology, Cambridge MA 02139*

Robin L. B. Selinger

*Catholic University, Washington DC, 20064*

Georg Kresse and Jürgen Hafner

*Institut für Theoretische Physik, Technische Universität Wien A-1040 Wien, Austria*

**Abstract.** Ionic model (SSCAD) calculations were performed for 36 different cation ordered supercell configurations in the pseudobinary perovskite related system $BaTaO_3 - BaZnO_3$. $Ba(Zn_{1/3}Ta_{2/3})O_3$ (BZT) in a $P\bar{3}m1$ ordered structure is the only single phase compound that is observed experimentally, and the one that SSCAD calculations predict as having lowest formation energy. It is therefore the presumed ground state at the BZT composition. The SSCAD results are supported by first principles VASP pseudopotential calculations which were performed for $BaTaO_3$, $BaZnO_3$, $Ba(Zn_{1/3}Ta_{2/3})O_3$, and $Ba(Zn_{1/2}Ta_{1/2})O_3$ with NaCl type ordering of $Zn$ and $Ta$. Finite temperature calculations that are based on the SSCAD results, predict a strongly first-order $P\bar{3}m1 \rightarrow Pm\bar{3}m$ transition, but the predicted $T_C$ appears to be about an order of magnitude too high.

## I INTRODUCTION

The state of B-site cation order in mixed $A(B', B'')O_3$ perovskites, such as $Ba(Zn_{1/3}Ta_{2/3})O_3$ (BZT) and $Pb(Mg_{1/3}Nb_{2/3})O_3$ (PMN), has a dramatic effect on their electronic properties. In BZT for example, changes in the degree of order are correlated with order of magnitude changes in the product of dielectric loss quality factor and frequency at microwave frequencies [1]. Similarly, the state of order strongly impacts dielectric dispersion in the relaxor ferroelectric PMN, which is widely used in multilayer capacitors [2]. BZT and PMN have analogous stoichiometries and B-site cation valences, but do not appear to adopt similar low-temperature ordered ground state (GS) structures, or electronic properties. The $P\bar{3}m1$ BZT structure (the apparent GS) is a 1:2 stacking modulation of Zn and Ta layers perpendicular to the $[111]_{cubic}$ direction of the $Pm\bar{3}m$ disordered (DIS) perovskite [3].

The structure of PMN is more controversial: Small angle resonant x-ray scattering studies [4] indicate that PMN adopts a NaCl-type 1:1 $Fm\bar{3}m$ ($a_{1:1} \sim 2a_0$) structure in which one B-site is occupied by Nb ions while the other has approximately 1/3 Nb plus 2/3 Mg (henceforth the 1:1 structure). A space-charge model has also been proposed [5–7]. It postulates that small regions with 1:1 Mg:Nb ordering are coherently modulated with disordered regions that are sufficiently Nb-enriched to maintain global charge balance. Both structural models imply significant configurational entropy which is unusual for a GS, and therefore, it seems likely that the observed state of PMN is a metastable intermediate temperature phase (or two-phase mixture) that fails to reach its true GS because of unfavorable kinetics.

The phase transitions 1:2 GS → 1:1, and 1:2 GS → DIS have been observed in various $A(B', B'')O_3$ perovskites (Table I) either in response to increased temperature or to alloying.

**TABLE 1.** Data on order-disorder transitions in $A(B', B'')O_3$ perovskites.

| System | Structure, x-, T-Range | Ref. |
|---|---|---|
| $Ca(Ca_{1/3}Nb_{2/3})O_3$ | 1:2 ⇌ 1:1<br>$1400 < T_C < 1500°C$ | [8] |
| $Ba(Ni_{1/3}Nb_{2/3})O_3$ | 1:2 ⇌ Disordered<br>$1350 < T_C < 1400°C$ | [9] |
| $Ba(Zn_{1/3}Nb_{2/3})O_3$ | 1:2 ⇌ Disordered<br>$1350 < T_C < 1400°C$ | [10] |
| $Ba(Co_{1/3}Nb_{2/3})O_3$ | 1:2 ⇌ Disordered<br>$T_C \approx 1500°C$ | [10] |
| $(1-x)Ba(Zn_{1/3}Ta_{2/3})O_3$-$(x)BaZrO_3$ | 1:2, $0 < x < 0.02$, $T = 1425°C$<br>1:1, $0.04 < x < 0.25$<br>DIS, $0.25 < x$ | [11] |
| $(1-x)Ba(Mg_{1/3}Nb_{2/3})O_3$-$(x)BaZrO_3$ | 1:2, $0 < x < 0.02$, $T = 1350°C$<br>1:1, $0.05 < x < 0.15$<br>DIS, $0.15 < x$ | [12] |
| $Ba_{1-x}La_x(Zn_{(1+x)/3}Ta_{(2-x)/3})O_3$ | 1:2, $0 < x < 0.02$, $T = 1500°C$<br>1:1, $0.02 < x < 0.20$ | [13] |
| $Ba_{1-x}K_x(Zn_{(1-x)/3}Ta_{(2+x)/3})O_3$ | 1:2, $0 < x \leq 0.10$, $T = 1500°C$ | [13] |

The experimental data in Table I indicate a competition between the 1:2 and 1:1 structures in $A(B'_{1/3}B''_{2/3})O_3$ perovskites. It follows that a sufficient model for order-disorder phenomena in these materials must yield the correct GS, and also permit the transition sequences: 1:2 GS → 1:1 → DIS and 1:2 GS → DIS. Previous statistical models of order-disorder phenomena in PMN [14,15] do not satisfy these conditions. The purpose of this paper is to determine what sort of Ising type model will be sufficient to treat order-disorder phenomena in these systems, and to make a first principles calculation for the specific example, $Ba(Zn_{1/3}, Ta_{2/3})O_3$.

## II   SIMPLE EMPIRICAL ISING MODELS

In a previous discussion [16] of the simple cubic GS problem [16–20] the first and second authors of this paper reported that a sufficient Ising Hamiltonian could be obtained by including the effective cluster interactions (ECI [21]) contained within the unit cube plus the linear triplet. This is actually a much more complicated Hamiltonian than is required. Ground state searches were performed for each of the 33 ECI's that that are contained in the unit cube plus the octahedron formed by a site plus its six nearest neighbors (nn + nn-octahedron, Figure 1).

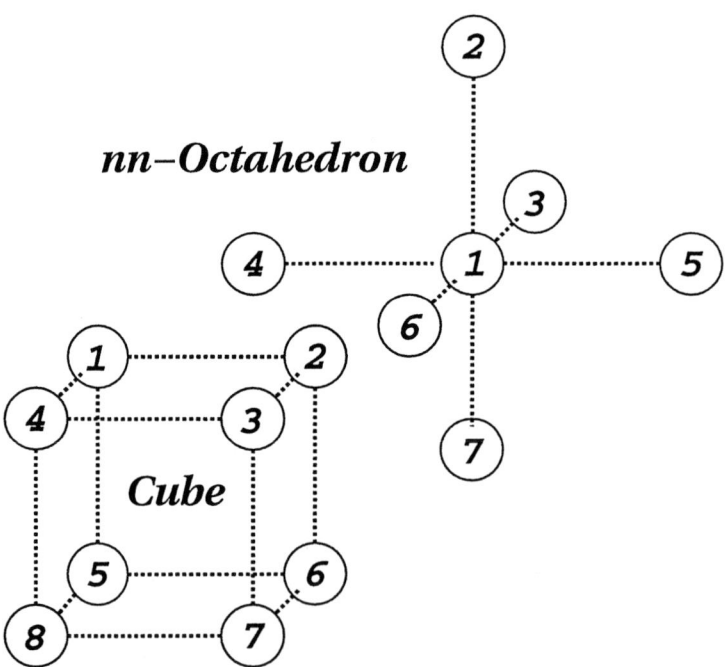

**FIGURE 1.** The unit cube and the nn octahedron clusters.

Within this set only the linear triplet interaction $[J(\circ - \circ - \circ)]$ is both necessary and sufficient to stabilize the 1:2 GS; i.e. only the linear triplet yields a GS at $x = 1/3$ for the 1:2 structure. Finite temperature behavior of the linear triplet model (Figure 2) was determined by Monte Carlo (MC) simulation [22] performed on a 30x30x30 simple cubic array of sites, in a canonical ensemble, via pairwise distant neighbor exchange (ions closer than five lattice spacings were not exchanged). At least 5000 iterations were performed at a given temperature before the calculations (order parameters) were considered to have converged. The linear triplet model yields a first-order 1:2 GS $\to$ DIS transition at $4.5183 < kT_C/[-J(\circ - \circ - \circ)] < 4.5185$.

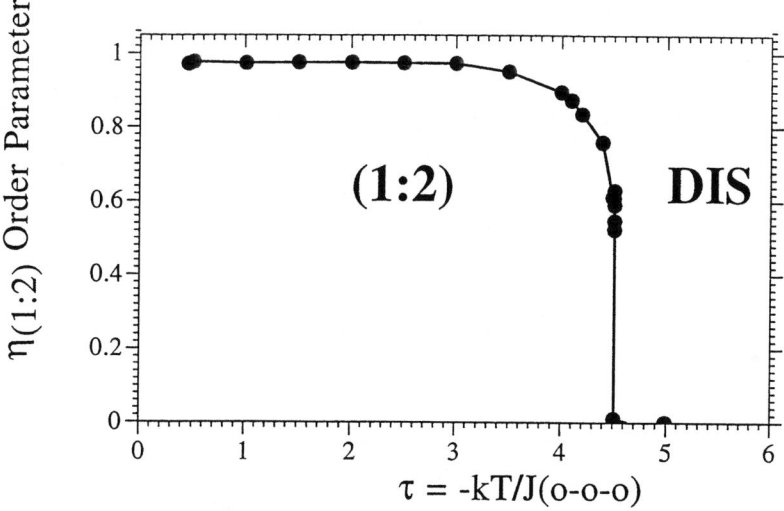

**FIGURE 2.** Phase diagram for the linear triplet model.

Obviously, the simplest Hamiltonian that yields a 1:2 GS $\to$ 1:1 $\to$ DIS transition sequence must be the linear triplet plus one additional ECI that stabilizes the 1:1 phase, e.g. the nn pair plus linear triplet, $(\circ-\circ-\circ) + (\circ-\circ)$. The phase diagram for this model (Figure 3) was determined by a combination of cluster variation method [23] calculations (cube + nn octahedron approximation, dashed lines), and MC simulations (triangular symbols). For values of $r = -J(\circ-\circ-\circ)/J(\circ-\circ) \leq 0.25$ this model yields the desired GS and transition sequence, but in this region of phase space in which the formation energy ($\Delta E$) for 1:2 phase is greater than that for the 1:1 phase, $\Delta E(1:2) > \Delta E(1:1)_{x=1/2}$, contrary to experiment as well as the results of both self consistent atomic deformation (SSCAD) [24] and pseudopotential [25] calculations of $\Delta E(1:2)$ and $\Delta E(1:1)_{x=1/2}$. Note the distinction between the 1:1 phase at $x = 1/3$, and the formation energy of a $Ba(Zn_{1/2}Ta_{1/2})O_3$ phase with NaCl-type ordering, $\Delta E(1:1)_{x=1/2}$.

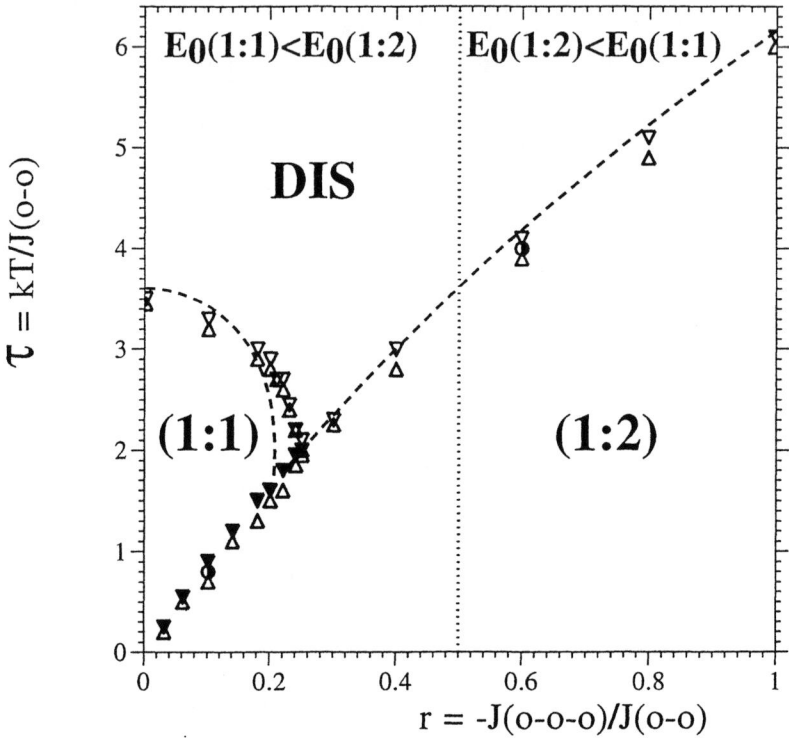

**FIGURE 3.** Phase diagram for the nn pair + linear triplet model. Dashed lines represent CVM results. Triangles that bracket phase boundaries were determined by Monte Carlo simulation. Half filled circles indicate two-phase mixtures in MC simulations.

A compound in a pseudobinary system, e.g. BZT in $BaTaO_3 - BaZnO_3$ or PMN in $PbNbO_3 - PbMgO_3$, that undergoes a strongly first-order 1:2 → DIS, or 1:2 → 1:1 transition **may** do so at a single temperature, or it may traverse a (1:2 + DIS) or (1:2 + 1:1) two-phase field with finite temperature range. Such two-phase fields occur in some of the MC simulations described in this paper, but they are quite narrow, and are typically not precisely resolvable. The presence of a (1:2 + 1:1) two-phase field would provide a thermodynamic basis for the composition modulations in PMN that Viehland et. al. [6] interpreted in terms of their space charge model. Note however that narrowness of the two-phase field implies a very limited difference between the 1:2 and 1:1 compositions.

## III  TOTAL ENERGY CALCULATIONS

### A  Ionic Model SSCAD Calculations

To perform a first-principles simulation one must calculate a set of ECI's. Several methods exist [26–30], and all use first principles total energy calculations to obtain differences in total energy between various atomic configurations (either for a series of different supercells, or for compositional modulations in a disordered medium). The method applied here is to calculate formation energies of ordered supercells with 40 or fewer atoms per cell.

Total energy calculations were performed for a series of perovskite based ordered superstructures in the composition range $(1-x)BaTaO_3 - (x)BaZnO_3$, $(1-x)BT - (x)BZ$. The vast majority of total energy calculations were performed with the SSCAD model code of Boyer et al. [24], which is based on Gordon-Kim [31] formulations. Atomic densities are assumed to be spherically-symmetric, although this approximation is relaxed in more recent versions of the code. [24] All structural degrees of freedom, cell internal and cell external, were optimized. An interatomic interaction cutoff of 0.79 $nm$ (15 $a.u.$) was used and total energies were converged to within 1.313 $kJ/mol$ (0.5 $mHartree/atom$), with many-body corrections applied. In BT, BZT$_{[111]}$, BZ, and a few of the simpler intermediate structures, such as $Ba(Zn_{1/2},Ta_{1/2})O_3$ with NaCl-type ordering, ionic charge densities were optimized as well. Optimized charge densities followed very regular systematics which were then assumed to hold for all additional structures:

- For $x = 1/3$, the ideal BZT composition, all ions adopt nominal valences: $Ba^{2+}$, $Mg^{2+}$, $Ta^{5+}$, $O^{2-}$.
- For $0 \leq x \leq 1/3$, excess Ta, $Ta^{5+} \rightarrow Ta^{(4+3x)+}$.
- For $1/3 \geq x \geq 1$, excess Zn, $O^{2-} \rightarrow O^{(x-7/3)}$.

The SSCAD results for total energies are plotted in Figures 4, and 5, calculated zero pressure equilibrium excess volumes are plotted in Figure 5. Typically, the reference state for such calculations is a mechanical mixture of end members (BT, and BZ in this case):

$$\Delta E_{Structure} = E_{Structure} - (x)E(BaTaO_3) - (1-x)E(BaZnO_3).$$

The values plotted in Figure 4 have been shifted by a constant value:

so that the BZT GS will have $\Delta E_{BZT} = 0$.

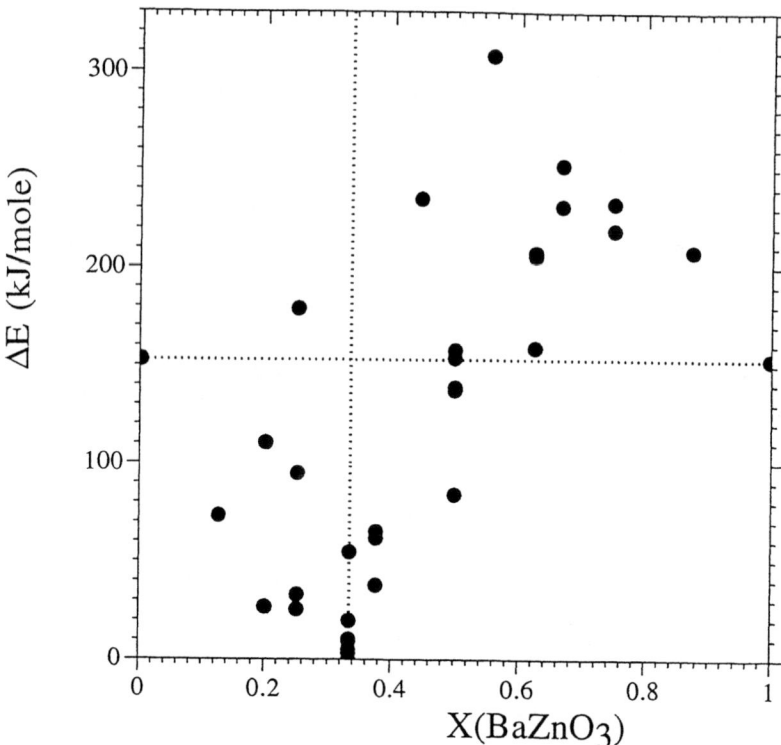

**FIGURE 4.** SSCAD calculations of $\Delta E_{formation}$ (per $ABO_3$ mole) for various supercells in $BaTaO_3 - BaZnO_3$; $\Delta E_{BZT}$ plots at $(x = 1/3, \Delta E = 0)$.

Experimentally, only the $P\bar{3}m1$ ordered BZT structure is observed, as single phase material, which implies that it is the minimum energy structure in BT-BZ, consistent with the results displayed in Figure 4, where $\Delta E_{BZT}$ plots at $(x = 1/3, E = 0)$. In general, $\Delta E$ increases as the composition departs from $x = 1/3$. Most calculated values for $\Delta V$ follow trends similar to those for $\Delta E$. Exceptions are the end members $BaTaO_3$ and $BaZnO_3$, and one structure at $x = 0.75$. That is, lower energy ordered structures typically exhibit smaller $\Delta V$.

The charge systematics described above constitute a crude approximation at best, but one that is expected to yield a qualitatively correct trend for compound formation energies. In general, one expects this approximation to work best for structures close to $x = 1/3$ and $\Delta E = 0$ (where all ions have their normal valences: $Ba^{2+}$, $Zn^{2+}$, $Ta^{5+}$, $O^{2-}$), and progressively worse for structures farther from this point. This is because structures that are farther from $(x = 1/3, \Delta E = 0)$ are more likely to be stabilized by changes in charge density distribution. In particular, some compounds will inevitably reduce their total energies via metallization, which

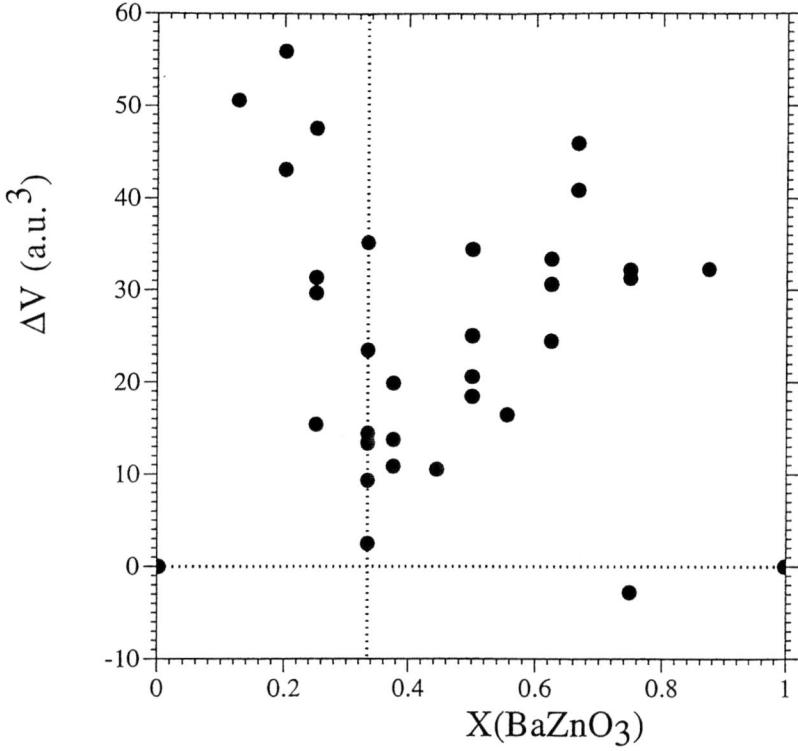

**FIGURE 5.** SSCAD results for $\Delta V$.

the SSCAD calculation will miss. This implies an expectation that "correct" $\Delta E$ are are expected to be systematically **lower** than the SSCAD results.

## B  VASP Pseudopotential Calculations

A limited number of formation energies were also calculated using the Vienna *ab initio* simulation program (VASP) [25]. These are plane-wave pseudopotential calculations that are performed in the local density approximation for exchange and correlation energies. Electronic degrees of freedom were optimized with a conjugate gradient algorithm, and both cell constant and ionic positions were fully relaxed. The results of these calculations are shown in Figure 6, where they are compared to the corresponding SSCAD results. As mentioned above, both the VASP and SSCAD results yield the same hierarchy of formation energies: $\Delta E(1:2) < \Delta E(1:1)_{x=1/2}$. Quantitatively however, there is a significant differ-

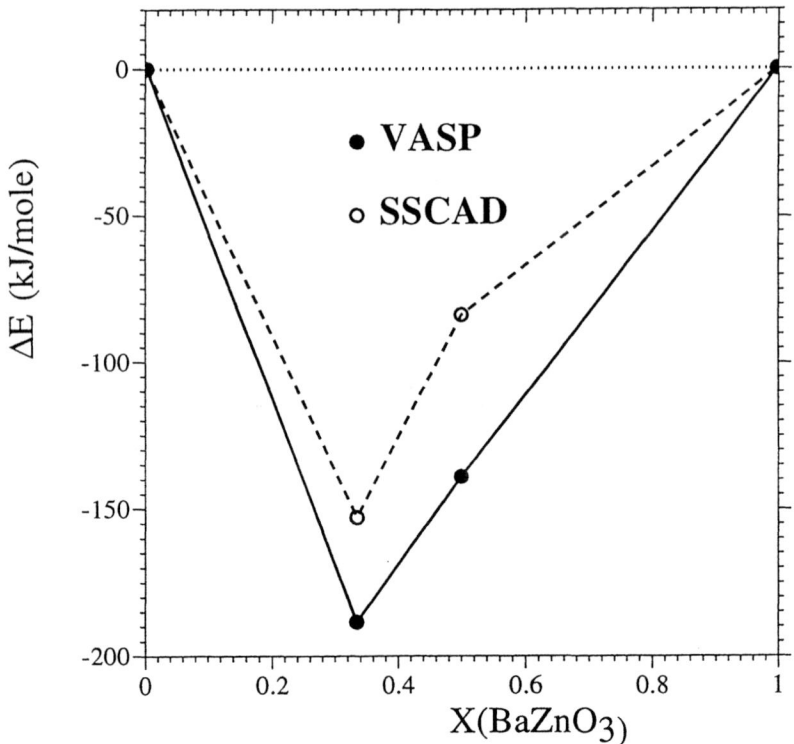

**FIGURE 6.** VASP and SSCAD results for the formation energies of $Ba(Zn_{1/3}Ta_{2/3})O_3$ ($x = 1/3$) and $Ba(Zn_{1/2}Ta_{1/2})O_3$ ($x = 1/2$).

ence between the VASP and SSCAD results.

## IV MONTE CARLO SIMULATIONS

The SSCAD results for $\Delta E$ were fit to a cluster expansion Hamiltonian that included all the interactions present in the cube plus the nn-octahedron (Table II). In spite of the large structural database, it was necessary to constrain the fit to avoid a Hamiltonian that predicts spurious GS's at $x \neq 1/3$, and this makes the simulation something less than a first principles calculation. Obviously, an absolute minimum $\Delta E$ at $x \neq 1/3$ would be highly surprising, and contradict experiment, because it implies unusual oxidation states for the various ions, and contradicts both the SSCAD results and experiment. Still, the requirement of constraints represents a significant approximation.

**TABLE 2.** Hamiltonian fit to SSCAD total energy data. Subcluster indicies in the Cube and nn-Octahedron are consistent with Figure 1.

| Sites per Subcluster | Subcluster Multiplicity | ECI kJ/mole | Subcluster in Cube | Subcluster in nn-Octahedron |
|---|---|---|---|---|
| 1 | 1 | 29.51 | 1 | 1 |
| 2 | 3 | 18.62 | 1,2 | 1,2 |
| 2 | 4 | 2.23 | 1,7 | |
| 2 | 6 | 0.58 | 1,3 | 2,3 |
| 2 | 3 | 21.92 | | 2,7 |
| 3 | 12 | -4.18 | 1,2,3 | 1,2,3 |
| 3 | 8 | -1.30 | 1,3,6 | 2,3,4 |
| 3 | 12 | 5.39 | | 2,3,7 |
| 3 | 24 | -2.34 | 1,2,7 | |
| 3 | 3 | 6.70 | | 2,1,7 |
| 4 | 24 | 0.32 | 1,2,3,7 | |
| 4 | 8 | -4.72 | 1,2,4,5 | 1,2,3,5 |
| 4 | 2 | -1.37 | 1,3,6,8 | |
| 4 | 6 | -0.51 | 1,3,5,7 | |
| 4 | 3 | -3.62 | | 2,3,6,7 |
| 4 | 24 | -1.33 | 1,7,6,8 | |
| 4 | 12 | 6.00 | | 2,3,4,7 |
| 4 | 12 | -1.91 | | 1,2,3,7 |
| 4 | 3 | 5.26 | 1,2,3,4 | |
| 5 | 8 | -3.74 | 1,3,6,8,5 | |
| 5 | 24 | 2.91 | 3,4,5,6,8 | |
| 5 | 24 | 2.43 | 4,5,6,7,8 | |
| 5 | 3 | -16.25 | | 1,3,4,5,6 |
| 5 | 6 | 5.50 | | 3,4,5,6,7 |
| 5 | 12 | 4.93 | | 1,2,3,4,5 |
| 6 | 1 | -14.59 | | 2,3,4,5,6,7 |
| 6 | 4 | 2.46 | 2,3,4,5,6,8 | |
| 6 | 12 | 0.41 | 2,4,5,6,7,8 | |
| 6 | 12 | 1.64 | 3,4,5,6,7,8 | |
| 6 | 6 | 5.00 | | 1,2,3,4,5,6 |
| 7 | 8 | 5.40 | 1,2,3,4,5,6,7 | |
| 7 | 1 | -47.31 | | 1,2,3,4,5,6,7 |
| 8 | 1 | 1.92 | 1,2,3,4,5,6,7,8 | |

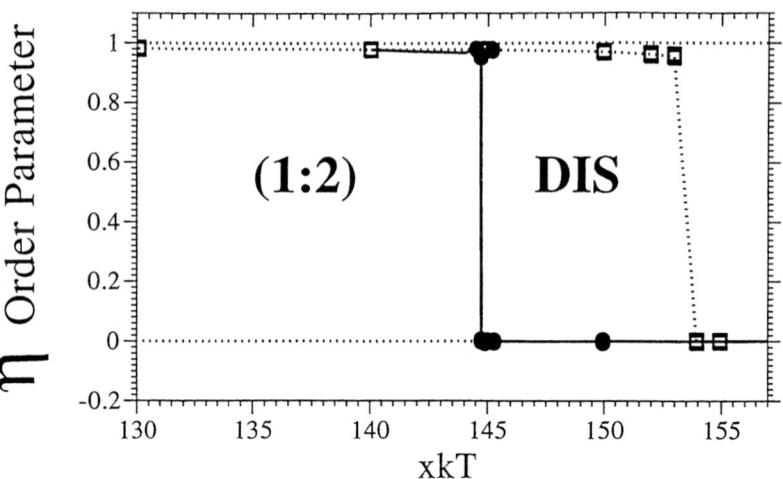

**FIGURE 7.** Monte Carlo simulation results for the temperature dependence of The 1:2 order parameter, $\eta$.

Finite temperature MC results (Figure 7) shows the temperature dependence of the 1:2 order parameter ($\eta_{(1:2)}$) The dotted line indicates calculations that started with a 1:2 ordered configuration, and solid lines indicate a mixture of (1:2 + DIS) phases. The latter is considered to yield the most accurate transition temperature for the 1:2 $\to$ DIS transition. Calculations that were started with a 1:1 ordered configuration always disordered in a few hundred iterations, both above and below $T_C$. A strongly first-order 1:2 $\to$ DIS transition is predicted. Given that $Ba(Zn_{1/3}Nb_{2/3})O_3$ (BZN) exhibits a 1:2 $\to$ DIS transition with 1350 < $T_C$ < 1400°C [10] (Table I), one expects similar behavior from $Ba(Zn_{1/3}Ta_{2/3})O_3$, but probably with a somewhat higher $T_C$; similar because $Nb^{5+}$ and $Ta^{5+}$ are very similar ions, and at slightly higher temperature because disordering of $Ba(Zn_{1/3}Ta_{2/3})O_3$ has not been observed experimentally. Quantitatively, the calculated value for $T_C \sim 15,362°C$ is probably much too high ($\sim 10$ times higher than what is observed in the analogous perovskites: $Ca(Ca_{1/3}Nb_{2/3})O_3, Ba(Ni_{1/3}Nb_{2/3})O_3, Ba(Zn_{1/3}Nb_{2/3})O_3, Ba(Co_{1/3}Nb_{2/3})O_3$ Table I).

## V  CONCLUSIONS

A sufficient Ising model to describe B-site order-disorder phenomena in $A(B'_{1/3}B''_{2/3})O_3$ perovskites must permit both the 1:2 GS and the transition sequence 1:2 GS $\to$ 1:1 $\to$ DIS. The linear triplet interaction, $J(\circ - \circ - \circ)$, is evidently both necessary and sufficient to stabilize the 1:2 GS, and a model that combines the linear triplet with the nn pair interaction is sufficient to generate the 1:2 GS → 1:1 → DIS transition sequence. It does so however, in a region of phase space (ECI space) in which $\Delta E(1:2) > \Delta E(1:1)_{x=1/2}$, contrary to experiment, and to SSCAD and VASP calculations. Finite-tempearture MC simulations that are based on SSCAD total energy calculations predict a strongly first order 1:2 GS → DIS transition, which is probably qualitatively correct; quantitatively however, the calculated value for $T_C$ appears to be much too high.

## VI  ACKNOWLEDGEMENTS

We thank H. T. Stokes, L. L. Boyer, and M. J. Mehl for the use of their computer program SSCAD, and A. Zunger for suggesting that we calculate formation energies at $x \neq 1/3$. RPM was supported by a National Research Council Postdoctoral Fellowship. This work supported by the Office of Naval Research under ONR contract No. N00014-94-F0017.

## REFERENCES

1. S. Kawashima, M. Nishida, I. Ueda, and H. Ouchi, J. Am. Ceram. Soc., **66**, 421 (1983)
2. S.G. Porter, Ferroelectrics, **33**, 193 (1981)
3. A. J. Jacobson, B.M. Collins and B. E. F. Fender, Acta Cryst. **B32**, 1083 (1976)
4. T. Egami, W. Dmowski, S. Teslic, P.K. Davies, I.-W. Chen, and H. Chen, in press, Ferroelectrics.
5. H.D. Rosenfeld and T. Egami, Ferroelectrics **158**, 351 (1994)
6. D. Viehland and J.-F. Li, J. Appl. Phys. **74**, 4121 (1993)
7. N. de Mathan, E. Husson, G. Calvarin, J.R. Gavarri, A. Hewatt, and A. Morell, J. Phys.: Cond. Matter **3**, 8159 (1991)
8. R.S. Roth, personal communication 1998.
9. I-T Kim, S.J. Chung and K.S. Hong, Ferroelectrics, **173**, 125- (1995).
10. U. Treiber and S. Kemmler-Sack, J. Solid State Chem., **43**, 51- (1982).
11. L. Chai and P.K. Davies, J. Am. Ceram. Soc. 80[12], 3193- (1997).
12. M. Akbas and P.K. Davies, submitted to: J. Amer. Ceram. Soc. (1997).
13. C-C. Lee, C-C Chou, and D-S. Tsai, In press, Ferroelectrics (1997).
14. A.G. Khachaturyan, Phys. Rev. B 48 2949- (1993).
15. L.A. Bursill, H. Qian, J.L. Peng, and X.D. Fan, Physica B, **216** 1 (1995).
16. R. McCormack and B.P. Burton, Computational Materials Science (in press)
17. S. Katsura and A. Narita, Progr. Theor. Phys., **50**, 1426 (1973).
18. M. Kaburagi and J. Kanamori, Progr. Theor. Phys., **54**, 30 (1975).
19. M.D. Lipkin, Physica A, **150**, 18 (1988).

# Effective Hamiltonian for the ferroelectric phase transitions in $KNbO_3$

U. V. Waghmare[1] and K. M. Rabe

*Department of Applied Physics, Yale University
P. O. Box 208284, New Haven, Connecticut, 06520-8284*

Henry Krakauer, Rici Yu,[2] and Cheng-Zhang Wang

*Department of Physics, College of William and Mary, Williamsburg, Virginia 23187-8795*

**Abstract.** An effective Hamiltonian for the ferroelectric transitions in $KNbO_3$ is constructed from first-principles density-functional-theory total-energy and linear-response calculations through the use of a localized, symmetrized basis set of "lattice Wannier functions." The detailed description provided here provides the foundation for the use of this effective Hamiltonian in classical Monte Carlo, classical molecular dynamics and quantum mechanical simulations, as well as guidance for the construction of effective Hamiltonians for related systems.

## INTRODUCTION

Many of the most interesting properties of perovskite ferroelectrics are dominated by the local polar distortions associated with the soft modes [1]. These properties include local structural fluctuations, the dielectric and piezoelectric responses, and the paraelectric-ferroelectric phase transition itself. With an effective Hamiltonian, we can efficiently focus on the behavior of these degrees of freedom to investigate these properties at a microscopic level. Studies of this kind include the perovskite ferroelectrics $PbTiO_3$ [2], $BaTiO_3$ [3] and $KNbO_3$ [4,5], as well as other perovskites with more complex structural transitions [6–8].

In this paper, we give a detailed description of an effective Hamiltonian for the perovskite ferroelectric $KNbO_3$ constructed from first-principles LAPW total-energy and linear response calculations. The finite-temperature behavior of this effective Hamiltonian has been studied in classical molecular dynamics simulations [4,5], yielding a phase diagram in reasonable agreement with available experiments

---

[1] Current address: Dept. of Physics, Harvard University, Cambridge, MA 02138.
[2] Current address: General Sciences Corporation, Silver Spring, MD

as well as microscopic insights into the differences between local and average structures. This effective Hamiltonian could also form the foundation of future investigations, including studies of the piezoelectric and dielectric response of $KNbO_3$ analogous to Refs. [10,11], dynamical properties of $KNbO_3$ at longer length scales, and the properties of solid solutions which have $KNbO_3$ as a end-member.

## METHOD

The general principles for the construction of a first-principles effective Hamiltonian based on lattice Wannier functions and their application to ferroelectric perovskites have been previously described in Ref. [2,12]. The following discussion of $KNbO_3$ closely parallels that of $PbTiO_3$ in Ref. [2], with one significant difference. For $KNbO_3$, in the construction of the lattice Wannier functions we use the eigenvectors of the dynamical matrix, in contrast to the use for $PbTiO_3$ of the eigenvectors of the force constant matrix. In $PbTiO_3$, tests showed that the difference in the resulting effective Hamiltonian subspace is rather small and both choices should give comparable results. The choice of the dynamical matrix eigenvectors for $KNbO_3$ is made to allow the use of the effective Hamiltonian in classical molecular dynamics simulations, since the form of the kinetic energy is thus greatly simplified.

The quantities needed to determine the parameters in the effective Hamiltonian are obtained from first-principles total-energy and linear-response calculations. Since the theoretical equilibrium lattice constant is 4.00 Å, just slightly smaller than the experimental value 4.016 Å [13], we have used the experimental value as the lattice constant of the cubic-perovskite high-symmetry reference structure. Results for the phonon dispersion, $Z^*$ and $\epsilon_\infty$ of this structure were presented in Ref. [14]. In the construction of the effective Hamiltonian, the better converged numerical values of $\Gamma$ point phonons, $Z^*$ and $\epsilon_\infty$ in Ref. [9] are used. In order to get the most accurate possible numerical values of the important anharmonic well-depths and the coupling of the polar distortions to homogeneous strain, we calculated the total energies of noninfinitesimal distortions with the all-electron LAPW+LO method [15], in which the pseudopotential approximation is not used, and the orthogonality is enforced for the semicore and valence orbitals by extending the LAPW basis functions to include local orbitals. With this method, the calculated well depths changed by about 0.2 mRy, significantly affecting some of the anharmonic parameters, particularly the soft-mode-strain coupling.

## CONSTRUCTION OF THE EFFECTIVE HAMILTONIAN

The construction of the effective Hamiltonian subspace begins with consideration of the calculated phonon eigenvalues and eigenvectors at $\Gamma$, $R$, $M$ and $X$ (Table

**TABLE 1.** Selected phonon frequencies ($cm^{-1}$) at high symmetry k-points calculated using DFT linear response. Symmetry labels follow the convention of Ref. [12].

| k-point | | frequencies | | frequencies | | frequencies |
|---|---|---|---|---|---|---|
| X | $X'_5$ | 148, 314 | $X'_2$ | 173, 777 | $X_1$ | 243, 456 |
| | $X_5$ | 107i, 151, 441 | $X_3$ | 284 | | |
| M | $M'_5$ | 146, 257, 398 | $M'_3$ | 101i, 375 | $M_4$ | 824 |
| | $M'_2$ | 148 | $M_5$ | 314 | $M_3$ | 150 |
| | $M_2$ | 476 | $M_1$ | 421 | | |
| | $R_{25}$ | 141 | $R_{15}$ | 163, 405 | $R'_2$ | 856 |
| R | $R'_{25}$ | 343 | $R'_{12}$ | 426 | | |
| $(111)\frac{\pi}{2a}$ | $\Lambda_1$ | 146, 250, 416, 815 | $\Lambda_2$ | 210 | $\Lambda_3$ | 137, 184, 228, 344, 473 |

1). The subspace has to include the unstable polar $\Gamma_{15}$ mode which freezes in to produce the low temperature rhombohedral structure. In addition, to achieve a good description of branches which emanate from this dominant unstable mode, the unstable endpoints of these branches $M_3'$ and $X_5$ are included. As can be seen from Table I of Ref. [12], the lattice Wannier functions which can build up this subset of modes transform like 3-dimensional vectors centered at Nb sites.

To include coupling of the relevant polar distortions ($\Gamma_{15}$) to local distortions of the unit cell (inhomogeneous strain), we expand the subspace to include the acoustic modes by choosing an additional set of lattice Wannier functions. Of the three possibilities (listed in Table I of Ref. [12]), K-centered 3-dimensional vectors are preferable to $O_{x,x}$ (1-dimensional vectors) and $O_{x,y}$ (2-dimensional vectors), since this choice corresponds to the smallest subspace expansion and highest site symmetry group. Furthermore, unlike $O_{x,x}$, the resulting subspace does not include the highest energy modes such as $R'_2$ and $M_4$.

Next, we obtain an explicit form for the Nb-centered LWF. This involves finding the symmetric coordination shells surrounding a Nb site and identifying the independent displacement patterns of each shell that transform according to the vector representation of the site symmetry group $O_h$. For a given shell there can be more than one pattern of displacements with a given transformation property. To each such pattern corresponds an independent amplitude parameter. By including the displacements of shells up through first neighbor K and second neighbor Nb shells, as well as selected displacements of O shells at first, second and fourth neighbors, we obtain a total of 12 parameters. The first shell of K atoms has 2 independent displacement patterns; there are 1, 2, 2 parameters for the zeroth, first and second shells of Nb atoms respectively, and 2, 2 and 1 parameters for the first, second and fourth shell of oxygen atoms respectively (displacements in the third shell were found not to be useful in the fit described below). These displacement patterns are shown in Fig. 1.

To determine the numerical values of these parameters for $KNbO_3$, we build up vectors $e_{\vec{q},\alpha}$ at high symmetry k-points in the Brillouin zone, namely $\Gamma$, X, M, and R, from the parametrized LWF using

**TABLE 2.** Values of the LWF parameters determined from first principles.

| $k_1$ | -0.00177645 | $k_2$ | -0.0557047 | $n_1$ | 0.800533 |
|---|---|---|---|---|---|
| $n_2$ | -0.056195 | $n_3$ | 0.0093515 | $n_4$ | -0.00773106 |
| $n_5$ | 0.00303931 | $o_1$ | -0.196627 | $o_2$ | -0.181273 |
| $o_3$ | -0.00449263 | $o_4$ | -0.0257974 | $o_5$ | 0.000845625 |

$$e_{\vec{q},\alpha} = \sum_{\vec{R}_i} exp(i\vec{q}\cdot\vec{R}_i)w_{i,\alpha} \qquad (1)$$

where $\vec{R}_i$ is a direct lattice vector and $w_{i,\alpha}$ is an LWF centered at the Nb site in the $i$th unit cell, $\alpha$ being its Cartesian component. This specifies each component of the vector as a linear function of the parameters, to be fit to the corresponding component of the appropriate eigenvector of the dynamical matrix calculated from first principles. The dynamical matrix eigenvectors are normalized so the sum over the five atom types of the squares of the components is equal to unity. As shown in Table 3, ten of the parameters listed above can be chosen exactly to reproduce the normalized eigenvectors of the unstable modes at $\Gamma_{15}$, $X_5$ and $M'_3$, as well as the normalized $R'_{25}$ eigenvector. The remaining two parameters, associated with Nb and K displacements, were used to fit to a normalized mode with maximum overlap with the Nb-dominated $M'_5$ mode at 398 cm$^{-1}$. Numerical values of these parameters, listed in Table 2, clearly show that the magnitude of the parameters

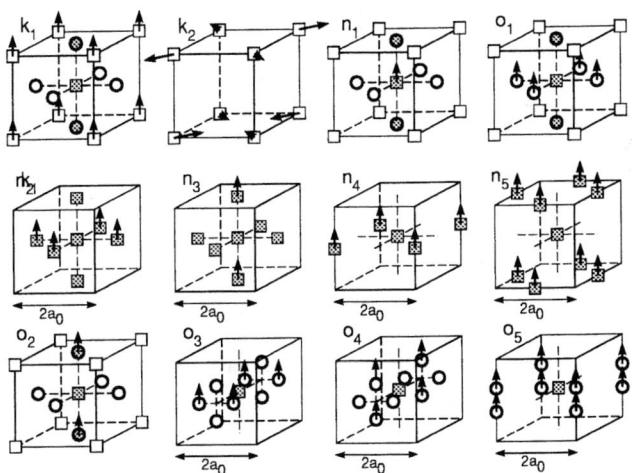

**FIGURE 1.** $z$ component of the vector-like Nb-centered lattice Wannier functions. $Nb$, $K$ and $O$ atoms are represented by solid squares, open squares and circles respectively. Parameters labeling the displacement patterns correspond to the length of the displacements (arrows) of atoms for the unit value of the LWF coordinate.

**TABLE 3.** Determination of LWF parameters. Linear combinations of these parameters for the modes at high symmetry k-points and the corresponding components of the normalized eigenvectors of the force constant matrix.

| Mode | combination of the parameters | component of the eigenvector |
|---|---|---|
| $\Gamma_{15}$ | $8k_1$ | -0.0142116 |
| | $n_1 + 4n_2 + 2n_3 + 4n_4 + 8n_5$ | 0.587846 |
| | $2o_1 + 4o_3$ | -0.411224 |
| | $2o_2 + 8o_4 + 8o_5$ | -0.56216 |
| $R'_{25}$ | $n_1 - 4n_2 - 2n_3 + 4n_4 + 8n_5$ | 1 |
| $M'_5$ | $8k_2$ | -0.445638 |
| | $n_1 - 2n_3 - 4n_4$ | 0.812754 |
| $M'_3$ | $n_1 - 4n_2 + 2n_3 + 4n_4 - 8n_5$ | 0.988777 |
| | $2o_2 - 8o_4 + 8o_5$ | -0.149402 |
| $X_5$ | $n_1 + 2n_3 - 4n_4$ | 0.85016 |
| | $2o_1 - 4o_3$ | -0.375283 |

decays rapidly with shell-radius, confirming the assumption of LWF localization. From these components, actual atomic displacements are obtained by multiplying by the factor $\frac{a_0}{\sqrt{m}}$, where $m$ is the mass of the atom type in a.m.u.

For the simplest treatment of inhomogeneous strain (the acoustic branches), an explicit expression for the K-centered LWF is not needed, since the goal is only to reproduce the long wavelength acoustic modes, whose dispersion is determined from the elastic constants. For a more refined treatment, an LWF could be parametrized as above and determined by fitting to a subset of the first principles eigenmodes $\Gamma_{15}$ (acoustic), $R_{15}$, $M'_2$, $M'_5$, $X'_2$ and $X'_5$.

Using the symmetry properties of the lattice Wannier basis for the effective Hamiltonian subspace, we write an explicit expression for $H_{eff}$ as a Taylor expansion in the lattice Wannier coordinates, invariant under the space group $Pm3m$. $\{\vec{\xi}_i\}$ and $\{\vec{u}_i\}$ denote the Nb-centered and K-centered lattice Wannier coordinates respectively. Each of these three dimensional vector degrees of freedom transforms according to the $\Gamma_{15}$ irrep of the point symmetry group $O_h$. Below, we organize the terms in the expansion of $H_{eff}$ into those acting exclusively in the Nb-centered subspace and the K-centered subspace, and those coupling the two.

In the Nb-centered subspace, we consider quadratic interactions up to third nearest neighbour with the most general form allowed by the space group symmetry:

$$\sum_i (A|\vec{\xi}_i|^2 + \sum_i \sum_{\vec{d}=nn1} [a_L(\vec{\xi}_i \cdot \hat{d})(\vec{\xi}_i(\vec{d}) \cdot \hat{d}) + a_T(\vec{\xi}_i \cdot \vec{\xi}_i(\vec{d}) - (\vec{\xi}_i \cdot \hat{d})(\vec{\xi}_i(\vec{d}) \cdot \hat{d}))]$$

$$+ \sum_i \sum_{\vec{d}=nn2} [b_L(\vec{\xi}_i \cdot \hat{d})(\vec{\xi}_i(\vec{d}) \cdot \hat{d}) + b_{T1}(\vec{\xi}_i \cdot \hat{d}_1)(\vec{\xi}_i(\vec{d}) \cdot \hat{d}_1) + b_{T2}(\vec{\xi}_i \cdot \hat{d}_2)(\vec{\xi}_i(\vec{d}) \cdot \hat{d}_2)]$$

$$+ \sum_i \sum_{\vec{d}=nn3} [c_L(\vec{\xi}_i \cdot \hat{d})(\vec{\xi}_i(\vec{d}) \cdot \hat{d}) + c_T(\vec{\xi}_i \cdot \vec{\xi}_i(\vec{d}) - (\vec{\xi}_i \cdot \hat{d})(\vec{\xi}_i(\vec{d}) \cdot \hat{d}))], \quad (2)$$

where $\vec{\xi}_i(\vec{d})$ denotes the LWF coordinate at a neighbor of site i at separation $\vec{d}$. Second neighbor sites are located along the diagonal of a square of side $a_0$. The unit vector $\hat{d}_1$ is in the plane of the square perpendicular to this diagonal, while $\hat{d}_2$ is perpendicular to the plane of the square. Beyond third neighbor we use a dipole-dipole form parametrized by the mode effective charge $\overline{Z}^*$ and the electronic dielectric constant $\epsilon_\infty$:

$$\sum_i \sum_{\vec{d}} \frac{(\overline{Z}^* e a_0)^2}{\epsilon_\infty} \frac{(\vec{\xi}_i \cdot \vec{\xi}_i(\vec{d}) - 3(\vec{\xi}_i \cdot \hat{d})(\vec{\xi}_i(\vec{d}) \cdot \hat{d}))}{|\vec{d}|^3}. \tag{3}$$

An important simplifying approximation is that the onsite potential, depending on the value of $\vec{\xi}_i$ at a single i, is the only set of terms including anharmonic interactions acting exclusively in the Nb-centered subspace. For $KNbO_3$, anharmonic terms only up to fourth order in $|\vec{\xi}_i|$ are necessary:

$$\sum_i (B|\vec{\xi}_i|^4 + C(\xi_{ix}^4 + \xi_{iy}^4 + \xi_{iz}^4)). \tag{4}$$

In the Nb-centered subspace, the parameters to be determined from first principles are $A, a_L, a_T, b_L, b_{T1}, b_{T2}, c_L, c_T, B, C$, and $\overline{Z}^*$. This determination relies on the explicit correspondence between the lattice Wannier coordinate $\{\vec{\xi}_i\}$ and the ionic displacements obtained above. This correspondence allows us to relate the first-principles total energies and the derivatives of total energies to various terms in $H_{eff}$. The parameters in the quadratic part of $H_{eff}$ are linearly related to linear combinations of elements of the force constant matrices obtained from density functional linear response at high symmetry k-points in the BZ. In Table 4 are given specific relations for modes at various k-points in the BZ including $\Gamma_{15}, X_1, X_5, M_5', M_3', R_{25}'$ and the $\Lambda_3$ modes at $(111)\pi/(6a_0)$ and $(111)\pi/(3a_0)$. The parameter $\overline{Z}^*$ is determined from the eigenvector of unstable $\Gamma_{15}$ and the Born effective charges $\{1.114, 9.664, -7.286, -1.746, -1.746\}$ calculated in Ref. [9], including charge neutrality corrections. Obtaining $\epsilon_\infty=6.63$ directly from DFT linear response [9] and solving the system of linear equations yields values for all the parameters in the quadratic part of $H_{eff}$, listed in Table 5. The resulting normal mode dispersion of $H_{eff}$ is shown in Fig. 2.

The parameters $B$ and $C$ appearing in the onsite anharmonic terms are determined from the total energies of uniformly distorted configurations $(\vec{\xi}_i = \vec{\xi})$, as shown in Fig. 3. The minimum energy configuration has rhombohedral symmetry ($\vec{\xi}$ along (111)). The difference among the energies of uniform distortions with different symmetries ((100), (110), (111)) is a reflection of the cubic anisotropy, which is described quite well by the fourth order terms. The resulting parameters are listed in Table 5.

To account for the effects of changes in lattice parameters at the structural phase transition, we include the lowest order terms in the homogeneous strain and its coupling to the Nb-centered subspace:

**TABLE 4.** Determination of coefficients in the quadratic part of $H_{eff}$. Linear combinations of these coefficients for the modes in the $H_{eff}$ subspace at high symmetry k-points are equated to the corresponding eigenvalues of the projected force constant matrix.

| k-point | Mode eigenvalue at k of the effective Hamiltonian | Value from LR ($eV Å^2$) |
|---|---|---|
| | $z = \overline{Z}^{*2}/\epsilon_\infty = 0.585269$ | |
| $\Gamma_{15}$ | $A + 2(a_L + 2a_T) + 4(b_L + b_{T1} + b_{T2}) + 8(c_L + 2c_T)/3 - 0.465649z$ | -0.0716693 |
| $X_1$ | $A - 2a_L + 4a_T - 4(b_L + b_{T1}) + 4b_{T2} - 8(c_L + 2c_T)/3 + 1.076919z$ | 0.362668 |
| $X_5$ | $A + 2a_L - 4b_{T2} - 8(c_L + 2c_T)/3 - 0.538455z$ | -0.0380006 |
| $M_5'$ | $A - 2a_L - 4b_{T2} + 8(c_L + 2c_T)/3 + 0.297567z$ | 0.259995 |
| $M_3'$ | $A + 2a_L - 4a_T - 4(b_L + b_{T1}) + 4b_{T2} + 8(c_L + 2c_T)/3 - 0.595134z$ | -0.0214604 |
| $R_{25}'$ | $A - 2a_L - 4a_T + 4(b_L + b_{T1}) + 4b_{T2} - 8(c_L + 2c_T)/3$ | 0.216625 |
| $(111)\frac{\pi}{3a}$ | $A + a_L + 2a_T + b_L + b_{T1} + b_{T2} + (c_L + 2c_T)/3$ $-(-1.5b_L + 1.5b_{T1} - c_L + c_T + 0.32784z)$ | 0.0307825 |
| $(111)\frac{\pi}{6a}$ | $A - a_L - 2a_T + b_L + b_{T1} + b_{T2} - (c_L + 2c_T)/3$ $-(-1.5b_L + 1.5b_{T1} + c_L - c_T + 0.0898486z)$ | 0.151074 |

**TABLE 5.** Parameters in the effective Hamiltonian (units of eV per unit cell, except for $\overline{Z}^{*2}/\epsilon_\infty$, which is dimensionless).

| | | | | | |
|---|---|---|---|---|---|
| $A$ | 2.425 | $a_L$ | 0.9693 | $C_{11}$ | 176.0 |
| $B$ | 4.107 | $a_T$ | -0.6151 | $C_{12}$ | 26.31 |
| $C$ | 21.71 | $b_L$ | 0.4396 | $C_{44}$ | 147.2 |
| | | $b_{T1}$ | -0.07914 | $g_0$ | 3.393 |
| | | $b_{T2}$ | -0.1254 | $g_1$ | -71.40 |
| $\overline{Z}^{*2}/\epsilon_\infty$ | 0.585269 | $c_L$ | -0.2923 | $g_2$ | 4.764 |
| | | $c_T$ | -0.07209 | | |

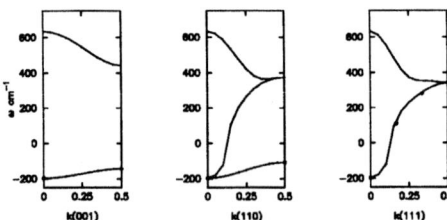

**FIGURE 2.** Normal mode dispersion of $H_{eff}$. Solid circles are the first principles mode eigenvalues used in the fitting.

$$\frac{N}{2}C_{11}\sum_\alpha e_{\alpha\alpha}^2 + \frac{N}{2}C_{12}\sum_{\alpha\neq\beta} e_{\alpha\alpha}e_{\beta\beta} + \frac{N}{4}C_{44}\sum_{\alpha\neq\beta} e_{\alpha\beta}^2 + Nf\sum_\alpha e_{\alpha\alpha}$$
$$+ g_0(\sum_\alpha e_{\alpha\alpha})\sum_i |\vec{\xi}_i|^2 + g_1\sum_\alpha (e_{\alpha\alpha}\sum_i \xi_{i\alpha}^2) + g_2\sum_{\alpha<\beta} e_{\alpha\beta}\sum_i \xi_{i\alpha}\xi_{i\beta}, \qquad (5)$$

where $e_{\alpha\beta}$ is a component of the strain tensor, $C_{11}, C_{12}, C_{44}$ are the elastic constants, and the parameters $g_0, g_1, g_2$ give the strength of coupling of strain with the local polar distortions $\xi_{i\alpha}$. All these parameters are determined from the single unit cell total-energy calculations for three independent types of unit cell distortions (isotropic, uniaxial and rhombohedral shear), with magnitudes of up to 2% of the experimental lattice constants. The total energies of these strained unit cells with no local polar distortion, shown in Fig. 4, give the three elastic constants $C_{11}, C_{12}$ and $C_{44}$. The linear term $f$ is set to zero in the simulations, to compensate for the first-principles underestimate of the lattice constant. For each of these unit-cell-strain types, we also compute the second derivative of energy with respect to uniform local polar distortions $\vec{\xi}_i = \vec{\xi}$, as shown in Fig. 4. These results yield the coupling parameters shown in Table 5.

In Ref. [2], the form of the terms in $H_{eff}$ acting in the K-centered subspace and the relation of the corresponding parameters to the elastic constants and strain-coupling constants included in Table 5 have been given explicitly. Because this "inhomogeneous strain" has been shown to have negligible effect on the phase transitions in the relatively small simulation cells used in Refs. [4,5], these terms were not included in the simulations described in those papers and will not be further discussed here.

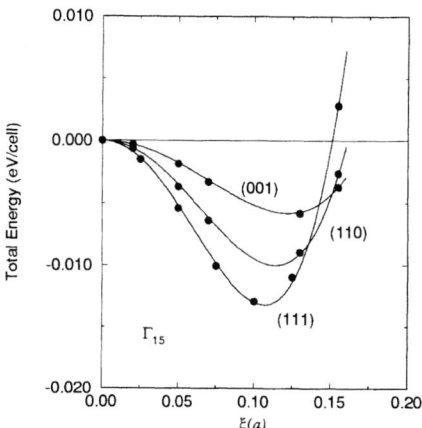

**FIGURE 3.** Total energies for uniformly distorted configurations ($\vec{\xi}_i = \vec{\xi}$) along directions (001), (110) and (111). Solid lines are the fit obtained with the parameters $B$ and $C$ in $H_{eff}$.

**TABLE 6.** Expansion parameters of the Hamiltonian for KNbO$_3$, following the conventions and table format of Ref. [3]. Energies are in Hartrees.

| On-site | $\kappa_2$ | 0.001547 | $\alpha$ | 0.000286 | $\gamma$ | −0.000481 |
|---|---|---|---|---|---|---|
| Intersite | $j_1$ | −0.000785 | $j_2$ | 0.001237 | | |
| | $j_3$ | 0.000230 | $j_4$ | −0.000160 | $j_5$ | 0.000331 |
| | $j_6$ | 0.000063 | $j_7$ | 0.000155 | | |
| Elastic | $B_{11}$ | 6.468 | $B_{12}$ | 0.967 | $B_{44}$ | 1.352 |
| Coupling | $B_{1xx}$ | −0.0868 | $B_{1yy}$ | 0.00433 | $B_{4yz}$ | 0.00152 |
| Dipole | $Z^*$ | 1.96986 | $\epsilon_\infty$ | 6.63 | | |

For the convenience of the reader, in Table 6 we give the first-principles effective Hamiltonian parameters for KNbO$_3$ following the conventions of Ref. [3]. The two forms are completely equivalent.

Having fully determined $H_{eff}$, we now explore the low energy surface of the model to confirm that it gives a correct ground state when compared with the real crystal. Since the anharmonic terms occur only in the Nb-centered subspace and are local (the anharmonicity is wavevector independent), it is easy to determine the ground state from the quadratic order terms. The lowest energy structure is obtained by freezing in the most unstable mode, which here is $\Gamma_{15}$. We consider changes in energy as this mode is frozen in with polarization along the (001), (110) and (111) directions. In Fig. 3, it can be seen that the rhombohedral state ((111)-

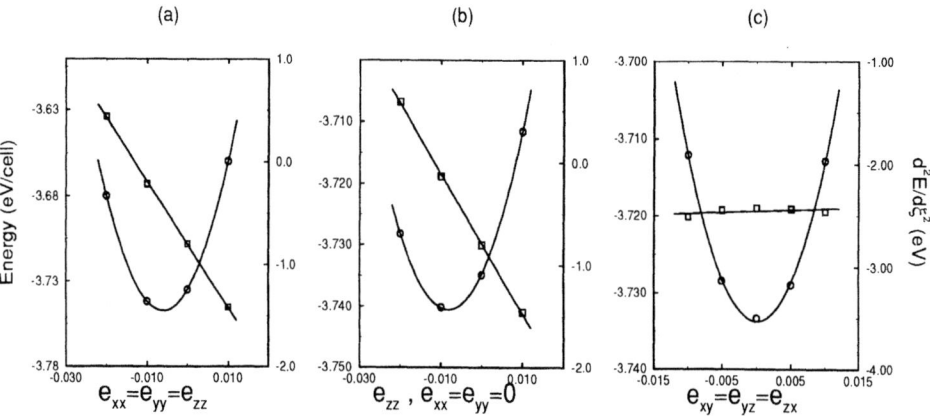

**FIGURE 4.** Energetics of the homogeneous strain ( (a) isotropic, (b) uniaxial and (c) shear) and its coupling to the uniform polar distortions. Circles are the total energies for the strained unit cell configurations with no polar distortions. Solid lines going through the circles are the fits obtained with the elastic constants $C_{11}$, $C_{12}$ and $C_{44}$. Squares correspond to the second derivative of the total energies with respect to uniform polar distortions for the strained unit cells. Solid lines going through the squares are the fits obtained with the coupling parameters $g_0$, $g_1$ and $g_2$.

distortion) has the lowest energy. If the unit cell is allowed to relax as the mode is frozen in, by minimizing over the homogeneous strain, we find an overall increase in distortion energy. While the tetragonal state is shifted the most, the ground state is still rhombohedral, as can be seen in Fig. 5. This is consistent both with previous first principles calculations [9,16] and experimental results [17].

## CONCLUSION

In conclusion, we have applied the method of lattice Wannier functions to construct an effective Hamiltonian for the ferroelectric phase transition in $KNbO_3$ completely from first principles. The detailed description provided here is intended to allow for the use of this effective Hamiltonian in classical Monte Carlo, classical molecular dynamics and quantum mechanical simulations, as well as to provide guidance for the construction of effective Hamiltonians for related systems.

## ACKNOWLEDGEMENTS

We are grateful for useful discussions with Christopher LaSota and Xuewen Wan. Work at William and Mary was supported by Office of Naval Research grant N00014-97-1-0049. Work at Yale was supported by ONR grant N00014-97-1-0047 and the Alfred P. Sloan Foundation. Computations were carried out at the Cornell Theory Center.

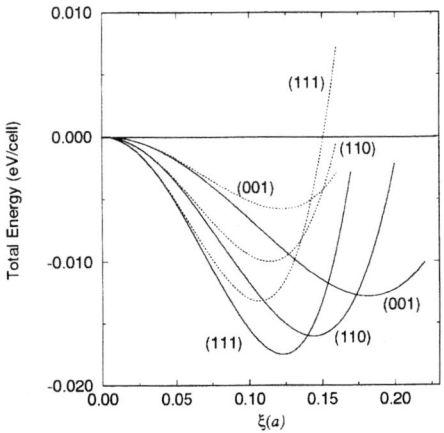

**FIGURE 5.** Model energetics of the uniform polar distortions along $(100), (110)$ and $(111)$. Dotted lines correspond to the polar distortions with the unstrained cubic unit cell, and solid lines to the distortions with unit cell allowed to relax with respect to homogeneous strain.

# REFERENCES

1. M. E. Lines and A. M. Glass, *Principles and Applications of Ferroelectrics and Related Materials* (Oxford, 1977), Chap. 8.
2. U. V. Waghmare and K. M. Rabe, Phys. Rev. **B55**, 6161 (1997).
3. W. Zhong, D. Vanderbilt, and K. M. Rabe, Phys. Rev. Lett. **73**, 1861 (1994); Phys. Rev. B **52**, 6301 (1995).
4. H. Krakauer, R. Yu, C.-Z. Wang, and C. LaSota, *Ferroelectrics*, in press.
5. H. Krakauer, R. Yu, C.-Z. Wang, K. M. Rabe and U. V. Waghmare, cond-mat/9710088.
6. U. V. Waghmare and K. M. Rabe, Ferroelectrics **194**, 135 (1997).
7. W. Zhong and D. Vanderbilt, Phys. Rev. Lett. **74**, 2587 (1995).
8. W. Zhong and D. Vanderbilt, cond-mat/9702198.
9. C.-Z. Wang, R. Yu, and H. Krakauer, Phys. Rev. **B54**, 11161 (1996).
10. K. M. Rabe and E. Cockayne, this proceedings.
11. A. Garcia and D. Vanderbilt, cond-mat/9801177.
12. K. M. Rabe and U. V. Waghmare, Phys. Rev. B **52**, 13236 (1995).
13. M. D. Fontana, G. Metrat, J. L. Servoin and F. Gervais, J. Phys. **C16**, 483 (1984).
14. R. Yu and H. Krakauer, *Phys. Rev. Lett.* **74**, 4067 (1995).
15. D. Singh, Phys. Rev. B **43**, 6388 (1991).
16. R. D. King-Smith and D. Vanderbilt, Phys. Rev. **B49**, 5828 (1994).
17. A. W. Hewat, J. Phys C6, 2559 (1973).

# First-Principles Study of Piezoelectricity in Tetragonal $PbTiO_3$ and $PbZr_{1/2}Ti_{1/2}O_3$

Gotthard Sághi-Szabó and Ronald E. Cohen

*Carnegie Institution of Washington*
*5251 Broad Branch Rd., N.W.*
*Washington, D.C. 20015-1305, USA*

Henry Krakauer

*Department of Physics, College of William and Mary*
*Williamsburg, VA 23187-8795, USA*

**Abstract.** Bulk spontaneous polarization, dynamical charges ($Z^*$) and piezoelectric stress tensor elements of tetragonal $PbZr_{1/2}Ti_{1/2}O_3$ (PZT 50/50) with B-site cations ordered along [001] and [111] directions and $PbTiO_3$ (PT) were determined from ground-state Berry's phase calculations using the all-electron Linearized Augmented Plane Wave (LAPW) method. Extensive internal strain contributions to the piezoelectric response in PT are due to large transverse effective charges and sensitivity of atomic positions to lattice strains. Theoretical proper moduli of PT, $e_{15} = 3.15 C/m^2$, $e_{31} = -0.93 C/m^2$ and $e_{33} = 3.23 C/m^2$, agree well with single crystal experimental data. Ordering along the polar [001] direction was found to enhance the $e_{33}$ piezoelectric stress modulus in chemically ordered PZT 50/50. Theoretical proper moduli in this latter material differ significantly from low temperature experimental data, indicating the possibility that the mechanism of piezoelectric response in PZT near the morphotropic phase boundary differs from conventional single crystal piezoelectricity found in PT.

## INTRODUCTION

Polycrystalline ceramics based on lead titanate and lead zirconate titanate $PbZr_{1-x}Ti_xO_3$ (PZT) are the materials of choice in a wide range of applications including actuators, ultrasonic transducers, piezoelectric transformers and acoustic scanners [1]. In contrast to newly discovered relaxor-PT materials with very high electromechanical coupling and low dielectric loss, represented by $PbZn_{1/3}Nb_{2/3}O_3$-$PbTiO_3$ (PZN-PT) and $PbMg_{1/3}Nb_{2/3}O_3$-$PbTiO_3$ (PMN-PT) [2], near morphotropic boundary compositions are necessary in PZT related materials to achieve high piezoelectric coefficients. However, the growth of good quality single crystals has not been successful for a significant part of the phase diagram [3]. Due to the lack of single crystal data, determination of bulk dielectric, piezoelectric and elastic properties usually involves indirect methods [4], or low temperature measurements to freeze out external domain wall and thermal effects [5]. Ab initio determination of piezoelectric properties can also provide us with the missing single crystal data for some of the important ferroelectric materials. In this paper, the piezoelectric response and dynamical charge tensor elements of the technologically important ferroelectric materials PZT 50/50 and PT are computed from first-principles within the density functional theory using the all-electron full-potential LAPW method [6]. This approach has already been successfully applied to modeling of polarization related properties of tetragonal ferroelectric $PbTiO_3$ [7,8], one of the end members of PZT based ceramics and relaxor-PT based PZN-PT, PMN-PT materials. This study is one of the first applications of modern polarization theory using the Berry's phase approach [9,10] to compute piezoelectric response in complex ferroelectrics.

$PbTiO_3$ is one of the simplest ferroelectric oxides, since it has a clearly established single transition with $T_c$=766K from a paraelectric cubic to a ferroelectric tetragonal phase, its electronic structure has been studied extensively [11–14], and is an obvious starting point for understanding piezoelectricity in ferroelectrics. In the more complex A(B'B")$O_3$ type ferroelectric perovskites, B-site chemical order might have an effect on polarization related properties. Long-range chemical ordering is very common in the subclass of Pb(B'B")$O_3$ perovskites, although it has not been observed experimentally in pure PZT [15]. Theoretical results showed, however, that PZT with compositions around the morphotropic phase boundary is likely a multiphase or metastable material, since the energetics of ordering, ferroelectricity, structural relaxations, and exsolution are all on the same low energy scale, comparable with thermal energies at or below synthesis temperatures [16].

## METHOD

### First-Principles Determination of Macroscopic Polarization

The total closed circuit (zero field) macroscopic polarization of a strained sample $\vec{P}^T$ can be expressed as $P_i^T = P_i^s + e_{i\nu}\epsilon_\nu$, where $P_i^s$ is the spontaneous polarization of the unstrained sample, $\epsilon_\nu$ is the strain tensor element, and $e_{i\nu}$ defines the piezoelectric tensor elements in Voigt notation. At low temperatures both PT and PZT 50/50 are tetragonal and belong to crystal class *4mm*. The three independent piezoelectric tensor components are $e_{31}=e_{32}$ and $e_{33}$, which describe the zero field polarization induced along the z axis when the crystal is uniformly strained in the basal $xy$ plane or along the z axis, respectively, and $e_{15}=e_{24}$ which measures the change of polarization perpendicular to the z axis induced by shear strain. This latter component is related to the induced polarization by $P_1 = e_{15}\epsilon_5$ and

$P_2 = e_{15}\epsilon_4$. The total induced polarization along the crystallographic c axis can be expressed as $P_3 = e_{33}\epsilon_3 + e_{31}(\epsilon_1 + \epsilon_2)$, where $\epsilon_1 = (a - a_0)/a_0$, $\epsilon_2 = (b - b_0)/b_0$ and $\epsilon_3 = (c - c_0)/c_0$ are strains along the a, b and c axis respectively, and $a_0$, $b_0$ and $c_0$ are lattice parameters of the unstrained reference structure. The change in total macroscopic polarization, containing both electronic and rigid ionic core contributions, is a well defined bulk property. Therefore the total piezoelectric constant can be calculated from finite differences of polarizations between crystals of different shapes and volumes.

The electronic part of the polarization was determined using the Berry's phase approach [9], a quantum mechanical theorem dealing with a system coupled to a slowly changing environment. One can calculate the polarization difference between two states of the same solid, under the necessary condition that the crystal remains an insulator along the path which transforms the two states into each other through an adiabatic variation of a crystal Hamiltonian parameter ($\lambda$). Expressing electronic polarization in terms of a dynamic current [17] instead of static charge separation provides well defined bulk quantities at the adiabatic limit. The magnitude of the electronic polarization of a system in state ($\lambda$) is defined only modulo $eR/\Omega$, where $\mathbf{R}$ is the shortest real-space lattice vector and $\Omega$ is the volume of the cell. In practice the $eR/\Omega$ factor can be eliminated by careful inspection, if the changes in polarization are such that $|\Delta \mathbf{P}| << |e\mathbf{R}/\Omega|$. The electronic polarization can be expressed as

$$P^{el}(\lambda) = -\frac{2e}{(2\pi)^3} \int_{BZ} d\mathbf{k} \frac{\partial}{\partial \mathbf{k}'} \phi^{(\lambda)}(\mathbf{k}, \mathbf{k}')\bigg|_{\mathbf{k}'=\mathbf{k}}, \quad (1)$$

where the integration domain is the reciprocal unit cell of the solid in state $\lambda$, and $\phi^{(\lambda)}$ are the geometric quantum phases, defined as phases of overlap-matrix determinants constructed from periodic parts of occupied valence Bloch states $v_n^{(\lambda)}(\mathbf{k})$ evaluated on a dense mesh of neighboring $\mathbf{k}$ points as $\phi^{(\lambda)}(\mathbf{k}, \mathbf{k}') = \mathrm{Im}\{\ln[\det\langle v_m^{(\lambda)}(\mathbf{k})| v_n^{(\lambda)}(\mathbf{k}')\rangle]\}$. The electronic polarization difference between two crystal states can be expressed as $P^{el} = P^{el}(\lambda_2) - P^{el}(\lambda_1)$. Common origins to determine electronic and core parts were arbitrarily assigned along the crystallographic axes. The individual terms in the sum do depend on the choice; however, the final results are independent of origin.

The elements of the macroscopic piezoelectric tensor can be further separated into two parts: a clamped-ion or homogeneous strain contribution evaluated at vanishing internal strain $u$ [18], and a term that is due to an internal microscopic strain, i.e. the relative displacements of differently charged sublattices:

$$e_{i\nu} = \frac{\partial P_i^T}{\partial \epsilon_\nu}\bigg|_u + \sum_k \frac{\partial P_i^T}{\partial u_{k,i}}\bigg|_\epsilon \frac{\partial u_{k,i}}{\partial \epsilon_\nu}, \quad (2)$$

where $P_i^T$ is the total induced polarization along the *ith* axis of the unit cell.

Eq. (2) can be rewritten in terms of the clamped-ion part and the diagonal (in PbTiO$_3$) elements of the Born (transverse) effective charge tensor as

$$e_{i\nu} = e_{i\nu}^{(0)} + \sum_k \frac{ea_i}{\Omega} Z_{k,ii}^* \frac{\partial u_{k,i}}{\partial \epsilon_\nu}, \qquad (3)$$

where $\Omega$ is the volume, $a_i$ is the lattice parameter, the clamped-ion term $e^{(0)}$ is the first term of Eq. (2), and is equal to the sum of rigid core $e^{(0),core}$ and valence electronic $e^{(0),el}$ contributions, and subscript $k$ corresponds to the atomic sublattices. $Z^*$ is the transverse (Born) effective charge:

$$Z_{k,i\nu}^* = Z_k^{core} + Z_{k,i\nu}^{*,el} = \frac{\Omega}{ea_i} \frac{\partial P_i}{\partial u_{k,\nu}}\bigg|_\epsilon. \qquad (4)$$

*Proper* macroscopic polarization changes ($P_i^P$) that do not include terms from the rotation or dilation of the spontaneous polarization ($P_i^s$) of a ferroelectric (and are apparently not observable [19]) are given by $P_i^P = P_i^T - \sum_j \left(\epsilon_{ij} P_j^s - \epsilon_{jj} P_i^s\right)$ [20]. Proper piezoelectric constants $e_{i\nu}^P$ can be expressed as

$$e_{31}^P = \frac{\partial P_3^T}{\partial \epsilon_1} + P_3^s, \ e_{15}^P = \frac{\partial P_1^T}{\partial \epsilon_5} - P_3^s, \qquad (5)$$

and $e_{33}^P = e_{33}^T$, that is the improper part of $e_{33}^T$ is zero. The difference between proper and total polarizations is due only to the homogeneous part, which can be expressed for $e_{31}$ ($e_{31}^{P,hom}$) as

$$e_{31}^{P,hom} = e_{31}^{hom} + P_3^s = \frac{\partial P_3^{el,T}}{\partial \epsilon_1} + P_3^{el,s}. \qquad (6)$$

One can derive a similar expression for $e_{15}^{P,hom}$. The homogeneous part appears as a pure electronic term only in the expression for the *proper* piezoelectric modulus. It is evident that the proper and total piezoelectric constants differ only in materials with non-zero polarization in the unstrained crystal (i.e. pyroelectrics).

The first term in Eq. (3) can be evaluated from polarization differences as a function of strain, with the internal parameters kept fixed at their values corresponding to zero strain. The second term, which arises from internal microscopic relaxation, [21,22] can be calculated after determining the elements of the dynamical transverse charge tensors and variations of internal coordinates $u_i$ as a function of strain. Transverse charges in general are the mixed second derivatives of a suitable thermodynamic potential with respect to atomic displacements and electric field. They measure the change in polarization induced by unit displacement of a given atom at zero electric field to linear order. In a polar insulator transverse charges indicate the extent of polarization changes induced by relative sublattice displacements. While many ionic oxides have Born effective charges close to their static value [23], ferroelectric perovskites display anomalously large dynamical charges [24,25,7].

**TABLE 1.** Structural parameters of tetragonal PbTiO$_3$. GGA parameters were obtained from constant volume optimization. Internal coordinates along the z direction ($u$) are given in terms of the c lattice constant.

|  | GGA [a] | LDA [b] | LDA [c] | Experiment [d] |
|---|---|---|---|---|
| a(a.u.) | 7.356 | 7.247 | 7.380 | 7.373 |
| c/a | 1.073 | 1.122 | 1.063 | 1.065 |
| $u_{Pb}$ | 0.000 | 0.000 | 0.000 | 0.000 |
| $u_{Ti}$ | 0.530 | 0.542 | 0.549 | 0.538 |
| $u_{O_1,O_2}$ | 0.610 | 0.634 | 0.630 | 0.612 |
| $u_{O_3}$ | 0.105 | 0.134 | 0.125 | 0.117 |

[a] experimental zero pressure volume Ref [40]
[b] optimized c/a at the experimental volume, Local Density Approximation (LDA), all-electron basis set.
[c] from Ref [14] (experimental volume and c/a, LDA, ultrasoft pseudopotentials with plane-wave basis).
[d] Ref [36], room temperature data.

## Computational Method

Total energy calculations were performed within the General Gradient Approximation (GGA) using the full-potential *ab initio* LAPW method with Local Orbital (LO) extension [6]. The Perdew-Burke-Ernzerhof 1996 (PBE96) [26] exchange-correlation parameterization was used in the calculations. The value of RK$_{max}$ was set to 8.3, LAPW sphere radii of 2.0, 1.7 and 1.6 a.u. were used for Pb, Ti and O, respectively. Pb *5d, 6s, 6p*, Ti *3s, 3p, 3d, 4s* and O *2s* and *2p* orbitals were treated as valence bands. Atomic core states were calculated relativistically, while valence states were treated semi-relativistically. The special points method [27-30] was applied for Brillouin-zone samplings with a 6 x 6 x 4 and 4 x 4 x 4 **k**-point mesh for the [001] and the [111] ordered structures, respectively. The **k**-space integrations in the Berry's phase calculations were made on a uniform 4 x 4 x 15 **k**-point mesh. Results were checked for convergence with respect to the number of **k**-points and the plane wave cutoff energy.

## RESULTS

The tetragonal structure of PbTiO$_3$ is completely defined by the *a* lattice constant, the *c/a* ratio and by three internal coordinates $u_i$. Table 1 contains the optimized internal parameters in the five atom unit cell at experimental volume and optimized *c/a* together with other theoretical and experimental results. Full, unconstrained optimization of structural parameters using the GGA resulted in an unreasonably large value of *c/a*=1.09 in PT. Therefore the volume of the PbTiO$_3$ unit cell was kept at its experimental value, and an optimum set of *c/a* and $u_i$ parameters which minimize the total energy was determined under this constraint.

**TABLE 2.** Structural parameters and $Z^*_{33}$ values of tetragonal PZT 50/50. Internal coordinates (u) are given in terms of the lattice constants of the P4mm unit cell.

|  | PZT P4mm | | | | PZT I4mm | | | |
| --- | --- | --- | --- | --- | --- | --- | --- | --- |
| Atom | x | y | z | $Z^*$ | x | y | z | $Z^*$ |
| $u_{Pb}$ | 0 | 0 | 0 | 3.46 | 0 | 0 | 0 | 3.23 |
| $u_{Pb}$ | 0 | 0 | 0.5300 | 2.83 | | | | |
| $u_{Zr}$ | 0.5000 | 0.5000 | 0.2435 | 6.06 | 0.5000 | 0.5000 | 0.2268 | 5.96 |
| $u_{Ti}$ | 0.5000 | 0.5000 | 0.7410 | 5.35 | 0.5000 | 0.5000 | 0.2255 | 5.69 |
| $u_O$ | 0.5000 | 0.5000 | 0.9555 | -4.44 | 0.5000 | 0.5000 | 0.9452 | -4.55 |
| $u_O$ | 0.5000 | 0.5000 | 0.4813 | -4.79 | 0.5000 | 0.5000 | 0.4613 | -4.95 |
| $u_O$ | 0.5000 | 0 | 0.1938 | -2.33 | 0.5000 | 0.0142 | 0.1922 | -2.16 |
| $u_O$ | 0.5000 | 0 | 0.7152 | -1.91 | | | | |

**TABLE 3.** Born effective charges of PbTiO$_3$.

| Atom | $Z^*_{xx}$ | $Z^*_{yy}$ | $Z^*_{zz}$ |
| --- | --- | --- | --- |
| Pb | 3.74/*3.90* [b] | 3.74 | 3.52 |
| Ti | 6.20/*7.06* [b] | 6.20 | 5.18 |
| O$_1$ | -2.61$_\perp$[a]/*-2.56* [b] | -5.18$_\|$[a]/*-5.83* [b] | -2.16 |
| O$_3$ | -2.15 | -2.15 | -4.38 |

[a] O$_1$ is on the xz face of the unit cell, $\perp$ marks the direction perpendicular to Ti-O bond in the xy plane, $\|$ indicates atomic displacement along the Ti-O bond
[b] Ref [24], LDA pseudopotential results at experimental structural parameters

The smallest ordered PZT unit cells, with B site cations ordered along the [001] or [111] directions, contain two formula units. LAPW total energies of the [001] ordered PZT structure were determined within *P4mm* symmetry, and the [111] tetragonal structure within the *I4mm* symmetry at the experimentally determined lattice parameters of the microcrystalline PZT 50/50 phase. Internal parameters were fully optimized (Table 2) using analytical forces [37]. These configurations served as reference states in further polarization calculations.

Born effective charges were obtained from changes of macroscopic polarization induced by small displacements of atomic sublattices (Tables 2 and 3). Ordering and chemical environment were found to have little effect on the magnitude of Born effective charges in these structures. Similarly to other perovskite type ferroelectric oxides, these charges were found significantly larger than their nominal ionic values.

The spontaneous polarization was computed in PT at the optimized external and internal structural parameters. Our theoretical value of 0.88 C/m$^2$ agrees well with the experimental value [35] of 0.75 C/m$^2$ (295 K). The agreement is even better if we take into account that experiments generally give reduced results due to cracking and charge leaking [36], and that spontaneous polarization usually does not saturate at room temperature in these oxides. In the [111] ordered PZT structure a theoretical spontaneous polarization value of 0.70 $C/m^2$ was computed, compared to a value of 0.74 $C/m^2$ in the P4mm structure. Theoretical polarization values of ordered PZT are in good agreement with low temperature experimental data [38,39] and are smaller than computed values in single crystal PT.

Piezoelectric constants were determined by using both the direct and the indirect Eq. (3) methods. In the direct approach, absolute macroscopic polarization values were computed at a reference structure $P^{ref}$ and at several strained structures $P^\epsilon$, with equilibrium internal parameters determined at each strain value. Applied strain values were typically in the ± 1% range. The slope of the $(P^\epsilon - P^{ref})$ vs. strain curve in the linear regime yielded directly the piezoelectric constants. The $\partial u_i/\partial \epsilon_\nu$ values were used together with the corresponding dynamical charge tensor elements to evaluate the individual microscopic strain contributions.

Clamped-ion contributions to the piezoelectric moduli in Eq. (3) were determined from slopes of polarization vs. strain curves. During this set of calculations, internal parameters were kept fixed at their values optimized in the unstrained reference structure. When combined with the effective charges and our displacement-strain derivatives we find no significant difference (less than 0.5%) between moduli calculated by direct and indirect methods, indicating that the linear approximation used to describe the piezoelectric response of PT and PZT 50/50 is valid for the applied magnitude of strains.

Table 4 contains the computed total and proper piezoelectric moduli of PT, and two sets of experimental data of the full piezoelectric stress tensor. Our theoretical value for the total $e_{33}$ modulus of PT is 3.23 $C/m^2$. The clamped ion contribution is -0.88 $C/m^2$, whereas the contribution from internal macroscopic strain is much larger with an opposite sign, 4.11 $C/m^2$. These values and the other theoretical *proper* $e_{15}$ and $e_{31}$ moduli are in good agreement with the single-crystal PT data reported by Li *et al.* [31] using the Brillouin scattering method. Methods used in this work give results for the intrinsic piezoelectric properties of an infinite single-domain perfect material, while values reported by Ikegami *et al.* [32] are based on experimental data obtained on a poled ceramic PT sample with small amounts of dopants. In the latter case, significant extrinsic contributions to piezoelectric constants can be expected in addition to the intrinsic contribution. It is however not uncommon, that extrinsic contributions are of the same magnitude as the intrinsic part [33].

Berlincourt et al. [34] measured the room temperature values of elastic compliances $s_{ij}^E$ and piezoelectric strain constants $d_{ij}$ for a number of poled ceramic PZT compositions using standard resonance methods. In tetragonal PZT, the piezoelectric stress constant $e_{ij}$ is related to the piezoelectric strain constants by

$$e_{33} = 2d_{31}C_{13}^E + d_{33}C_{33}^E, \tag{7}$$

where

$$C_{13}^E = -s_{13}^E/s, C_{33}^E = (s_{11}^E + s_{12}^E)/s, s = s_{33}^E(s_{11}^E + s_{12}^E) - 2(s_{13}^E)^2 \tag{8}$$

Their measurements give the values $C_{13}^E = 5.60 \times 10^{10}$ N/m$^2$, $C_{33}^E = 11.07 \times 10^{10}$ N/m$^2$, and $e_{33} = 27.0$ C/m$^2$ in the ceramic material.

**TABLE 4.** Piezoelectric stress tensor elements (C/m$^2$) of tetragonal PbTiO$_3$ and ordered PZT 50/50.

|  | PbTiO$_3$ | | | PZT 50/50 (P4mm) | PZT 50/50 (I4mm) |
|---|---|---|---|---|---|
|  | $e_{15}$ | $e_{31}$ | $e_{33}$ | $e_{33}$ | $e_{33}$ |
| Homogeneous | 2.87 | -2.60 | -0.88 | -0.65 | -0.65 |
| Proper Homogeneous | 1.99 | -1.72 | -0.88 | -0.65 | -0.65 |
| Internal Strain | 1.16 | 0.79 | 4.11 | 5.46 | 4.25 |
| Total | 4.03 | -1.81 | 3.23 | 4.81 | 3.60 |
| Proper Total | 3.15 | -0.93 | 3.23 | 4.81 | 3.60 |
| Experiment | 3.92[a] | -0.98[a] | 3.35[a] | 27.0[c] |  |
|  | 2.96[b] | 0.46[b] | 6.50 [b] | 11.9[d] |  |
|  | 4.4[e] | 2.1[e] | 5.0 [e] |  |  |
|  | 4.8[e] | -0.67 [e] | 4.1[e] |  |  |

[a] Ref. [31]
[b] Ref. [32]
[c] Ref. [34], RT data
[d] Ref. [5], low T data
[e] Ref. [41], two sets of velocity RMS deviations

Our theoretical values of 3.60 C/m$^2$ (I4mm) and 4.81 C/m$^2$ (P4mm) differ significantly from these values. However, room temperature measurements do include external contributions to the piezoelectric modulus, such as domain wall and thermal defect motions. At very low temperatures these contributions can be eliminated from the measurements of $d_{ij}$ piezoelectric strain coefficients [5]. Using the low temperature data of standard piezoelectric resonance measurements obtained for poled, pure, ceramic PZT 50/50, the calculated value of $e_{33} = 11.9$ C/m$^2$ is still more than twice as big as the theoretical value.

Further investigation of various contributions to the $e_{33}$ modulus of PZT 50/50 reveals several important details about the nature of piezoelectric response in PZT 50/50. Homogeneous contributions (-0.65 C/m$^2$) are independent of ordering and are slightly smaller in magnitude than the contribution found in tetragonal PbTiO$_3$ (-0.88 C/m$^2$). However, ordering along the polarized [001] direction enhances the internal strain part of the $e_{33}$ modulus, resulting in larger piezoelectric response.

# CONCLUSIONS

We have used the Berry's phase approach within the all-electron LAPW+LO formalism to compute polarization related properties of single crystal PT and chemically ordered PZT 50/50 ordered single crystal materials. Chemical B-site ordering along the [001] direction couples better with [001] strain, which results in larger $e_{33}$ piezoelectric stress modulus in the P4mm PZT structure than in the [111] ordered material. Theoretical intrinsic piezoelectric moduli in ordered PZT 50/50 were found to be somewhat larger, but relatively close to the $e_{33}$ modulus of single crystal PbTiO$_3$. This indicates that at the atomic level, no significant difference was found between PZT 50/50 and PT. Spontaneous polarization, dynamical charge and piezoelectric moduli all support this similarity. Based on our computations, PZT with compositions close to the morphotropic boundary should behave very similarly to PT, which lacks the excellent piezoelectric properties of PZT. Since the computed piezoelectric moduli in both ordered PZT phases differ significantly from experimental data, even after experimentally "freezing out" certain extrinsic contributions, we conclude that the large piezoelectric response in PZT 50/50, unlike in PT, cannot be described in simple terms of conventional single crystal piezoelectric response.

## ACKNOWLEDGMENTS

This work was supported by the Office of Naval Research under ONR grants N00014-97-1-0052 (REC) and N00014-97-1-0049 (HK). Computations were performed on the CRAY J90/16-4096 computer at the Geophysical Laboratory, Carnegie Institution of Washington supported by grant NSF EAR-9304624. The authors would like to thank Michel Posternak, Andrea Dal Corso, David Singh, Raffaele Resta ard Richard Martin for helpful discussions.

## REFERENCES

1. Uchino, K., *Piezoelectric Actuators and Ultrasonic Motors*, Boston: Kluwer Academic Publ., 1996.
2. Park, S.-E., and Shrout, T. R., J. Appl. Phys. **82**, 1804 (1997).
3. Clarke,R., and WHatmore, R.W., J. Cryst. Growth, **33**, 29 (1976).
4. Amin, A., Haun, M.J., Badger, B., McKinstry, H.A., and Cross, L.E., Ferroelectrics , **65**, 107 (1985).
5. Zhuang, Z.Q., Haun, M.J., Jang, S.J., and Cross, L.E., in *Proceedings of the 6th IEEE International Symposium on the Applications of Ferroelectrics, Lehigh, PA, Institute of Electrical and Electronics Engineers, New York*, 1986, pp. 394.
6. Singh, D.J., *Planewaves, Pseudopotentials and the LAPW Method*, Boston: Kluwer Academic Publishers, 1994.
7. Sághi-Szabó, G., Cohen, R.E., and Krakauer, H., Ferroelectrics , in press.
8. Sághi-Szabó, G., Cohen, R.E., and Krakauer, H., submitted.
9. King-Smith, R.D., and Vanderbilt, D., Phys. Rev. B **47**, 1651 (1993).
10. Resta, R., Rev. Mod. Phys. **66**, 899 (1994).
11. Cohen, R.E., and Krakauer, H., Phys. Rev. B **42**, 6416 (1990).
12. Cohen, R.E., Nature **358**, 136 (1992).
13. Rabe, K.M., and Waghmare, U.V., J. Phys. Chem. Solids **57**, 1397 (1996).
14. Garcia, A., Vanderbilt, D., Phys. Rev. B **54**, 3817 (1996).
15. Randall, C.A., Bhalla, A.S., Shrout, T.R., and Cross, L.E., Ferroelec. Lett., **11**, 103 (1990).

16. Sághi-Szabó, G., and Cohen, R.E., Mat. Res. Soc. Symp. Proc. **453**, 191 (1997).
17. Resta R.,Ferroelectrics bf 136, 51 (1992).
18. de Gironcoli, S., Baroni, S., and Resta, R., Phys. Rev. Lett. **62**, 2853 (1989).
19. Nelson, D.F., *Electric, Optic and Acoustic Interactions in Dielectrics* (Wiley, New York, 1979).
20. Martin, R. M., Phys. Rev. B **4**, 1607 (1972).
21. Dal Corso, A., Posternak, M., Resta, R., and Baldereschi, A., Phys. Rev. B **50**, 10715 (1994).
22. Posternak, M., Baldereschi, A., Catellani, A., and Resta, R., Phys. Rev. Lett. **64** 1777 (1990); Tagantsev, A. K., *ibid.* **69**, 389 (1992); Baldereschi, A., Posternak, M., and Resta, R., *ibid.* **69**, 390 (1992).
23. Posternak, M., Baldereschi, A., Krakauer, H., and Resta, R., Phys. Rev. B, in press.
24. Zhong, W., King-Smith, R.D., and Vanderbilt, D., Phys. Rev. Lett. bf 72, 3618 (1994).
25. Resta, R., Posternak, M., and Baldereschi, A.,Phys. Rev. Lett. **70**, 1010 (1993).
26. Perdew, J.P., Burke, K., and Ernzerhof, M., Phys. Rev. Lett. **77**, 3865 (1996).
27. Baldereschi, A., Phys. Rev. B **7**, 5212 (1973).
28. Chadi, D.J., and Cohen, M.L., Phys. Rev. B **8**, 5747, (1973).
29. Monkhorst, H.J.,and Pack, J.D., Phys. Rev. B **13**, 5188 (1976).
30. Monkhorst, H.J., and Pack, J.D., Phys. Rev. B **16**, 1748 (1977).
31. Li, Z., Grimsditch, M., Xu, X., and Chan, S.-K., Ferroelectrics, **141**, 313 (1993).
32. Ikegami, S., Ueda, T., and Nagata, T., J. Accoust. Soc. Am. **50**, 1060 (1971).
33. Zhang, Q.M., Wang, H., Kim, N., and Cross, L.E., J. Appl. Phys. **75**, 454 (1993).
34. Berlincourt, D.A., Cmolik, C., and Jaffe, H., Proc. IRE, **48**, 220 (1960).
35. Gavrilyachenko V.G., *et al.*, Sov. Phys. Solid State **12**, 1203 (1970).
36. *Ferroelectrics: Oxides, Landolt-Bornstein Numerical Data and Functional Relationships in Science and Technology, Group III, Vol 28, Pt. a*, Berlin: Springer-Verlag, 1996.
37. Yu, R., Singh, D., and Krakauer, H., Phys. Rev. B **43**, 6411 (1991).
38. Berlincourt, D., and Krueger, H.A., J. Appl. Phys. **30**, 1804 (1959).
39. Tsuzuki, K., Sakata, K., and Wada, M., Ferroelectrics bf 8, 501 (1974).
40. Shirane, G., and Pepinsky, R., Acta Cryst. **9**, 131 (1956).
41. Kalinichev, A.G., Bass, J.D., Sun, B.N., and Payne, D.A., J. Mater. Res. **12**, 2623 (1997).

# Temperature-dependent dielectric response of BaTiO$_3$ from first principles

## Alberto García[*] and David Vanderbilt[†]

[*]*Departamento de Física Aplicada II, Universidad del País Vasco Apdo. 644, 48080 Bilbao, Spain, email: wdpgaara@lg.ehu.es*
[†]*Department of Physics and Astronomy, Rutgers University 136 Frelinghuysen Road, Piscataway, NJ 08854-8019, USA, email: dhv@physics.rutgers.edu*

**Abstract.** Monte Carlo simulations with an effective Hamiltonian parametrized from first principles are performed to study the dielectric response of BaTiO$_3$ as a function of temperature, with particular emphasis on the behavior of the dielectric constant near the transition from the ferroelectric tetragonal phase to the paraelectric cubic phase.

The peculiar dielectric properties of ferroelectric materials stem from the coupling of the electric field to polar distortions of the crystal lattice. In one common scenario, the progressive softening of a lattice vibrational mode in the neighborhood of a phase transition brings about a dramatic rise in the value of the dielectric constant. The resulting very large relative permitivities (in the thousands range) have found important technological applications. From a practical point of view, however, the development of materials with a desired response faces the difficulty of trying to separate experimentally the influences of many effects: composition, structure, domain configuration, etc. There are also open theoretical issues such as whether the relevant phase transitions are indeed associated to the above-mentioned "soft-mode" (and are thus of "displacive" character) or exhibit "order-disorder" characteristics. The use of first-principles calculations can help in both fronts. They allow the study of the response of a system under "controlled" conditions that would be very difficult or impossible to realize in the laboratory. Besides, they provide a microscopic view of the materials which is simply not available experimentally and, unlike simplified models, they are tailored to the detailed chemical composition of the system.

In the last few years much progress has been made in the computational study of ferroelectric materials through the use of effective Hamiltonians which contain the physically relevant degrees of freedom of the structure. The effective Hamiltonians are constructed on the basis of first-principles calculations, and the statistical mechanics of the system is then studied by Monte Carlo simulation. Calculations of the phase transition sequence [1] and ferroelectric domain walls [2] in BaTiO$_3$, and of the ferroelectric transition in PbTiO$_3$ [3], have proved the usefulness of this approach.

In this work we present the first calculations of the temperature-dependent dielectric response of the perovskite BaTiO$_3$ from first principles, with particular emphasis on the behavior of the dielectric tensor in the vicinity of the phase transition from the tetragonal ferroelectric to the cubic paraelectric phase.

The basic ingredient of the Metropolis Monte Carlo algorithm is the generation of a sequence of states which are distributed according to the Boltzmann probability Prob$_j = \exp(-\beta U^j)$, where $U^j$ is the energy of the state $j$ and $\beta = (kT)^{-1}$. A state $j_{n+1}$ is added to the sequence after state $j_n$ on the basis of a transition probability $\pi(n, n+1) = \min\{1, \exp[\beta(U^{j_n} - U^{j_{n+1}})]\}$. In our method the energy $U$ of the system is represented by means of an effective Hamiltonian $H_{\text{eff}}$ which is basically a Taylor expansion of the energy surface around the high-symmetry cubic perovskite structure. $H_{\text{eff}}$ is written in terms of the dynamical variables which are relevant to the low-energy distortions: the amplitudes $\{\mathbf{u}\}$ of the local modes (three degrees of freedom per unit cell) which represent the "soft" transverse optical phonon and are directly related to the polarization $\mathbf{P}$ of the crystal [4] [$\mathbf{P} = (Z^*/V)\sum \mathbf{u}$, where $Z^*$ is the mode effective charge and $V$ is the cell volume]; a set $\{\mathbf{v}\}$ of displacement variables representing the acoustic modes; and the six components of the homogeneous strain $\eta$. The parameters of the energy expansion, including those for the on-site local-mode self-energy, the interaction between local modes (both short-range and dipole-dipole), the elastic energy, and the local mode-elastic coupling, are computed using highly accurate first-principles LDA calculations with Vanderbilt ultrasoft pseudopotentials [5]. More details about the construction of the effective Hamiltonian can be found in Ref. [1]. The extension of the standard Metropolis Monte Carlo algorithm to include the effects of stress $\sigma$ and electric field $\mathbf{E}$ involves replacing the Boltzmann probability factor $\exp(-\beta U^j)$ by $\exp[-\beta(U^j - V_0\sigma_\nu\eta_\nu^j - E_i\mathcal{P}_i^j)]$ in the acceptance criterion for state $j$ (here $\mathcal{P}_i = VP_i$ is the $i$th component of the net dipole moment of the crystal and $V_0$ is the volume for zero strain). For a given temperature, stress, and field, the strain $\eta$ and the mode variables are allowed to fluctuate, their average values determining the strain and net polarization of the system. This extended framework has recently been used to study the piezoelectric response of BaTiO$_3$ as a function of temperature, and to illustrate the influence of electric fields on the phase diagram of this material [6].

Here, we use this approach to compute the dielectric response of the cubic and the tetragonal (ferroelectric) phases of BaTiO$_3$. The tetragonal phase is stable from approximately 278K to 403K and exhibits a spontaneous polarization that we take to be along the $z$ axis. The linear dielectric response coefficients (dielectric tensor) are given, in S.I. units, by $\varepsilon_{ij} = \varepsilon_0(1+\chi_{ij})$, where $\chi_{ij}$ is the dimensionless dielectric susceptibility defined by

$$\chi_{ij} = \frac{1}{\varepsilon_0}\left(\frac{\partial P_i}{\partial E_j}\right)_{\sigma,T} \simeq \frac{1}{V\varepsilon_0}\left(\frac{\partial \mathcal{P}_i}{\partial E_j}\right)_{\sigma,T} = \frac{1}{V\varepsilon_0}\beta\left(\langle\mathcal{P}_i\mathcal{P}_j\rangle - \langle\mathcal{P}_i\rangle\langle\mathcal{P}_j\rangle\right). \tag{1}$$

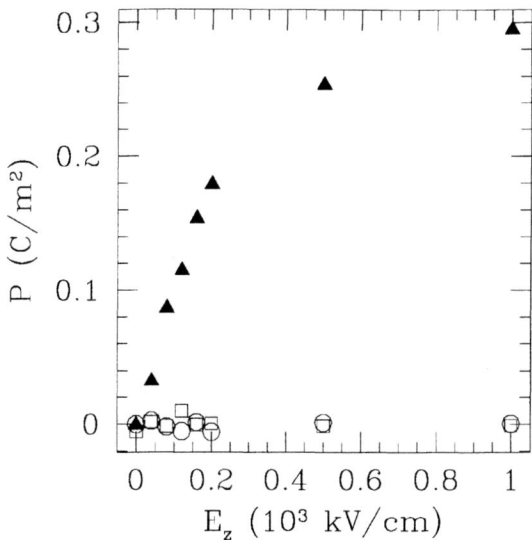

**FIGURE 1.** Average polarization vs. electric field for the cubic phase of BaTiO$_3$ at a rescaled temperature of 500K. Solid triangles and open circles and squares represent the $z$, $x$, and $y$ components of the polarization, respectively.

The approximate equality reflects the neglect of the field derivative of the volume [7]. Here the averages are to be computed using the extended Boltzmann factor defined above [8]. According to these equations, one could either compute the linear dielectric response from direct calculations of the average polarization as a function of electric field ("direct approach"), or from an analysis of the statistical correlation between polarization components ("correlation approach").

As an example of the direct approach, we show in Fig. 1 the results of a series of simulations for the cubic phase in which an electric field of progressively greater magnitude is applied along the $z$ direction. For each field value, the simulation box (a cube with $10 \times 10 \times 10 = 1000$ unit cells) was allowed to equilibrate for $2 \times 10^4$ Monte Carlo sweeps (MCS) [9] and polarization averages were taken over another $2 \times 10^4$ MCS. A fit to the $P_z$ vs. $E_z$ curve in the linear region corresponding to field strengths up to approximately 150 kV/cm can be used to extract the dielectric susceptibility. For higher fields, nonlinear effects are clearly present. It is important to realize that the nonlinearity is not put in explicitly, as the only extra term in the simulation is $-E_i \Delta \mathcal{P}_i$, which is linear in the field. The nonlinearity appears through the terms of higher order in the local mode variables $\{\mathbf{u}\}$ (and their coupling to the strain) in the effective Hamiltonian. A closer analysis of these effects could form the starting point of an investigation of the nonlinear dielectric response in this material.

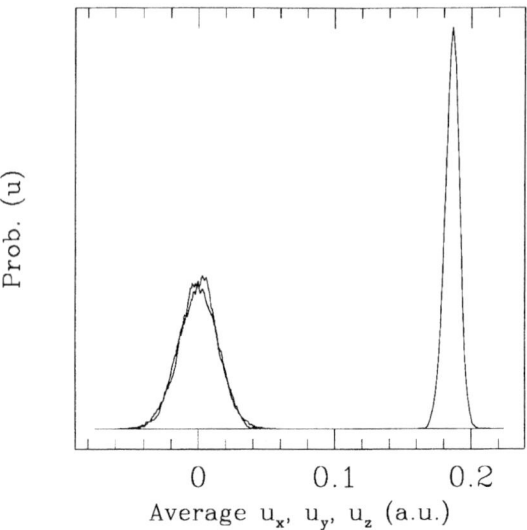

**FIGURE 2.** Probability histograms for the local mode components in the tetragonal phase, obtained from a run with $3 \times 10^5$ MCS and an $L = 14$ simulation box (rescaled temperature: 355 K). The vertical scale is arbitrary.

The computation of linear response coefficients from the correlation approach requires only one simulation at zero field, although relatively long runs (of at least $10^5$ MCS) are needed to obtain good statistics. Also, the quality of the calculated correlations should improve with the size of the simulation box (recall that the relations in Eq. 1 are strictly valid only in the thermodynamic limit). We have therefore performed our calculations using larger boxes, with $12 \times 12 \times 12 = 1728$ ($L = 12$) and $14 \times 14 \times 14 = 2744$ ($L = 14$) unit cells. Figure 2 shows histograms for the averages of the three components of the local mode in the tetragonal phase, computed with an $L = 14$ box and $3 \times 10^5$ MCS. The profiles are quite clean and gaussian-looking. The diagonal components of $\chi$ (which are the relevant ones for the tetragonal and cubic phases) are related to the width of the statistical distribution of the corresponding component of **P**, and thus to the standard deviation of the system-average of the local mode amplitudes. Wider distributions (such as those for the $x$ and $y$ components in Fig. 2) indicate larger values for the corresponding components of the dielectric tensor.

Note that the correlation method provides information about the whole tensor from a single run, which is particularly economical when dealing with lower-symmetry phases. In these cases the use of the direct method becomes more cumbersome. For example, the determination of $\chi_{33}$ and $\chi_{11}$ for the tetragonal phase would involve two separate series of direct calculations, for varying $E_z$ and $E_x$,

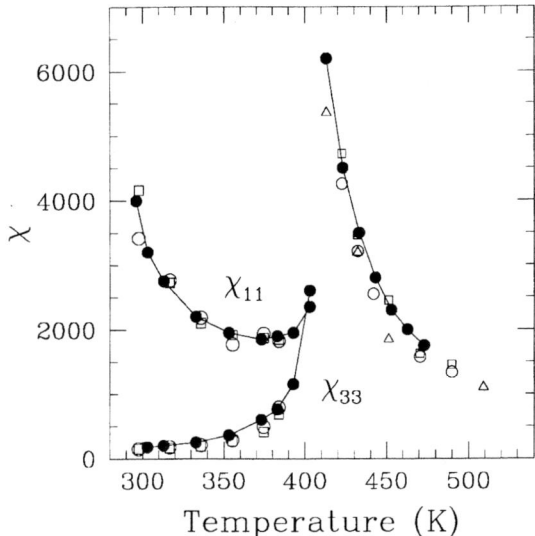

**FIGURE 3.** Constant-stress dielectric susceptibility for the tetragonal and cubic phases of BaTiO$_3$. Solid circles are experimental data from Ref. [12]. Open symbols represent computed values, plotted vs. rescaled temperature (see text). Triangles are from direct P vs. E simulations, and circles and squares are from correlation analysis with $L = 12$ and $L = 14$ boxes, respectively.

respectively. In the latter, extra care would be needed at temperatures close to the transition to ensure that a transverse field does not cause a switch of the spontaneous polarization from the $z$ to the $x$ direction.

Before presenting our results we must discuss an important point regarding the temperature scale. An effective Hamiltonian based on a *finite* Taylor expansion of the energy should not be expected to reproduce perfectly the behavior of the material at relatively high temperatures. In particular (see Ref. [1]) the theoretical transition temperatures are progressively shifted downwards with respect to the true ones [10], the agreement worsening as the temperature increases. This shift might be related to the neglect of higher order terms in the interaction between local modes in $H_{\text{eff}}$, as transition temperatures depend basically on the details of the interaction (as a simple example, recall the Ising model, for which $T_c$ is proportional to the spin-spin coupling $J$) [11]. In order to provide a better comparison of our results to experiment, we have therefore linearly rescaled the theoretical temperatures so that the end points of the range of stability of the tetragonal phase coincide with the experimental $T_c$'s. The rescaling is also extrapolated into the range of stability of the cubic phase. By fixing the points at which phase transitions occur, we are able to focus on the consequences of lattice instability for the dielectric response. The more important, low-energy regions of the energy surface,

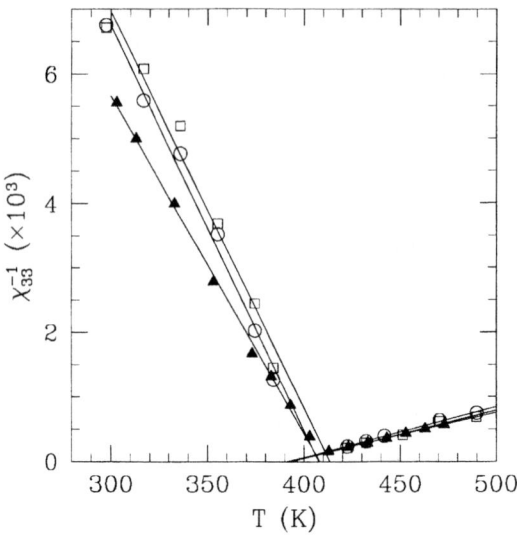

**FIGURE 4.** Fit of the inverse susceptibility to a Curie-Weiss form. Solid triangles are experimental data from Ref. [12]. Open circles and squares represent values computed from correlation analysis with $L = 12$ and $L = 14$ boxes, respectively.

which are correctly parametrized by our $H_{\text{eff}}$, presumably play the most important role in determining this response.

Figure 3 shows the computed constant-stress dielectric susceptibility for the cubic and tetragonal phases, together with recent experimental data [12] for the low-frequency dielectric response of $BaTiO_3$ [13]. The agreement is excellent, with the simulations reproducing in detail all the features of the observed behavior, including the large anisotropy of the dielectric tensor in the tetragonal phase. Our data from the direct simulation of the **P** vs. **E** curves (cubic phase only) and statistical correlation with various box sizes are comparable within the aforementioned limitations of a finite simulation box. Near the transition temperature $T_c$ the pseudo-divergent behavior of $\chi_{33}$ can be approximated to a Curie-Weiss form $\chi_{33}^{-1} = C(T-T_0)$, where $T_0$ is a temperature close but not identical to $T_c$ (Figure 4). The fitted values of the Curie temperature $T_0$ extracted from the tetragonal and cubic phases are very similar, reflecting the fact that the transition is only weakly first-order. The values of the Curie-Weiss constants above and below the transition are not related in a simple way in a first-order transition, but are predicted from mean-field arguments (Landau theory [14]) to satisfy $C_{below} = -2C_{above}$ for a second-order transition. Simulations with a three-dimensional $\phi^4$ model [15] (which could be seen as a simplified form of our effective Hamiltonian) have shown that this mean-field relationship between the $C$ constants is only valid in the limit of a pure displacive phase transition, and that $|C_{below}|/2C_{above}$ increases with the degree of order-disorder character. Our data (and experiment) show that $|C_{below}|/2C_{above} \simeq 4$. Assuming the deviations from mean-field behavior are similar for our $H_{\text{eff}}$ and the $\phi^4$ model, this result would be consistent with the analysis in Ref. [1] pointing to a relatively strong order-disorder character of the cubic-tetragonal transition in $BaTiO_3$.

In conclusion, we have shown how the temperature-dependent dielectric response of a system can be computed from first principles using an effective Hamiltonian and Monte Carlo simulations. As an application, we have presented the first calculations of the dielectric response of $BaTiO_3$ as a function of temperature. The analysis of the behavior of the dielectric susceptibility in the vicinity of the cubic-tetragonal transformation gives evidence for a certain degree of order-disorder character in the transition.

This work was supported in part by the UPV research grant 060.310-EA149/95 and by the ONR Grant N00014-97-1-0048. We thank J.M. Perez-Mato and Karin Rabe for useful comments.

## REFERENCES

1. W. Zhong, D. Vanderbilt and K. M. Rabe, Phys. Rev. Lett. **73**, 1861 (1994); Phys. Rev. B **52**, 6301 (1995).
2. J. Padilla, W. Zhong, and D. Vanderbilt, Phys. Rev. B **53**, R5969 (1996)
3. U. Waghmare and K. Rabe, Phys. Rev. B **55**, 6161 (1997).
4. To relate the net macroscopic polarization to the soft-mode average, thus excluding the contributions from other polar modes, is equivalent to assuming that there are no anharmonic effects in the dynamics of the other modes, so that their average amplitude is zero.
5. D. Vanderbilt, Phys. Rev. B **41**, 7892 (1990).
6. A. Garcia and D. Vanderbilt, preprint:cond-mat/9712312 (1997).
7. The linear change in volume is strictly zero by symmetry in the cubic phase, and in the tetragonal phase for fields perpendicular to the ferroelectric axis. If an electric field is applied along the ferroelectric axis, the longitudinal elongation and the transverse contraction are such that the net change in volume is very nearly zero (the piezoelectric coefficients satisfy $d_{33} \simeq -2d_{31}$).
8. The relationship between $\chi$ and the polarization correlations can be simply obtained by differentiation of the expression $\Lambda\langle X \rangle = \sum_j X^j \text{Prob}_j$, where $\text{Prob}_j$ is the extended Boltzmann factor, $\Lambda$ is the extended partition function, and $X$ is any component of the net polarization.
9. A Monte Carlo sweep is completed after each local variable is considered for a "flip attempt" and each component of the homogeneous strain suffers $2L + 1$ attempted changes, where $L$ is the linear size of the simulation box.
10. For the rhombohedral to orthorhombic to tetragonal to cubic transition sequence, the experimental $T_c$'s are respectively 187K, 278K, and 403K, while the theoretical ones are 197K, 230K, and 295K.

11. Another possible source of the error in the computed values of transition temperatures is the underestimation by the LDA of the equilibrium lattice constant. The simulations are run at a negative effective pressure to compensate for this effect (see Ref. [1]).
12. Z. Li, M. Grimsditch, C. M. Foster, and S.-K. Chan, J. Phys. Chem. Solids **57**, 1433 (1996).
13. At high frequencies the strain is not able to follow the polarization and the experiment measures the *constant strain* dielectric susceptibility.
14. L. Landau and E. M. Lifshitz, *Electrodynamics of Continuous Media* (Pergamon, Oxford, 1960).
15. S. Radescu, I. Etxebarria, and J.M. Perez-Mato, J. Phys. Cond. Matter **7**, 585-95 (1995)

# Temperature-dependent dielectric and piezoelectric response of ferroelectrics from first principles

K. M. Rabe and E. Cockayne

*Department of Applied Physics, Yale University*
*P. O. Box 208284, New Haven, Connecticut, 06520-8284*

**Abstract.** A method for the calculation of the temperature dependence of dielectric and piezoelectric responses, based on the use of a first-principles effective Hamiltonian, is described. Results are presented for the ferroelectric perovskite $PbTiO_3$. While the method includes only the soft-mode contributions to the responses, it is argued to give a good description of the divergences or near-divergences of the response functions near the cubic-tetragonal transition. The expression of the response functions in terms of correlation functions is used to provide a real-space interpretation of the responses which clearly distinguishes between $PbTiO_3$ and the related materials $BaTiO_3$ and $KNbO_3$.

## INTRODUCTION

Systems with large dielectric and piezoelectric responses are of interest both from a fundamental and a technological point of view [1]. Understanding the microsopic origin of a high sensitivity of the polarization of a ferroelectric oxide to applied field, in the case of the dielectric constant, or stress-induced changes in strain, in the case of piezoelectricity, can lead to the optimization of these properties through the appropriate choice of materials. Empirically, large responses are observed in stoichiometric compounds such as $PbTiO_3$ and $BaTiO_3$ for temperatures near the ferroelectric phase transition [2–5]. In solid solutions, such as PZT, enhanced responses are observed over a wide temperature range [6].

In this paper, we discuss the modeling of the piezoelectric and dielectric response of ferroelectric perovskites, specifically $PbTiO_3$ and $BaTiO_3$, though the use of first-principles effective Hamiltonians previously constructed in Refs. [7–10]. The calculations give reasonable agreement with experiment, considering that this modeling gives only the contribution of the soft modes to the responses, and not other contributions such as thermal expansion and the response of other polar modes. However, for temperatures near the ferroelectric phase transition, the soft modes are expected to dominate and thus this approach should capture the essential physics of the large observed responses. In addition, it should correctly describe the trends between different materials. Finally, with this microscopic approach, we are able directly to relate the calculated response functions to the characteristic correlations of local unit cell polarizations, which should yield further insight into the nature of the large responses of interest.

## FIRST-PRINCIPLES EFFECTIVE HAMILTONIANS

The construction of first-principles effective Hamiltonians for $PbTiO_3$ and $BaTiO_3$ has previously been described in detail [8–10]. Briefly, the model consists of one vector per five-atom unit cell which represents the local polarization associated with the ferroelectric distortion. The potential $\mathcal{H}_{eff}(\{\vec{\xi}_i\}, e_{\alpha\beta})$ for these vector degrees of freedom is expanded for the high-symmetry cubic perovskite reference structure, including local anharmonic terms, quadratic intersite interactions which are assumed to be dipolar beyond third neighbors, and lowest order coupling to homogeneous strain.

The extension of this model to include the effects of electric field is accomplished by writing the dependence of the polarization on the model degrees of freedom to lowest order:

$$P(\{\vec{\xi}_i\}, e_{\alpha\beta}) = \sum_i \overline{Z}^* ea_0 \vec{\xi}_i \tag{1}$$

and adding to the potential the coupling term $-\vec{P} \cdot \vec{E}$ where $\vec{E}$ is the macroscopic electric field [11]. The dielectric response is then simply obtained from $\frac{d<P_\alpha>}{dE_\beta}$, while the piezoelectric response can be expressed as $\frac{d<e_{\alpha\beta}>}{dE_\gamma}$, where the brackets are used to denote the thermal expectation value. The effects of macroscopic stress $\sigma_{ij}$ can be included by adding to the potential the coupling term $-\sigma_{ij} \cdot e_{ij}$. With these couplings, the thermodynamic identity $\frac{de_{jk}}{dE_i} = \frac{dP_i}{d\sigma_{jk}}$ can be obtained (the well-known equality of the direct and converse piezoelectric effects [11]). In the rest of this paper, we work at zero external stress and evaluate the piezoelectric response using $\frac{d<e_{\alpha\beta}>}{dE_\gamma}$.

Calculations of the temperature dependent properties of the system are carried out using classical Monte Carlo calculations, as previously described [8,9]. The calculations of the response functions presented here involved 7x7x7 simulation cells, 10,000 Monte Carlo sweeps (MCS) for thermalization, 200,000 MCS for the computation of thermal expectation values, and several runs at each temperature with different random number seeds to estimate statistical error. The calculations of the real-space correlation functions presented here involved 10x10x10 simulation cells, 10,000 Monte Carlo sweeps (MCS) for thermalization, and 100,000 MCS for the computation of thermal expectation values. In order to compute the temperature-dependence of the order parameter and thus identify the cubic-tetragonal transition, a small symmetry-breaking field $E_z$ of magnitude $\frac{20}{\epsilon_\infty}$ kV/cm was applied. The response function results include this nonzero field.

# CORRELATION FUNCTION EXPRESSIONS FOR RESPONSE FUNCTIONS

In the Monte Carlo simulations described above, the response functions are computed using the fact that they can be expressed as correlation functions. This allows the appropriate derivatives to be calculated in one Monte Carlo run.

Specifically, we consider the dielectric tensor:

$$\epsilon_{\alpha\beta} = \epsilon_\infty + 4\pi\chi_{\alpha\beta}$$

where

$$\chi_{\alpha\beta} = \frac{1}{\Omega}\frac{\partial <P_\alpha>}{\partial E_\beta}.$$

Here, $\Omega$ is the volume of the unit cell and $\vec{P}$ is the polarization per unit cell, specifically

$$<P_{i\alpha}> = \frac{\int\{de_{\alpha\beta}\}\{\prod_j d\vec{\xi}_j\}(\frac{Z^*ea_0}{N}\sum_i \xi_{i\alpha})exp(-\beta(H_{eff} - Z^*ea_0 \sum_i \xi_{i\alpha} \cdot \vec{E}))}{\int\{de_{\alpha\beta}\}\{\prod_j d\vec{\xi}_j\}exp(-\beta(H_{eff} - Z^*ea_0 \sum_i \xi_{i\alpha} \cdot \vec{E}))} \quad (2)$$

or

$$<P_{i\alpha}> = (\frac{1}{N\beta})\frac{\partial ln Z(\beta,\vec{E})}{\partial E_\alpha}$$

where we have defined

$$Z(\beta,\vec{E}) = \int\{de_{\alpha\beta}\}\{\prod_j d\vec{\xi}_j\}exp(-\beta(H_{eff} - Z^*ea_0 \sum_i \xi_{i\alpha} \cdot \vec{E})).$$

Differentiating Equation 2 with respect to $E_\alpha$, we readily find the correlation function expression:

$$\chi_{\alpha\beta} = \frac{\beta(Z^*ea_0)^2}{\Omega}(<\sum_i \xi_{i\alpha} \frac{1}{N}\sum_j \xi_{j\beta}> - N<\xi_\alpha><\xi_\beta>) \quad (3)$$

where

$$<\xi_\alpha> = \frac{1}{N}<\sum_i \xi_{i\alpha}>.$$

The piezoelectric tensor can be expressed in terms of correlation functions in an completely analogous way:

$$d_{\gamma\alpha\beta} = \frac{\partial <e_{\alpha\beta}>}{\partial E_\gamma}$$

where

$$<e_{\alpha\beta}> = \frac{\int\{de_{\alpha'\beta'}\}\{\prod_j d\vec{\xi}_j\}e_{\alpha\beta}exp(-\beta(H_{eff} - Z^*ea_0\sum_i \vec{\xi}_i \cdot \vec{E}))}{\int\{de_{\alpha'\beta'}\}\{\prod_j d\vec{\xi}_j\}exp(-\beta(H_{eff} - Z^*ea_0\sum_i \vec{\xi}_i \cdot \vec{E}))}. \quad (4)$$

Differentiating Equation 4 with respect to $E_\gamma$, we find the correlation function expression:

$$d_{\gamma\alpha\beta} = \beta(Z^*ea_0)(<e_{\alpha\beta}\sum_j \xi_{j\gamma}> - N<e_{\alpha\beta}><\xi_\gamma>). \quad (5)$$

When the homogeneous strain appears only up to quadratic order in the effective Hamiltonian, as is the case for the effective Hamiltonians available for PbTiO$_3$ [8,9], BaTiO$_3$ [10] and KNbO$_3$ [12], the piezoelectric tensor can be reexpressed in terms of correlations of local polar distortions. Because the integral over strain is Gaussian, completely equivalent expressions for the thermal expectation values involving $e_{\alpha\beta}$ can be obtained by making the following substitutions (and cyclic permutations thereof):

$$e_{xx} \to -\frac{1}{N}(\frac{(g_0 + \frac{g_1}{3})}{(C_{11} + 2C_{12})}\sum_i |\vec{\xi}_i|^2 + \frac{g_1}{3(C_{11} - C_{12})}\sum_i (2\xi_{ix}^2 - \xi_{iy}^2 - \xi_{iz}^2)) \quad (6)$$

$$e_{xy} \to -\frac{1}{N}\frac{g_2}{C_{44}}\sum_i \xi_{ix}\xi_{iy} \quad (7)$$

For the example of $d_{33}$, this yields

$$d_{33} = -\frac{\beta(Z^*ea_0)}{N}(\frac{(g_0 + \frac{g_1}{3})}{(C_{11} + 2C_{12})}\sum_{i,j}(<|\vec{\xi}_i|^2\xi_{jz}> - <|\vec{\xi}_i|^2><\xi_{jz}>)$$

$$+\frac{g_1}{3(C_{11} - C_{12})}\sum_{i,j}(2(<\xi_{ix}^2\xi_{jz}> - <\xi_{ix}^2><\xi_{jz}>)$$

$$-(<\xi_{iy}^2\xi_{jz}> - <\xi_{iy}^2><\xi_{jz}>)$$

$$-(<\xi_{iz}^2\xi_{jz}> - <\xi_{iz}^2><\xi_{jz}>)) \quad (8)$$

While Equation 5 proves more convenient for the actual computation of $d_{\gamma\alpha\beta}$, it is Equation 8 that naturally leads to a microscopic interpretation of the soft-mode response, as we will discuss further below.

## $\epsilon_{zz}$ AND $d_{33}$ FOR PbTiO$_3$

The results of the Monte Carlo calculations for $\epsilon_{zz}$ are shown in Figure 1. At room temperature, the calculated value is 66, which should be compared to the single-crystal measurement $\epsilon_{zz} = 80$ [13]. The agreement is surprisingly good, considering that the calculation includes only the soft-mode contribution. The room temperature value is already considerably larger than the zero-temperature calculated value of 42. The overall temperature dependence is very similar to that observed experimentally [2,13], with a slow increase with temperature below the transition, a near-divergence both above and below $T_c$ which is cut off by the first-order character of the cubic-tetragonal transition, and an enhanced value of the dielectric response in the high-temperature cubic phase.

The results of the Monte Carlo calculations for $d_{33}$ are shown in Figure 2. At room temperature, the calculated value is 76, which should be compared to the single-crystal measurement $d_{33} = 83.7$ [13]. As for the dielectric response, the agreement is surprisingly good, considering that the calculation includes only the soft-mode contribution. The room temperature value is already considerably larger than the zero-temperature calculated value of 48. The overall temperature dependence is very similar to that observed experimentally [2,4], with a slow increase with temperature below the transition, a near-divergence below $T_c$ which is cut off by the first-order character of the cubic-tetragonal transition, and a drop to zero in the high-temperature cubic phase, as required by symmetry.

These results can be compared with the closely analogous calculation of the piezoelectric coefficients of $BaTiO_3$ (Figure 1 in [14]), noting that for this system as well, remarkably good quantitative agreement with the experimentally measured values is obtained.

# LANDAU THEORY FOR THE DIVERGENT RESPONSES

The near-divergences observed in the results described above can be already understood within a simple Landau theory which describes coupling of a scalar

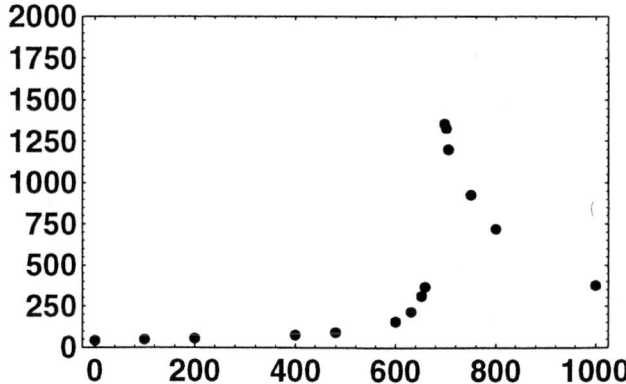

**FIGURE 1.** Dielectric response $\epsilon_{zz}$ as a function of temperature, in K.

polar mode $\xi$ with a one-component strain $e$ and electric field $E$:

$$F(\xi, e; T, E) = A\xi^2 + B\xi^4 - E\xi + ce^2 + ge\xi^2 \qquad (9)$$

where the temperature dependence enters through $A = A_0(T - T_c)$. If we first eliminate $e$, which is accomplished by the substitution $e \to \frac{g}{c}\xi^2$, the following effective fourth-order free energy results:

$$F(\xi, e; T, E) = A\xi^2 + (B - \frac{g^2}{4c})\xi^4 - E\xi \qquad (10)$$

Minimizing with respect to $\xi$ yields the following expressions for the dielectric susceptibility

$$\chi = \frac{d\xi_{min}}{dE} = \frac{-1}{4A} \quad (T < T_c)$$
$$= \frac{1}{2A} \quad (T > T_c) \qquad (11)$$

and piezoelectric coefficient

$$d = \frac{de_{min}}{dE} = -\frac{g}{4Ac}\sqrt{\frac{-A}{2(B - \frac{g^2}{4c})}} = -\frac{g}{c}\xi_{min}\chi \quad (T < T_c)$$
$$= 0 \quad (T > T_c) \qquad (12)$$

As observed in the full simulations, $d$ diverges as $T_c$ is approached from below, while $\chi$ diverges on both the low and high temperature sides, and is larger above $T_c$. In PbTiO$_3$, these divergences are cut off due to the first-order character of the transition, so that, for example, a Curie-Weiss fit to $\chi$ above $T_c$ yields a value of $T_0$ about 80 K below $T_c$.

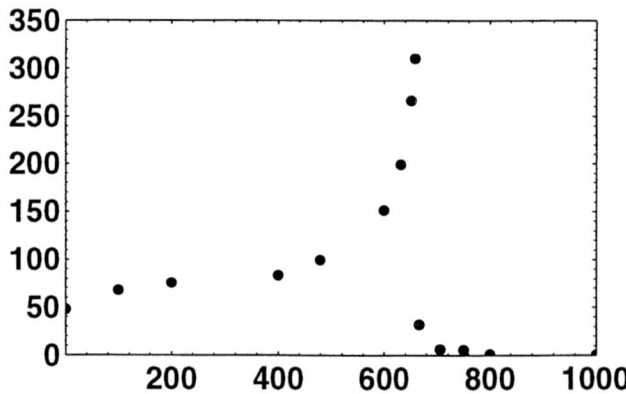

**FIGURE 2.** Piezoelectric response $d_{33}$ as a function of temperature, in K.

# REAL-SPACE DECOMPOSITION OF THE CORRELATION FUNCTIONS

The correlation function expressions for the response functions allow the interpretation of the responses in terms of the characteristic intersite correlations between the local polar distortions. Some of the correlation functions of interest have been previously calculated in effective Hamiltonian studies of $KNbO_3$ [12] and $BaTiO_3$ [15], allowing us to connect to and build upon the insights derived in those investigations to develop a microscopic understanding of the dielectric and piezoelectric responses.

For the dielectric response, the first term inside the parentheses of Equation 3 can be reorganized:

$$\frac{1}{N}\sum_i (<\xi_{i\alpha}\xi_{i\beta}> + \sum_{j,nn} <\xi_{i\alpha}\xi_{j\beta}> + \sum_{j,nnn} <\xi_{i\alpha}\xi_{j\beta}> +...)$$

If correlations become infinite ranged, as at a second order transition, this will scale like N, leading to a divergence in the thermodynamic limit.

For the piezoelectric tensor, we can similarly reorganize the correlation functions which appear in Equation 5, for example:

$$\frac{1}{N}\sum_i (<\xi_{i\alpha}\xi_{i\alpha}\xi_{i\gamma}> + \sum_{j,nn} <\xi_{i\alpha}\xi_{i\alpha}\xi_{j\gamma}> + \sum_{j,nnn} <\xi_{i\alpha}\xi_{i\alpha}\xi_{j\gamma}> +...)$$

The real-space correlation functions appearing in the expressions above are computed using fast fourier transforms of $\vec{\xi}_i$ and $\xi_{i\alpha}\xi_{i\beta}$. For the dielectric susceptibility, for example, the real space expressions can be written in terms of $\chi_{\alpha\beta}(\vec{q})$ as follows (taking $\vec{R}_j = \vec{R}_i + \vec{d}$):

$$\frac{1}{N}\sum_i <\xi_{i\alpha}\xi_{j\beta}> = \frac{1}{N}\sum_i \frac{1}{N^2}\sum_{\vec{q},\vec{q}\,'} <\xi_\alpha(\vec{q})\xi_\beta(\vec{q}\,')> exp(-i\vec{q}\cdot\vec{R}_i - i\vec{q}\,'\cdot(\vec{R}_i+\vec{d}))$$

$$= \frac{1}{N^2}\sum_{\vec{q}} <\xi_\alpha(\vec{q})\xi_\beta(\vec{q})^*> exp(i\vec{q}\cdot\vec{d})$$

It can be checked that this gives the correct relation to $\chi_{\alpha\beta}$ when $<\vec{\xi}>= 0$:

$$\chi_{\alpha\beta} = \frac{\beta(Z^*ea_0)^2}{\Omega}\sum_{\vec{d}}\frac{1}{N}\sum_i <\xi_{i\alpha}\xi_{j\beta}> = \sum_{\vec{d}}\frac{1}{N}\sum_{\vec{q}}\chi_{\alpha\beta}(\vec{q})exp(i\vec{q}\cdot\vec{d}) = \chi_{\alpha\beta}(\vec{q}=0)$$

In Figure 3, we show the real-space decomposition in the $xz$ plane of the high-T dielectric susceptibility correlation function $<\xi_{iz}\xi_{i+\vec{d},z}>$ for $PbTiO_3$ at two temperatures above $T_c$. The correlation is rather isotropic, and the range of the correlations increases as expected as the temperature decreases towards $T_c$. For comparison, the same correlation function for BaTiO$_3$ at two temperatures above the cubic-tetragonal transition temperature is also shown in Figure 3. The highly anisotropic "chain-like" correlations discussed in previous work [12,15] are clearly evident, and the correlation length transverse to the chains increases as the temperature decreases towards T$_c$. Similar behavior is expected for KNbO$_3$. The qualitative difference in the nature of the correlations can be directly attributed to the difference in the dispersion relation of the soft mode along R. The implications of the different character of these correlations for the dielectric behavior of PbTiO$_3$ and BaTiO$_3$ are currently under investigation.

As can be seen from Equation 5, the piezoelectric response depends on third-order two-site correlations of the form $< \xi_{i\alpha}^2 \xi_{j\beta} > - < \xi_{i\alpha}^2 >< \xi_{j\beta} >$. These have not to our knowledge been considered in previous studies. Calculations of these real-space correlations for PbTiO$_3$ and BaTiO$_3$ above and below the transition temperature are currently in progress.

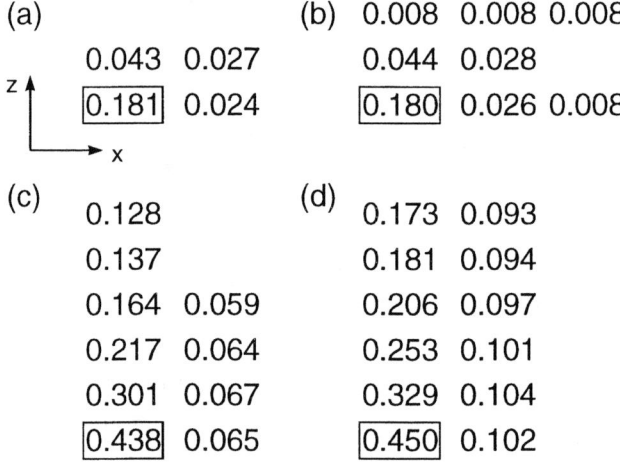

**FIGURE 3.** Real-space decomposition in the $xz$ plane of the high-T dielectric susceptibility correlation function $< \xi_{iz}\xi_{i+\vec{d},z} >$ for (a) PbTiO$_3$ at T = 706 K, (b) PbTiO$_3$ at T = 690 K, (c) BaTiO$_3$ at T = 343 K, and (d) BaTiO$_3$ at T = 308 K. The numerical value in the box indicates the on-site correlation $< \xi_{iz}\xi_{iz} >$, while the numbers above and to the right of the box give the correlations for the corresponding values of $\vec{d}$. All correlations have been normalized by dividing by the square of the ground state value of $\xi_z$ (0.08$^2$ for PbTiO$_3$ and 0.03$^2$ for BaTiO$_3$). Only the largest correlations are shown (above 0.0078 for PbTiO$_3$ and above 0.056 for BaTiO$_3$).

# DISCUSSION AND CONCLUSIONS

As already discussed, the use of an effective Hamiltonian for calculating the dielectric and piezoelectric response involves a number of approximations. The effects of thermal expansion and the neglect of other polar modes cannot be incorporated without adding additional degrees of freedom to the model (unless one is willing to accept semi-empirical input). However, the low-order expansion around the cubic reference structure can be improved, which in particular will yield better model results for the properties of the $PbTiO_3$ tetragonal ground state [18]. This could involve both $\mathcal{H}_{eff}$ and $P$. In the case of the polarization $P$, the next order terms involve including strain dependence and $\xi$ dependence of the effective charges. First-principles calculations [16,17] suggest that the former corrections are rather small compared to the latter. Calculations for these refinements for $PbTiO_3$ are currently in progress.

Since the effective Hamiltonian approach is limited to the description of soft-mode contributions to the responses, we focus on the near-divergent responses where these contributions strongly dominate. Fortunately, this regime is, in any case, the one of greatest interest for understanding the origin of and engineering large responses in a variety of systems characterized by proximity to a lattice instability. Further insight can be gained from the real-space decomposition of the divergent responses based on their expressions in terms of correlation functions.

Finally, these results for pure systems suggest an interpretation for the origin of large responses in solid solutions. For such systems, there is a distribution of local environments, and for a given temperature some subset of lattice degrees of freedom will be marginally stable. This idea has been investigated using first principles results in $Pb_{1-x}Ge_xTe$, and more details are presented in Refs. [19,20].

# ACKNOWLEDGMENTS

We thank R. E. Cohen, A. Garcia, R. M. Martin, D. Vanderbilt, and U. V. Waghmare for helpful discussions. This work was supported by ONR grant N00014-97-1-0047 and the Alfred P. Sloan Foundation, and was in part carried out at the Aspen Center for Physics.

# REFERENCES

1. J. M. Herbert, *Ferroelectric Transducers and Sensors*, New York: Gordon and Breach, 1982.
2. J. P. Remeika and A. M. Glass, Mater. Res. Bull. **5**, 37 (1970).
3. K. Kakuta, T. Tsurumi and O. Fukunaga, Jpn. J. Appl. Phys. **34**, 5341 (1995).
4. Z. Li, M. Grimsditch, C. M. Foster and S.-K. Chan, J. Phys. Chem. Solids **57**, 1433 (1996).
5. C. J. Johnson, Appl. Phys. Lett. **7**, 221 (1965).

6. F. Jona and G. Shirane, *Ferroelectric Crystals*, New York: Macmillan, 1962.
7. K. M. Rabe and U. V. Waghmare, Phys. Rev. **B52**, 13236 (1995).
8. K. M. Rabe and U. V. Waghmare, J. Phys. Chem. Solids **57**, 1397 (1997).
9. U. V. Waghmare and K. M. Rabe, Phys. Rev. **B55**, 6161 (1997).
10. W. Zhong, D. Vanderbilt, and K. M. Rabe, Phys. Rev. Lett. **73**, 1861 (1994); Phys. Rev. B **52**, 6301 (1995).
11. J. F. Nye, *Physical Properties of Crystals*, Oxford: Clarendon, 1964.
12. H. Krakauer, R. Yu, C.-Z. Wang, and C. LaSota, *Ferroelectrics*, in press; H. Krakauer, R. Yu, C.-Z. Wang, K. M. Rabe and U. V. Waghmare, cond-mat/9710088.
13. Z. Li, M. Grimsditch, X. Xu and S.-K. Chan, Ferroelectrics **141**, 313 (1993).
14. A. Garcia and D. Vanderbilt, cond-mat/9801177.
15. J. Padilla, W. Zhong and D. Vanderbilt, mtrl-th/9509005.
16. C.-Z. Wang, R. Yu, and H. Krakauer, Phys. Rev. **B54**, 11161 (1996).
17. Ph. Ghosez, X. Gonze, Ph. Lambin and J.-P. Michenaud, Phys. Rev. **B51**, R6765 (1995).
18. U. V. Waghmare, unpublished.
19. E. Cockayne and K. M. Rabe, cond-mat/9712232.
20. E. Cockayne and K. M. Rabe, these proceedings.

# Configuration Dependence of Physical Properties of a Ferroelectric Solid Solution

Eric Cockayne and Karin M. Rabe

*Department of Applied Physics, Yale University, P.O. Box 208284, New Haven, CT 06520-8284*

**Abstract.** In this article, we motivate the detailed comparison of the physical properties of individual configurations of a ferroelectric solid solution as a means toward developing first principles models for these systems. We compare energies, dielectric constants $\epsilon_\infty$, mode effective charges of local polar distortions, and the zero temperature piezoelectric behavior of several ordered $Pb_3GeTe_4$ supercells. Cluster expansions of these properties show the importance of second-neighbor effects, which can be related to symmetry-preserving relaxation and its effect on the symmetry breaking polar instabilities.

## INTRODUCTION

Ferroelectric solid solutions are of great technological importance. For example, the largest piezoelectric response are found in mixed ferroelectrics, such as $Pb(Zr_{1-x}Ti_x)O_3$. Recently, giant piezoelectricity was discovered in the relaxor ferroelectric systems $Pb(A_{1/3}Nb_{2/3})O_3$–$PbTiO_3$, (A = Zn, Mg). [1] Understanding the physics of ferroelectric solid solutions on the microscopic level would be of great theoretical interest and could also point to new ways to tune their piezoelectric response and other properties.

Ab initio calculations have proved successful in relating properties of stoichiometric ferroelectrics to phenomena on the atomic level [2]. Structural parameters [2,3], dielectric constants, effective charges, phonon dispersion relations [4,5], and polarizations [6] have been calculated from first principles. To predict the behavior at finite temperature, first-principles models have been constructed. [7–11]. Generally, these models are based on a vector representation of the local polar distortions responsible for the ferroelectric phase transition. An electric dipole moment is associated with each local distortion. At long distance, the interaction between local distortions is dipole-dipole; significant corrections appear at short range. Anharmonic terms, elastic constants and strain coupling to local distortion also appear, determining the ground state and the nature of the phase transitions in the models. [12] The models obtained allow for simulation of ferroelectric phase transitions, where the Curie point is generally in good agreement with experiment. The models also allow piezoelectricity and dielectric functions to be computed [13,14]. In the following, we discuss the modeling of ferroelectric and piezoelectric behavior in solid solutions, and lay part of the groundwork for a $Pb_{1-x}Ge_xTe$ model by exploring the configuration dependence of certain quantities essential for constructing the model.

## WHY COMPARE CONFIGURATIONS?

Consider a ferroelectric solid solution. The physical properties of this solid solution are the ensemble averages of the properties of the individual ensemble members. Thus one needs to be able to compute these physical properties for individual ensemble members. A given ensemble member (configuration) can be treated as if it were an actual crystal structure and modeled in the same manner as for a stoichiometric ferroelectric.

While models for a few chosen small unit cell configurations are obtained using this approach, some model for the general configuration is necessary in order to properly take ensemble averages. Given the nature of the model described in the Introduction, we expect that in a model for the general configuration, there will again be polar instabilities, a local basis for these distortions, and interactions between the local distortions. However, now the local polar distortions themselves and their interactions will be site-dependent. In principle, the site-dependence of any quantity can be described by a cluster expansion. The models obtained for the series of individual small unit cell configurations can be used to obtain the cluster expansions for a model that is valid for all configurations.

In modeling stoichiometric ferroelectrics, two kinds of truncation in the models are necessary to prevent an explosion of terms: the order of the expansion of the energy in powers of the local distortion variable and the range of the local interactions. The cluster expansion approach necessitates a third kind of physically motivated truncation: the range of the cluster expansion. Where possible, terms should be found whose configuration dependence is unimportant and then kept constant for all configurations. For those terms where a cluster expansion is necessary, it should be truncated at the shortest possible range. Toward this end, it is thus very important to establish and understand the configuration dependence of those quantities that determine the model parameters.

$Pb_{1-x}Ge_xTe$ provides an excellent prototype system for investigating the form of first-principles models in solid solutions using the approach just outlined. For all compositions above a critical composition $x \approx 0.005$ [15], $Pb_{1-x}Ge_xTe$ undergoes a transition from a cubic phase for $T > T_c(x)$ to a rhombohedral phase at $T < T_c(x)$ [16]. Because the endmembers of the solid solution series have the rocksalt structure with only 2 atoms per primitive cell, a large number of different mixed configurations can be investigated without the need for excessively large supercells. To simplify the problem further, we focus on configuration dependence at a fixed composition, $Pb_{0.75}Ge_{0.25}Te$. This composition was chosen to be near the low Ge concentration regime of physical interest, while allowing for a variety of supercells with only 8 atoms. We present results on the five 8-atom supercells of highest symmetry, shown in Figure 1. In each configuration studied, all Ge ions are translationally equivalent and therefore have the same environment, greatly simplifying the analysis.

## FIRST PRINCIPLES METHODS

In our study of the structural phase transition in the cP8 configuration [10], we give full details of the first principles calculations. Briefly, we performed all calculations using density functional theory within the local density approximation (LDA). Ab initio pseudopotentials were used for the ions and a plane wave basis set with cutoff energy 300 eV was used for the Kohn-Sham eigenfunctions. Total energies were calculated via conjugate gradients optimization using the CASTEP 2.1 package [17]. The CASTEP 2.1 package was modified [18] to do variational linear response [5] calculations of force constants, Born effective charges and dielectric constants. Brillouin zone averages for the cP8 configuration were performed using a $4 \times 4 \times 4$ Monkhorst-Pack set. For the other configurations, the same **k** point grid was used, folded into the corresponding Brillouin zone.

## SYMMETRY PRESERVING RELAXATION

We found two distinct ways in which the total energies of the configurations that we studied could be lowered with respect to the energy of the structures with all atoms fixed on rocksalt positions: symmetry-preserving relaxation and symmetry-breaking polar instabilities. The existence of symmetry-preserving relaxation follows from group theory. Except for the cP8 configuration, all of the

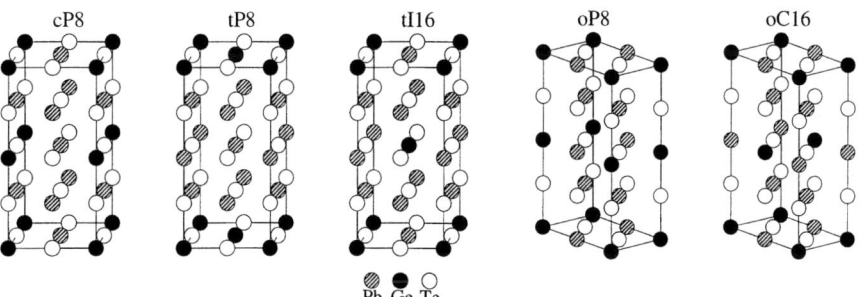

**FIGURE 1.** $Pb_3GeTe_4$ configurations investigated. Each has 8 atoms per primitive unit cell. (The Pearson symbols specify the number of atoms in the conventional unit cell.)

configurations have one or more identity irreducible representations (irreps) among its normal modes. An identity irrep implies an energy term that is linear in the corresponding mode and thus a lowering of the energy by relaxation in the ionic displacement subspace spanned by these modes. This relaxation neither lowers the symmetry nor causes net polarization in the configurations studied. The other normal modes in the systems have non-identity irreps; thus to lowest order, the total energy is quadratic in these modes. The existence of instabilities (normal modes whose harmonic coefficient is negative) with their concurrent symmetry breaking is a system-specific phenomenon, as is the polar nature of these instabilities. This section focuses on symmetry-preserving relaxation; the next section focuses on symmetry-breaking polar instabilities.

Before investigating symmetry-preserving relaxation, we relaxed each configuration of Figure 1 with respect to strain, holding all ions fixed at ideal rocksalt positions. At this point, the total energies all differed by less than 10 meV/(8 atom cell) and the (pseudocubic) lattice parameters all differed by less than 0.06%. Had all subsequent calculations been performed with respect to the same pseudocubic lattice parameters, e. g. the cP8 value of $a = 6.275$ Å, the strain energies involved would been negligible.

Each configuration was allowed to undergo symmetry preserving relaxation, with its strain held fixed. The relaxed atomic coordinates can be simply described (to within 0.03 Å) by displacement of the Te ions in each Pb-Te-Ge chain segment 0.12 Å toward to the Ge ion, with all other ions held fixed. [19] Qualitatively, relaxation in $Pb_{1-x}Ge_xTe$ can be regarded as the result of the size mismatch between the smaller Ge and the larger Pb ions. The energy of the cP8 configuration, for which no symmetry-preserving relaxation occurs, was 171 meV/cell higher than for the relaxed oC16 configuration, in which every Te is displaced.

The energies of the relaxed configurations were fit to a pair-only cluster expansion out to fourth cation-cation neighbors. In meV/cell, the total energy of the configurations tested is given by:

$$E = E_0 - 2.9\overline{N}_1 + 26.4\overline{N}_2 - 1.5\overline{N}_3 + 1.3\overline{N}_4, \tag{1}$$

where $\overline{N}_i$ is the average number of $i$'th neighbor Ge ions per Ge ion. The cation sublattice in $Pb_{1-x}GeTe$ is fcc; each cation has 12 first neighbors cations at $(a/2, a/2, 0)$, etc., 6 second neighbors at $(0, 0, a)$, etc., 24 third neighbors at $(a/2, a/2, a)$, etc., and 12 fourth neighbors at $(a, a, 0)$, etc., where $a$ is the conventional fcc lattice parameter. The sensitivity of energy to $\overline{N}_2$ is an order of magnitude larger than for the other terms, illustrating the importance of the above-mentioned relaxation. The Ge and the Pb in a linear Ge-Te-Pb segment where the Te undergoes significant relaxation are second neighbors; the more Ge-Te-Pb segments, the fewer Ge-Te-Ge segments, the lower $\overline{N}_2$, and the lower the relaxed energy.

The expression 1 is a fit of 5 relaxed cell energies to five parameters. It is also important to get an estimate of the predictive power of a cluster expansion truncated at a smaller number of terms. We did a series of cluster expansions in terms of $\overline{N}_1$, $\overline{N}_2$ and $\overline{N}_3$ to each subset of 4 configurations [20] in Figure 1, and used the resulting expression to calculate the energy of the configuration that was left out. The rms error was 16 meV/cell and the maximum error was 20 meV/cell. This compares with the root mean square variance of 57 meV/cell for the relaxed energies of the set of configurations studied, showing modest but not excellent predictability for a cluster expansion out to third neighbor. In what follows, we present results based on three or fewer configurations and the cluster expansions are truncated where the number of parameters matches the number of unknowns. Calculations on additional configurations would be necessary to make any meaningful test of predictability.

## SYMMETRY BREAKING INSTABILITIES

Next, we looked at energy lowering via symmetry-breaking distortions. For each configuration, we then calculated the normal modes at the zone center. We performed linear response force constant calculations at $\mathbf{q} = 0$ for each relaxed configuration and then diagonalized the corresponding dynamical matrix to obtain the normal modes. For each configuration, there were symmetry-breaking polar instabilities, with a three-dimensional vector representation centered on the Ge ion. [21] Further $\mathbf{q} \neq 0$ calculations on cP8 and tP8 showed that instabilities throughout the Brillouin zone could be well-represented by a local (lattice Wannier function [24]) basis with one vector on each Ge ion. This supports the Ge off-centering model for ferroelectricity in $Pb_{1-x}Ge_xTe$ [22] and gives strong evidence that the correct form for a first-principles model for arbitrary configurations will have one vector per Ge ion.

In order to appropriately model the long-range physics within our models, both the electronic dielectric tensor $\epsilon_\infty$ and the polarization associated with the local instabilities must be determined. These are linear response functions of the relaxed high-symmetry structures. We have calculated the dielectric tensors $\epsilon_\infty$ and the Born effective charges for the cP8, tP8, and tI16 configurations. The Born effective charge tensors $\mathbf{Z}_j^\star$ and the normal mode ionic displacement pattern $\mathbf{u}_{j\beta}$ associated with the zone-center instability which transforms like the vector component $\alpha$ were used to calculate the so-called mode effective charge tensor through

$$\overline{Z}^\star_{\alpha\beta} = \sum_{j\gamma} (Z_j^\star)_{\alpha\gamma} (u_{j\beta})_\gamma \qquad (2)$$

The normal modes were all normalized such that $\sum_j |\mathbf{u}_{j\beta}|^2$ was 1 Å$^2$.

The results for the $\epsilon_\infty$ and $\overline{Z}^\star$ tensors are shown in Table 1. Fitting to a pairwise cluster expansion out to second neighbor, we obtain:

$$(\epsilon_\infty)_{\alpha\alpha} = 42.5 - 0.175\overline{N}_{1\|} - 0.1\overline{N}_{1\perp} + 1.7\overline{N}_{2\|} + 0.2\overline{N}_{2\perp}, \qquad (3)$$

where $\overline{N}_{i\perp}$ is the mean number of $i$'th neighbor Ge ions per Ge ion in a direction perpendicular to $\alpha$ and $\overline{N}_{i\|}$ is the mean number of $i$'th neighbor Ge ions per Ge ion in a direction having nonzero component along $\alpha$. Likewise, the mode effective charge is given (in units of eÅ) by

$$\overline{Z}^*_{\alpha\alpha} = 17.31 - 0.14\overline{N}_{1\|} - 0.12\overline{N}_{1\perp} - 2.17\overline{N}_{2\|} - 0.22\overline{N}_{2\perp}. \quad (4)$$

The relative configuration dependence of $\overline{Z}^*$ is larger than that of $\epsilon_\infty$. The three configurations all have diagonal dielectric and mode effective charge tensors; it should be noted that off-diagonal terms will be nonzero in the general (asymmetric) configuration. For the mode effective charge tensors, the more Ge-Te-Pb or Pb-Te-Ge segments along $z$, the lower $\overline{N}_{2\|}$ and the higher $\overline{Z}^*_{zz}$. We have previously shown how this increase can be explained in terms of local bonding [23]. The differences in the Born effective charges of individual ions between configurations is relatively small; it is primarily the difference in the normal mode eigenvector between configurations that leads to the increase in $\overline{Z}^*$. In particular, in Ge-Te-Pb chain segments, relaxation of the Te away from the Pb leads effectively to an off-centering instability of the Pb ion that does not occur when the Pb-Te distance is smaller. [23] The active participation of the Pb ion helps to increase $\overline{Z}^*$.

**TABLE 1.** Electronic dielectric constant and mode effective charges in three $Pb_3GeTe_4$ configurations.

| Component | cP8 | tP8 | tI16 |
|---|---|---|---|
| $(\epsilon_\infty)_{xx}$ | 46.7 | 45.6 | 46.3 |
| $(\epsilon_\infty)_{yy}$ | 46.7 | 45.6 | 46.3 |
| $(\epsilon_\infty)_{zz}$ | 46.7 | 42.9 | 43.3 |
| $\overline{Z}^*_{xx}$ | 12.10 | 11.97 | 12.54 |
| $\overline{Z}^*_{yy}$ | 12.10 | 11.97 | 12.54 |
| $\overline{Z}^*_{zz}$ | 12.10 | 15.95 | 16.44 |

# PIEZOELECTRICITY AT ZERO TEMPERATURE

The above calculations were for relaxed high symmetry states. We now turn our attention to properties of the fully distorted LDA ground states. We have previously reported the piezoelectric tensors at zero temperature for the cP8 and tP8 configurations [23]. In this section, we give further results for the tI16 configuration, write a simple cluster expansion for the zero-temperature piezoelectricity in $Pb_{0.75}Ge_{0.25}Te$, and use this expansion to estimate the piezoelectric tensor of the disordered system.

The LDA ground state for the tI16 configuration is given in Table 2. In Table 3, we give the full piezoelectric tensors for the LDA ground states of the cP8, tP8 and tI16 configurations. In each case, the orientation is with respect to the axes of Figure 1, and the symmetry has been broken such that the Ge displacement and the polarization lie in the $+\{111\}$ quadrant.

**TABLE 2.** LDA ground state of $Pb_3GeTe_4$-tI16. Monoclinic, space group Cm, $a = 8.984$ Å, $b = 8.876$ Å, $c = 7.738$ Å, $\beta = 54.08°$.

| Atom | Wyckoff position | x | y | z |
|---|---|---|---|---|
| Pb | 2(a) | 0.5034 | 0 | 0.0061 |
| Pb | 4(b) | 0.5020 | 0.2489 | 0.5000 |
| Ge | 2(a) | 0.0291 | 0 | 0.0069 |
| Te | 2(a) | 0.2355 | 0 | 0.5158 |
| Te | 2(a) | 0.7578 | 0 | 0.4753 |
| Te | 4(b) | 0.2391 | 0.2383 | 0.9976 |

**TABLE 3.** Comparison of piezoelectric strain tensors for the ground states of three $Pb_3GeTe_4$ configurations (in $C/m^2$). Components in parentheses are equal to other components via symmetry. Components which do not appear in the table are related by symmetry to those that do; e.g. $e_{22} = e_{11}$ for each configuration

| Component | cP8 | tP8 | tI16 |
|---|---|---|---|
| $e_{11}$ | 1.8 | 2.5 | 2.4 |
| $e_{12}$ | -0.6 | -1.2 | -1.2 |
| $e_{13}$ | (-0.6) | -0.9 | -0.6 |
| $e_{14}$ | -0.2 | -0.6 | -0.5 |
| $e_{15}$ | 1.0 | 1.2 | 0.2 |
| $e_{16}$ | (1.0) | 1.1 | 1.1 |
| $e_{31}$ | (-0.6) | -0.3 | -0.5 |
| $e_{33}$ | (1.8) | 5.1 | 6.5 |
| $e_{34}$ | (1.0) | 2.2 | 6.4 |
| $e_{36}$ | (-0.2) | -0.5 | -0.7 |

If only pair terms are included in the cluster expansion for piezoelectricity, then the piezoelectric tensor is given by

$$\mathbf{e} = \mathbf{e}_0 + \sum_\mathbf{d} \overline{N}_\mathbf{d} \mathbf{e}_\mathbf{d}, \qquad (5)$$

where $\mathbf{e}$ is the piezoelectric tensor, $\mathbf{e}_0$ is a configuration-independent term, $\{\mathbf{d}\}$ is the set of cation-cation separations, $\overline{N}_\mathbf{d}$ is the average number of Ge neighbors per Ge ion at separation $\mathbf{d}$, and $\mathbf{e}_\mathbf{d}$ is the correction to the piezoelectric tensor due to neighbors at separation $\mathbf{d}$. We are now dealing with a property of $Pb_{0.75}Ge_{0.25}Te$ at zero temperature; the discussion of piezoelectric tensor in this section only applies to the subensemble of zero temperature structures that have been poled by a field in the +111 direction. The subensemble has rhombohedral symmetry; the average piezoelectric tensor has only four independent components: $e_{11}$, $e_{12}$, $e_{14}$ and $e_{15}$.

Symmetry constrains the form of the individual $\mathbf{e}_\mathbf{d}$. Consider for example the effect of adding one Ge-Ge pair separated by $(a/2, a/2, 0)$ to a perfectly rhombohedral configuration. The global symmetry is broken to monoclinic. By writing the appropriate symmetrized form of the piezoelectric tensor for a monoclinic system and subtracting that for a rhombohedral system, the correct symmetry for $\mathbf{e}_{(a/2,a/2,0)}$ is obtained. It is straightforward to show that the tensors $\mathbf{e_d}$ for two different $\mathbf{d}$ are related if and only if the two $\mathbf{d}$ are equivalent by symmetry under the rhombohedral group. In the cation sublattice, the 12 first neighbors break under rhombohedral distortion into 2 groups of 6, while the 6 second neighbors are all equivalent.

The cP8, tP8 and tI16 configurations are sufficient to separate the first neighbor and second neighbor contributions to the piezoelectric tensor, but insufficient for determining the independent contributions of the two kinds of first neighbor. In Table 4, we give $\mathbf{e}_0$ and the contributions $\mathbf{e}_{(a/2,a/2,0)} + \mathbf{e}_{(a/2,-a/2,0)}$ and $\mathbf{e}_{(0,0,a)}$. Both the tensors $\mathbf{e}_{(a/2,a/2,0)} + \mathbf{e}_{(a/2,-a/2,0)}$ and $\mathbf{e}_{(0,0,a)}$ have the same symmetry as that for the tP8 and tI16 ground state piezoelectric tensors.

The dominant terms reflect the piezoelectric components that change the most from configuration to configuration. The lack of second neighbor Ge pairs along $(0,0,a)$ tends to increase the value of $e_{33}$, while the lack of Ge pairs separated by $(\pm a/2, \pm a/2, 0)$ and $(0, 0, a)$ both increase the value of $e_{34}$. The physics of the $(0, 0, a)$ pairs is clear: the presence of a Ge-Te-Pb chain segment means that there is relaxation of the Te atom joining them and thus a weakening of the instability that transforms like $z$. A weak instability implies large response [23,14]. The $e_{33}$ and $e_{34}$ components are the ones enhanced because the instability along $\hat{z}$ is effectively coupled most strongly to these components. [10]. The exact source of the large influence of first neighbor pairs on $e_{34}$ has yet to be determined.

**TABLE 4.** Terms in cluster expansion of $Pb_{0.75}Ge_{0.25}Te$ piezoelectric tensor (in $C/m^2$)

| Component | $e_0$ | $e_{(a/2,a/2,0)} + e_{(a/2,-a/2,0)}$ | $e_{(0,0,a)}$ |
|---|---|---|---|
| $e_{11}$ | 7.7 | 0.1 | -0.3 |
| $e_{12}$ | -1.0 | 0.2 | 0.3 |
| $e_{13}$ | (-1.0) | -0.1 | 0.0 |
| $e_{14}$ | -1.2 | 0.0 | 0.1 |
| $e_{15}$ | 5.6 | 0.5 | 0.4 |
| $e_{16}$ | (5.6) | 0.0 | 0.0 |
| $e_{31}$ | (-1.0) | 0.1 | -0.1 |
| $e_{33}$ | (7.7) | -0.7 | -2.4 |
| $e_{34}$ | (5.6) | -2.2 | -2.7 |
| $e_{36}$ | (-1.2) | 0.1 | 0.2 |

Given the data in Table 4, it is possible to estimate the piezoelectric tensor of disordered $Pb_{0.75}Ge_{0.25}Te$. We assume for present purposes that all members of the poled subensemble are equiprobable. Then there are no spatial correlations and $\overline{N}_\mathbf{d} = 0.25$ for all $\mathbf{d}$. Using the values in Table 4 and Eq. 5, we estimate for disordered $Pb_{0.75}Ge_{0.25}Te$ that $e_{11} = 5.9$, $e_{12} = -0.9$, $e_{14} = -1.0$, and $e_{15} = 3.7$, reproducing the tensor form expected for rhombohedral symmetry. The estimated values of the components suggest that the piezoelectric response of the cP8 configuration is particularly unrepresentative for the disordered system. Because of the strong dependence of piezoelectric response on the magnitude of instability in a system and the configuration dependence of the magnitude of local instability, piezoelectric results on single high symmetry supercells of mixed ferroelectrics should be treated with caution.

## CONCLUSIONS

This work describes the necessity of cluster expansions in first principles models for piezoelectricity and ferroelectricity in solid solutions and the importance of comparing the properties of different configurations. In the specific example of $Pb_{0.75}Ge_{0.25}Te$, strong configuration dependence is found for relaxed energies, the nature of the local polar instabilities, and the polarization associated with these instabilities. A good model for ferroelectricity in a mixed system must be able to account for this variability. We show how results of the zero temperature piezoelectricity on several ordered supercells can be used to obtain an estimate of the zero temperature piezoelectricity of a completely disordered cell.

## ACKNOWLEDGEMENTS

This work was supported by ONR N00014-97-J-0047.

## REFERENCES

1. S.-E. Park and T. R. Strout, in *1996 IEEE Ultrasonics Symposium Proceedings* (New York: IEEE, 1996), v.2, p. 935; *J. Appl. Phys.* **82**, 1804 (1997), and references therein.
2. R. E. Cohen, *Nature* **358**, 136 (1992).
3. D. J. Singh, *Phys. Rev. B* **52**, 12559 (1995).
4. S. Baroni, P. Giannozzi and A. Testa, *Phys. Rev. Lett.* **58**, 1861 (1987).
5. X. Gonze, D. C. Allan and M. P. Teter, *Phys. Rev. Lett.* **68**, 3603 (1992).
6. R. D. King-Smith and D. Vanderbilt, *Phys. Rev. B* **47**, 1651 (1993).
7. K. M. Rabe and J. D. Joannopoulos, *Phys. Rev. Lett.* **59**, 570 (1987).
8. W. Zhong, D. Vanderbilt, and K. M. Rabe, *Phys. Rev. Lett.* **73**, 1861 (1994); *Phys. Rev. B* **52**, 6301 (1995).
9. K. M. Rabe and U. V. Waghmare, *Ferroelectrics* **164**, 15 (1995); U. V. Waghmare and K. M. Rabe, *Phys. Rev. B* **55**, 6161 (1997).
10. E. Cockayne and K. M. Rabe, *Phys. Rev. B* **56**, 7947 (1997).

11. H. Krakauer, R. Yu, C.-Z. Wang, K. M. Rabe and U. V. Waghmare, unpublished (cond-mat/9710088); H, Krakauer, R. Yu, C. Z. Wang, and C. Lasota, *Ferroelectrics* (to be published).
12. K. M. Rabe and U. V. Waghmare, *Phil. Trans. R. Soc. Lond.* **A 354**, 2897 (1996).
13. A. Garcia and D. Vanderbilt, unpublished (cond-mat/9712312).
14. K. M. Rabe, these Proceedings.
15. S. Takaoka and K. Murase, *Phys. Rev. B* **20**, 2823 (1979).
16. D. K. Hohnke, H. Holloway and S. Kaiser, *J. Phys. Chem. Solids* **33**, 2053 (1972).
17. M. C. Payne et al., "CASTEP 2.1", Cavendish Laboratory, University of Cambridge (1991).
18. U. Waghmare, Ph. D. thesis, Yale University (1996).
19. If a Te atom belongs to Pb-Te-Ge chain segments in more than one Cartesian direction, as for example the $z = 0$ Te atoms in the oP8 and oC16 configurations, then its relaxation is given approximately by the vector sum of the appopriate 0.12 Å relaxations for the individual chains.
20. For technical reasons, the subset of four configurations with cP8 missing could not be used in this analysis. The particular values of $\overline{N}_1$, $\overline{N}_2$, and $\overline{N}_3$ for that set of four configurations gives an indeterminate set of linear equations.
21. The vector representation breaks into irreps according to the point group of the Ge site in the given configuration.
22. Yu. A. Logachev and B. Ya. Moǐzhes, *Sov. Phys. Solid State* **19**, 1635 (1977). (*Fiz. Tverd. Tela* (Leningrad) **19**, 2793 (1977)).
23. E. Cockayne and K. M. Rabe, unpublished (cond-mat/9712232).
24. K. M. Rabe and U. V. Waghmare, *Phys. Rev. B* **52**, 13236 (1995).

# Local Lattice and Precursor Effects in $SrTiO_3$ and $PbTiO_3$

A. Bussmann-Holder

*Max-Planck-Institut für Festkörperforschung*
*Heisenbergstr. 1, D-70569 Stuttgart, Germany*

**Abstract.** The lattice dynamics of $SrTiO_3$ and $PbTiO_3$ have been reinvestigated within the two-dimensional anharmonic polarizability model. Opposite to former calculations, which were based on the self-consistent phonon approximation, in the present calculations the nonlinearly coupled equations of motion are solved exactly for arbitrary q for the dynamical displacement coordinates and associated frequencies. It is found that in both compounds the oxygen ion dynamics are highly anharmonic already far away from the the structural instability. The displacement patterns can be related to dynamical stripe and domain formations. The potential associated with these dynamics is calculated and shows periodically fluctuating areas with attractive and repulsive regions.

Various new experimental techniques like, e.g. EXAFS, PDF, NMR [1] have recently revealed that many perovskite ferroelectrics show locally polarized clusters which are not compatible with the undistorted high-temperature paraelectric phase. These precursor domains have been evidenced as order/disorder effects dominating the phase transition dynamics [2]. As in addition to these findings, soft mode behavior is observed corresponding to a displacive phase transition, a rather controversial situation emerged. Theoretically it has recently been shown that order/disorder and displacive effects may well coexist, as both effects obey different time scales [2,3]. In the following the previous calculations are extended to search for the origin of polarized local regimes which exist well above $T_c$. The lattice dynamics of $SrTiO_3$ and $PbTiO_3$ have previously been calculated within the nonlinear polarizability model [4]. The anharmonicity is attributed in that model to the local nonlinear polarizability of the oxygen ion [5] which induces a double-well potential. The temperature dependence of the soft ferroelectric mode, Raman scattering data, dielectric constant have been calculated by using the self-consistent phonon approximation (SPA) which corresponds to replacing higher-order displacement coordinates by thermal averages [1,6]. This in turn defines self-consistently temperature-dependent renormalized local couplings, which become zero at $T_c$ and lead to the structural instability. Even though very good agreement with various exerimental data could be achieved, local effects are not describable within the pseudo-harmonic SPA.

In the following the formerly introduced two-dimensional polarizability model [2] is solved exactly for arbitrary $q$ from which the real space dynamical displacement coordinates are obtained. Even though PbTiO$_3$ and SrTiO$_3$ are distinctly different with respect to ferroelectricity, their local dynamics show an astonishing resemblance. The model has been introduced in Ref. 2 and consists of a two-dimensional projection of the perovskite ABO$_3$ structure on the $(11\frac{1}{2})$ plane. The BO$_3$ unit is treated as a highly polarizable cluster with mass $m_1$ and harmonic ($g_2$) and fourth-order ($g_4$) core-shell coupling. The coupling to the A-sublattice (mass $m_2$) is indirect through the BO$_3$ shells. Next nearest neighbor BO$_3$ core-core coupling $f'$ ensures stability of the lattice in the limit of compensations of harmonic and fourth-order couplings.

The already Fourier transformed equations of motion read:

$$\left\{\omega^2(q) - \omega_1^2(q)\right\} \xi_1(q) = g_2/m_1 W(q) \left[1 + g_4/g_2 W(q)^2\right]$$
$$\left\{\omega^2(q) - \omega_2^2(q)\right\} \xi_2(q) = m_2 \omega^2 c_x c_y \left\{\xi_1(q) + W(q)\right\}$$
$$0 = -m_2 \omega_2^2 (\xi_1(q) + W)q) \left\{1 + g_4/g_2 W(q)^2\right\} + m_2 \omega_2 c_x c_y \xi_2(q) \qquad (1)$$

where $\xi_1(q)$ is the q-dependent displacement coordinate of ion $i$ with mass $m_1$ and W the relative core-shell coordinate of $m_1$, which can be associated with a q-dependent fluctuating dipole $q_0 eW(q)$. $\omega_1^2 = 4f'/m_1(s_x + s_y)$ and $\omega_2^2 = 8f/m_2$ are acoustic and zone boundary optic mode frequencies which are taken from experimental data; $c_{x,y} s_{x,y}$ are x,y-dependent direction *cosine* and *sine* functions. The parameters used are listed in Table 1. Equation 1 is solved numerically for $\xi_1(q)$,

TABLE 1. Parameters used in the calculations.

| | SrTiO$_3$ | PbTiO$_3$ |
|---|---|---|
| $\omega_1$ [THz] | 1.81 | 1.49 |
| $\omega_2$ [THz] | 8.89 | 5.5 |
| $g_4$ [$10^{22} g_s^{-2}$cm$^2$] | 0.904 | 4.606 |

W(q) and frequencies $\omega$(q) by starting from a trial set of displacement coordinates to solve for $\omega^2$(q), solve with these $\omega$(q) for $\xi_1$(q), W(q) and iterate until convergence is achieved. Temperature dependent effects are included by varying the height of the onsite double well potential, i.e. $g_2$, calculate $g = g_2 + 3g_4 W(q)^2$ which corresponds to the pseudoharmonic limit, solve for the q = 0 optic mode and compare $g$ and $\omega^2$ (q=0) with results from previoius SPA calculations. Interestingly, the agreement is extremely good.

The q-dependent dispersion of the lowest transverse acoustic and optic mode in $\langle 11 \rangle$ and $\langle 10 \rangle$ direction agree also, on the average, with previous results, but a refined resolution at small q-values reveals that in both SrTiO$_3$ and PbTiO$_3$ pronounced anomalies in the acoustic mode dispersion appear. In SrTiO$_3$ the

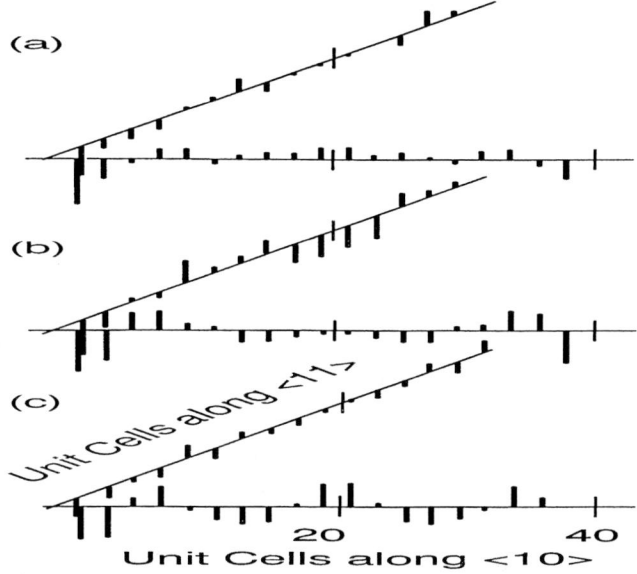

**FIGURE 1.** Real space TiO$_3$-core displacement $\xi_1$ along $\langle 11 \rangle$ and $\langle 10 \rangle$ direction for PbTiO$_3$ at T = 550K (Fig.1a), T = 510K (Fig.1b), T = 480K (Fig.1c).

origin of this anomaly has been attributed to a new quantum coherernt state [7] and/or mode-mode crossing [8] and coupling. As in the present calculation, similar anomalies are observed also for PbTiO$_3$ it is argued that these are intrinsic to ferroelectric perovskites [9].

The calculated $\xi_1$ and W show anomalous q-dependencies in both compounds. In order to interpret these anomalies, they have been Fourier transformed to real space. In Fig.1 $\xi_1$ of PbTiO$_3$ is shown for three different temperatures. At T = 550K (Fig. 1a), the PO$_4$-core dynamical displacements in the $\langle 11 \rangle$ direction are small and not translationally invariant. In the $\langle 10 \rangle$ direction evidence for the formation of a periodic pattern appears with a period of 8 lattice constants [10]. With decreasing temperature (Fig.1b) the development of this pattern becomes more pronounced, while the displacement in $\langle 11 \rangle$ direction become larger and highly anharmonic. Below the transition temperature (T = 480K, Fig.1c), stripe-like features appear along $\langle 10 \rangle$, and due to the very close vicinity to the structural instability, the displacments along $\langle 11 \rangle$ slow down again. The results for SrTiO$_3$ for $\xi_1$ are shown in Fig. 2 and resemble closely the previous results. For all three temperatures (T = 85K, Fig.2a; T = 40K, Fig.2b; T = 10K, Fig.2c), the displacement along $\langle 11 \rangle$ is highly anharmonic but decreases with decreasing temperature, as is expected from the softening of the optic mode. Along $\langle 10 \rangle$ a similar reduction in $\xi_1$ with temperature appears, but in addition, a periodic pattern develops similar to PbTiO$_3$. The q-dependence of the dipole eq$_0$W associated with W(q) is shown in Fig.3 for

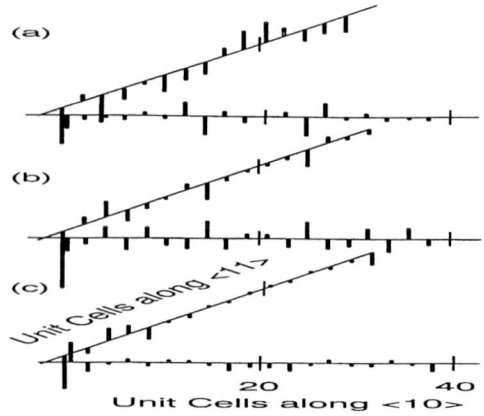

**FIGURE 2.** Real space TiO$_3$-core displacement $\xi_1$ along $\langle 11 \rangle$ and $\langle 10 \rangle$ direction for SrTiO$_3$ at T = 85K (Fig.2a), T = 40K (Fig.2b) and T = 10K (Fig.2c).

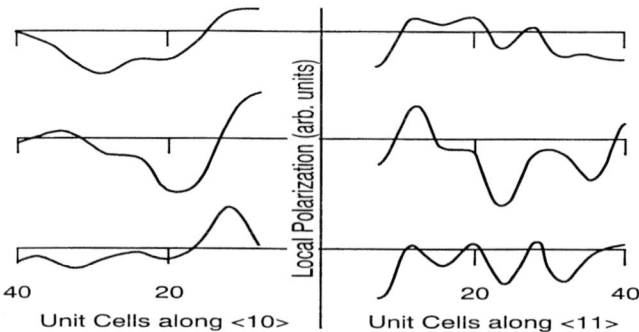

**FIGURE 3.** Real space polarization fluctuations in PbTiO$_3$ along $\langle 11 \rangle$ and $\langle 10 \rangle$ direction for temperatures T = 550K (top), T = 510K (middle) and T = 480K (bottom).

PbTiO$_3$ for the same temperatures as the displacements. At all three temperatures strong polarization fluctuations are observed along the $\langle 11 \rangle$ direction which tend to a uniquely polarized state below $T_c$, but are still exhibiting oscillations along the $\langle 10 \rangle$ direction. The fluctuation period is large and consequently broad uniquely polarized domain states are expected even far above $T_c$. The same behavior has also been found in SrTiO$_3$, but in this compound the dipole oscillations exhibit a longer lattice periodicity. It is important to note that when the temperature is about 100K above the critical temperature, the above described anomalous displacements and dipole oscilations are negligibly small.

The calculated displacements, dipole moments, frequencies have been reinserted into the Hamitonian in order to solve for the dynamical potential associated with these anharmonic and anomalous dynamics. The real space potential is shown in

Fig.4 for PbTiO$_3$ for temperatures T ≈ 600K, 550K, 510K, and 480K, respectively. As indicated above, for the highest temperature (Fig.4a), practically no contri-

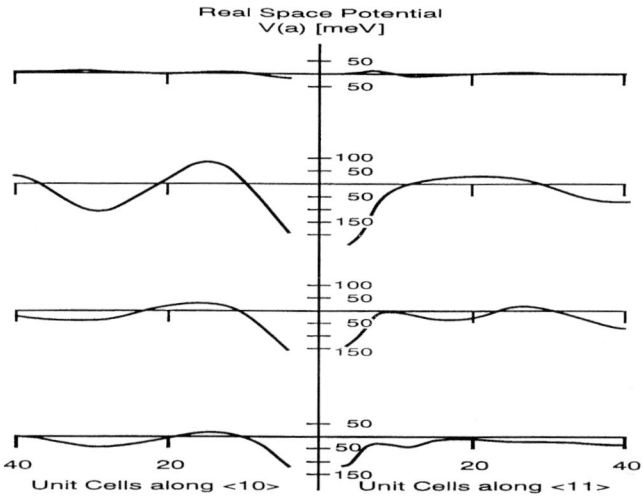

**FIGURE 4.** Real space dynamical potential of PbTiO$_3$ along $\langle 11 \rangle$ and $\langle 10 \rangle$ direction for temperatures T = 590K (Fig.4a), T = 550K (Fig.4b), T = 510K (Fig.4c) and T = 480K (Fig.4d).

butions to the total potential arise, as the system is nearly harmonic. At T = 550K (Fig.4b) and 510K (Fig.4c), strong fluctuations in the dynamical potential appear with broad repulsive areas in the $\langle 11 \rangle$ direction and fluctuating repulsive and attractive regions along $\langle 10 \rangle$ which have an approximate periodicity 16 lattice constants even for T < T$_c$ (Fig.4c). In the ferroelectric phase, T = 480K, the lattice is stable along $\langle 11 \rangle$, the potential is nearly constant and always attractive.

In conclusion, the lattice dynamics of SrTiO$_3$ and PbTiO$_3$ have been reinvestigated by studying exact solutions of the two-dimensional analogue of the polarizability model. It is found that both compounds behave pseudo-harmonically at sufficiently high temperatures, but already far above the freezing temperature of the soft optic mode, the local dynamics of the TiO$_3$ cluster start to deviate from this behavior. The displacement patterns show correlated areas on the scale of some few lattice constants and precursor-type polarized domain patterns appear. The origin of these structures can be attributed to dynamical potential fluctuations which exhibit repulsive areas on a similar length scale.

# REFERENCES

1. N. Sicron, et al., Phys. Rev. B **50**, 13168 (1994); S. Rod, F. Borsa, J.J. van der Link, Phys. Rev. B **38**, 2267 (1988); Y. Yacoby, Y. Girshberg, E.A. Stern, Z. Phys. B **104**, 725 (1997).

2. For a review see: M. Stachiotti, A. Dobry, R. Migoni, A. Bussmann-Holder, Phys. Rev. B **47**, 2473 (1993), and references therein.
3. A. Bussmann-Holder, A.R. Bishop, G. Benedek, Phys. Rev. B **53**, 11521 (1996).
4. R. Migoni, H. Bilz, D. Bäuerle, Phys. Rev. Lett. **37**, 1155 (1976); A. Bussmann-Holder, H. Bilz, G. Benedek, Phys. Rev. B **39**, 9214 (1989).
5. A. Bussmann-Holder, H. Bilz, R. Roenspiess, K. Schwarz, Ferroelectrics **25**, 343 (1980).
6. C. Perry, R. Currat, H. Buhay, R. Migoni, W. Stirling, J. Axe Phys. Rev. B **39**, 8666 (1989); R. Migoni, K.-H. Rieder, K. Fisher, H. Bilz, Ferroelectrics **13**, 377 (1976); R. Migoni, thesis, University of Stuttgart, 1976; D. Khatib, R. Migoni, G.E. Kugel, L. Godefroz, J. Phys. Cond. Matter **1**, 9811 (1991); G.E. Kugel, M.D. Fontana, W. Kress, Phys. Rev. B **35**, 813 (1987); M. Stachiotti, R. Migoni, J. Phys. Cond. Matt. **2**, 4341 (1990); M. Stachiotti, R. Migoni, U. Höchli, J. Phys. Cond. Matt. **3**, 3689 (1991).
7. E. Courtens, B. Hahlen G. Coddens, B. Hemion, Physica B **219** & **220**, 577 (1986); E. Courtens, B. Hehlen, E. Farki, A.K. Tagantsev, Z. Phys. B **104**, 641 (1987).
8. J.F. Scott, H. Ledbetter, Z. Phys. B ]bf 104, 635 (1997).
9. A. Bussmann-Holder, Phys. Rev. B **56**, 10762 (1997).
10. A. Bussmann-Holder, A.R. Bishop, Phys. Rev. B **56**, 5297 (1997).

# Self-ordered second-component clusters in solid solutions on the basis of ferroelectric perovskites: Nb clusters in KTaO$_3$.

R. I. Eglitis[1,2], V.S.Vikhnin[3], P.A.Markovin[3], G.Borstel[1]

[1] Universität Osnabrück – Fachbereich Physik, D-49069 Osnabrück, Germany
[2] Institute of Solid State Physics, University of Latvia, 8 Kengaraga, Riga LV-1063, Latvia
[3] A.F.Ioffe Physical Technical Institute, 194021 St. Petersburg, Russia

**Abstract.** Semi-empirical Hartree-Fock calculations using the intermediate neglect of differential overlap (INDO) method, for self-ordered cubic symmetry clusters of seven Nb ions in KTaO$_3$ are performed with the aim of verifying the cluster model [1] of second component-induced phase transitions in ferroelectric perovskite matrices. It is shown that such a seven-particle cluster in KTaO$_3$:Nb has two types of quasidegenerate states of different nature. Namely, the state with the equilibrium full-symmetric dilatation and off-center displacements in [111]-directions of the central Nb-ion in the cluster, and the state with the equilibrium full-symmetric compression without any off-center effect. The consequences of such cluster structures on the multi-well potential are discussed. In particular, an avalanche-like behaviour of the temperature dependence of the ferroelectric order parameter is obtained. The results of semi-empirical INDO calculations of a seven-impurities cluster of Nb-ions in KTaO$_3$ confirm the key assumptions of the cluster model of induced phase transitions.

*Keywords:* perovskites, ferroelectric phase transitions, clusters, solid solutions, $KTaO_3$, semi-empirical Hartree-Fock calculations, phase transitions

## I INTRODUCTION

Non-empirical *ab initio* methods are still cumbersome and time-consuming in the treatment of the electronic and spatial structure of complex systems, especially with partialy covalent chemical bonding, like perovskites. In order to be able to study the relatively complicated case of perovskite solid solutions, there is a need to close the gap between numerically exact but time consuming *ab initio* methods, on one hand, and broadly used, simple but not so reliable *ad hoc* parameter-dependent phenomenological approaches, such as the shell model, on the other hand. One possible compromise is to use a semiempirical quantum chemical method that is parameter-dependent, but with parameters which are more or less transferable, for the specific chemical constituents given, and not subject to adjustment for each new compound in question. An example of such a method is the updated Intermediate Neglect of the Differential Overlap (INDO) method [2–6], which is a semi-empirical version of the Hartree-Fock method. In recent years, the INDO method has been applied very succesfully for the study of defects, both in the bulk and on the surface, in many oxide materials [3–5,7–9], as well as semiconductors [10,11]. We recently applied this method to the study of phase transitions and frozen phonons in $KNbO_3$ [6], pure and Li-doped $KTaO_3$ [12], and $F$ centers in $KNbO_3$ [13].

Taking into account the increasing number of successful INDO method applications during recent years, we have chosen this method for the study of self-ordered cubic symetry clusters of seven Nb ions in $KTaO_3$ with the aim of verifying the cluster model [1] using an $4\times4\times4$ times extended large unit cell (LUC) containing 320 atoms.

A solution to problems in the interpretation of the impurity-induced ferroelectric phase transition by on-center impurities (like $Ba^{2+}$ impurities in incipient ferroelectric $SrTiO_3$) or by weak off-center impurities (like $Nb^{5+}$ impurities in incipient ferroelectric $KTaO_3$, and $Ca^{2+}$ impurities in $SrTiO_3$) had been proposed [1] on the basis of the self-ordered cluster model. The main assumption of this model is the following: The clusters of the second component atoms are formed and self-order. Self-ordered clusters of a second component in solid solutions on the basis of ferroelectric perovskites have their own degrees of freedom (of order-disorder as well as of displacive types). The percolation of the corresponding local order parameters as well as the dynamical percolation of soft, low-frequency local vibrations leads here to cooperative behaviour which finally (at a specific concentration of a second component) induces the ferroelectric phase transition. The actual mechanisms of the existence of self-ordered high symmetry clusters, as well as their soft internal degrees of freedom had been developed in Ref. [1] on the basis of concrete models of impurity ions interaction with the lattice. However it would be very useful to use first-principles calculations to verify the two main features of the model: (i) the existence of self-ordered high-symmetry clusters with rather high bonding energy in spite of the absence of a threshold of soluability for the solid solutions under consideration, and (ii) the existence of soft internal degrees of freedom of such clusters. Such an investigation is performed in the present article. The seven-impurities Nb clusters in $KTaO_3$ of cubic point symmetry are considered as an example.

## II THE METHOD OF CLUSTER PROPERTIES CALCULATIONS

The Fock matrix elements in the modified INDO approximation [3–5] contain a number of semiempirical parameters. The orbital exponent $\zeta$ enters the radial part of Slater-type atomic orbitals:

$$R_{nl}(r) = (2\zeta)^{n+\frac{1}{2}} \left[(2n)!\right]^{-\frac{1}{2}} r^{n-1} \exp(-\zeta r), \tag{1}$$

where $n$ is the main quantum number of the valence shell. We used a valence basis set including $4s$, $4p$ atomic orbitals (AO) for K, $2s$, $2p$ for O, $5s$, $5p$, $4d$ for Nb and $6s$, $6p$, $5d$ for the Ta atom. The diagonal matrix elements of the interaction of an electron occupying the $\mu$-th valence AO on atom A with its own core are taken as

$$U_{\mu\mu}^A = -E_{\text{neg}}^A(\mu) - \sum_{\nu \in A} (P_{\nu\nu}^{(0)A} \gamma_{\mu\nu} - \frac{1}{2} P_{\nu\nu}^{(0)A} K_{\mu\nu}), \tag{2}$$

where $P_{\mu\mu}^{(0)A}$ are the diagonal elements (initial guess) of the density matrix, $\gamma_{\mu\nu}$ and $K_{\mu\nu}$ are one-centre Coulomb and exchange integrals, respectively. $E_{\text{neg}}^A(\mu)$ is the initial guess of the $\mu$-th AO energy. The interaction of an electron on the $\mu$-th AO belonging to the atom $A$ with the core of another atom $B$ reads:

$$V_\mu^B = Z_B \left\{ 1/R_{AB} + [\langle \mu\mu | \nu\nu \rangle - 1/R_{AB}] \exp(-\alpha_{AB} R_{AB}) \right\}, \tag{3}$$

where $R_{AB}$ is the distance between atoms $A$ and $B$, $Z_B$ is the core charge of atom $B$, and parameter $\alpha_{AB}$ describes the non-point character of this interaction. The resonance-integral parameter $\beta_{\mu\nu}$ enters the off-diagonal Fock matrix elements for the spin component $u$:

$$F_{\mu\nu}^u = \beta_{\mu\nu} S_{\mu\nu} - P_{\mu\nu}^u \langle \mu\mu | \nu\nu \rangle, \tag{4}$$

where the $\mu$th and $\nu$th AO are centered at different atoms, $S_{\mu\nu}$ is the overlap matrix between them, and $\langle \ | \ \rangle$ are two-electron integrals.

Thus the INDO parametrization scheme contains the following set of parameters per atom: $\zeta$, $E_{neg}$, $P^{(0)}$, $\alpha$, $\beta$. INDO parameters for Nb and Ta were optimized by us earlier, fitting our INDO parameters to the results of *ab initio* calculations, as well as available experimental data for $KNbO_3$ and $KTaO_3$, and tabulated [6,12]. In the present work we used the same set of INDO parameters as in Ref [6,12].

## III  THE RESULTS: MULTI-WELL CLUSTER POTENTIAL STRUCTURE

The exploration of the Nb-clusters in $KTaO_3$ was performed using the periodic LUC method, as it is implemented in the updated CLUSTERD [3-6] computer code. We extended our primitive $KTaO_3$ unit cell $4 \times 4 \times 4$, i.e. 64 times, which is equivalent to a bandstructure calculation at 64 **k** points in the Brillouin zone. In order to calculate the Nb clusters in $KTaO_3$, we replaced in our $4 \times 4 \times 4$ times extended unit cell, containing 320 atoms, seven Ta atoms by seven Nb atoms, as is shown schematically in Fig. 1. After this, in order to find the energy minima of Nb clusters in $KTaO_3$, we allowed to six Nb atoms to relax towards the central Nb atom (see Fig. 1). The positions of K and O atoms were kept fixed. The results

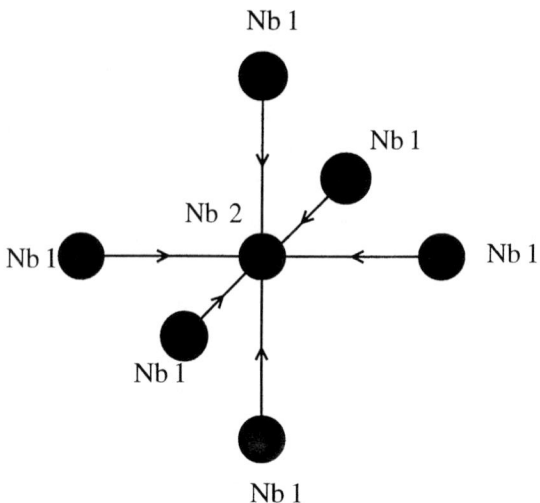

**FIGURE 1.** Nb - cluster containing seven Nb atoms inside $KTaO_3$.

of our calculations show that for six Nb atoms shifted symmetrically towards the central Nb atom by 0.187 Å, the total energy per LUC is lowered by 0.088 eV. Moreover, the outward uniform displacement of six Nb atoms from the central Nb (see Fig.2) by 0.073 Å is also favorable and lowers the energy by approximately 0.03 eV (see Fig. 3.). In the case, when 6 Nb atoms are shifted outwards from the central Nb atom, the central Nb-atom moves off-center from central position in the [111] direction by 0.27 Å, further lowering the total energy of system by 0.09 eV - to give a total energy reduction of approximately 0.12 eV. These configurations correspond to a ground-state vibronic multiplet. The Nb atom reveals also an instability in the [100] direction. The shift of the central Nb ion along the [100] direction by 0.192 Å lowers the cluster energy additionally by 0.056 eV in the same case when six Nb atoms are shifted outwards from the central Nb atom. The total cluster-structure induced energy lowering in the ground state, which coresponds to the situation when six Nb atoms are symetrically shifted in the outward direction from the center of the Nb cluster, and the central Nb atom is off-center in the [111] direction is approximately 0.12 eV (see Fig. 3). According to our calculations in the case when 6 Nb atoms are in the state of energy minima arising from the symetrical shift of six Nb atoms towards the central Nb atom (excited state), the central Nb atom exhibits on-center behaviour (see Fig. 3). It should be emphasized that the polar distortion dependence of the cluster energy in the latter excited state is very smooth. This circumstance can lead to a rather soft polar quasilocal cluster frequency on the one hand, and to the possibility of a defect-induced dipole instability and weak off-center displacements on the other.

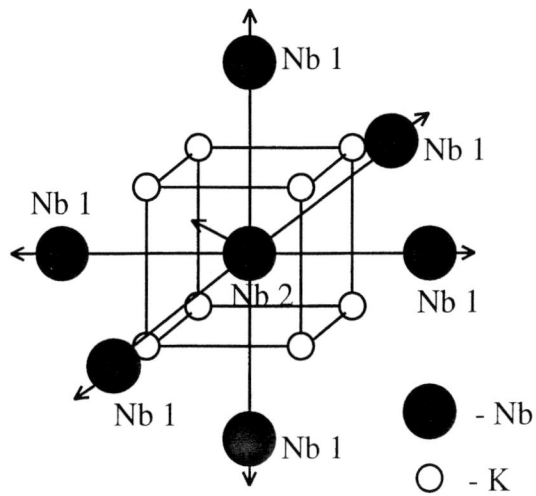

**FIGURE 2.** Sketch of the symmetric repulsion of six Nb atoms outwards from the central Nb atom, which, as a consequence, is off-center then both in [100] and [111] directions.

## IV THE PECULIARITIES OF THE SECOND-COMPONENT INDUCED FERROELECTRIC PHASE TRANSITION IN SOLID SOLUTION KTaO$_3$:Nb. EFFECT OF Nb-CLUSTERS.

The first important peculiarity of clusters under consideration in a high polarizable matrix such as KTaO$_3$ ($\varepsilon \gg 1$) is the appearance of avalanche-like behaviour of induced polarisation in the temperature region of the cluster-induced ferroelectric phase. The physical reason for this phenomenon in the ferroelectric phase is connected with the temperature dependence of the resulting energy splitting between the dipole ground state of cluster (with equilibrium full-symmetric dilatation and off-center displacement along [111] direction and along the order parameter field) and excited cluster state (with equilibrium full symmetric compression and with on-center position of the central Nb-ion in the cluster). These states are split by the energy $\Delta$ in the paraelectric phase. But in the impurity cluster-induced ferroelectric phase, the main field of such clusters order parameter field acts along the [111] direction. This field is proportional to the occupation of the above mentioned dipole state oriented along [111] direction. That is why the real splitting $\tilde{\Delta}$ between the ground dipole state and the excited non-dipole state equals

$$\tilde{\Delta} = \Delta + \frac{yA}{T} \tag{5}$$

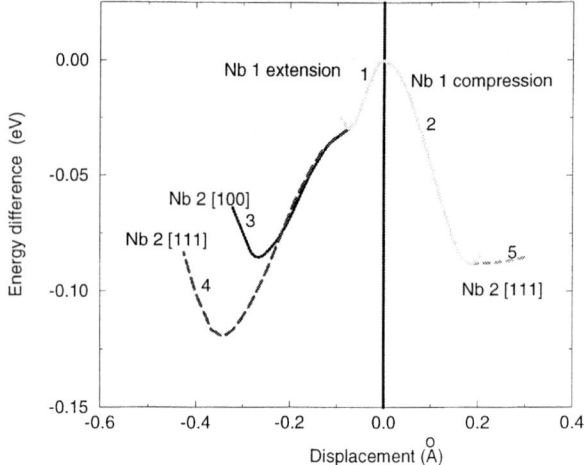

**FIGURE 3.** Displacement energy of six Nb atoms relaxed outwards (curve 1) and inwards (curve 2) as it is shown schematically in Fig. 2 and Fig. 1, respectively. The central Nb atom is off-center in [100] and [111] directions, curves (3 and 4), in the case of the symmetric dilatation of the cluster, but is on-center (curve 5), in the case of compression of the cluster.

where the dipole mean field-induced splitting $yA/T$ is evaluated in the framework of linear interaction of the electric dipole moment of the central Nb ion in the cluster with the polarisation of the soft TO mode via the local field [14,15] under the assumption that the critical temperature in pure KTaO$_3$ equals zero. Here $y = y(T)$ is the occupation of the ground dipole state in the cluster, $A$ is a constant, $A \sim (ex_0)^2\gamma^2$, $\gamma$ is the local field factor, $x_0$ is the equilibrium off-center displacement. Let us consider, for simplicity, the three-well model case, where two off-center wells have minima in [111] and [$\bar{1}\bar{1}\bar{1}$] directions, and the excited state (with the energy $\Delta$) is a non-dipole one. For this case the inequalities

$$1 > e^{-\frac{\tilde{\Delta}}{kT}}, 1 \gg e^{-\frac{2yA}{kT^2}} \qquad (6)$$

are fulfilled; then the ground dipole state occupation approximately equals

$$y^( = \frac{n_0}{(1 + e^{-\frac{\tilde{\Delta}}{kT}})} \qquad (7)$$

$n_0$ is the total concentration of clusters under consideration. On the basis of expressions (5) and (7) we can obtain the equation for the critical dipole state occupation $y_c$, for which the first derivative $dy/dT$ tends to infinity. As a result, at this occupation the first derivative of the cluster-induced polarization satisfies $\mid dP/dT \mid_{y_c} \to \infty$. The equation for $y_c$ is transcendental and has the following form

$$(\Delta^2 + 4k[\ln(\frac{n_0}{y}-1)^{-1}]yn_0A)^{-1/2}\left\{[\ln(\frac{n_0}{y}-1)^{-1}]kn_0A + kyn_0A(\frac{n_0}{y}-1)\frac{n_0}{(n_0-y)^2}\right\} -$$
$$-\frac{(\Delta + \{\Delta^2 + 4kyn_0A[\ln(\frac{n_0}{y}-1)^{-1}]\}^{1/2})(\frac{n_0}{y}-1)\frac{n_0}{(n_0-y)^2}}{(\ln(\frac{n_0}{y}-1)^{-1})} = 0 \quad (8)$$

where $n_0 < 2y$ is satisfied. The $y_c$ value can be considered as a concentration threshold for the appearance of a well defined cluster-induced ferroelectric phase (where the condition $\tilde{\Delta} \gg M_2^{1/2}$, $M_2$ is the second moment of fluctuations of internal cluster induced fields, is fulfilled). The existence of a concentration threshold of a second component-induced ferroelectric phase transition has been confirmed experimentally for the $KTaO_3$:Nb system (see, for example, Ref. [16]). We analyze now the solution of this equation (8).

Let us consider the actual case of $y_c \leq n_0^{(c)}$. Here $n_0^{(c)}$ is the threshold concentration of clusters, which can be connected with the critical temperature $T_c(n_0)$ near the ferroelectric phase transition threshold in accordance with the experimental approximation [16] $T_c \sim (x - x^c)^{1/2}$ with $x$ and $x^c$ as the total concentrations of second component ions and corresponding critical concentrations, respectively, which satisfied the equalities $x = n_0W$, $x^c = n_0^{(c)}W$ in the cluster model ($W$ is number of second component ions within the clusters).

Let us also take into account the possible realizing of the following inequality:

$$\Delta^2 \gg 4kyn_0A[\ln(\frac{n_0}{y}-1)^{-1}] \quad (9)$$

as well as of the inequality

$$\frac{n_0}{n_0-y} \gg [\ln(\frac{n_0}{y}-1)^{-1}] \quad (10)$$

which corresponds to $y_c \leq n_0^{(c)}$ action. Using Eq. (8) - (10) for the case $y_c \leq n_0^{(c)}$, the solution of (8) under the above mentioned condition is

$$y_c = \frac{n_0^{(c)}}{1+e^{-\frac{2\Delta^2}{k(n_0^{(c)})^2A}}} \quad (11)$$

Following from Eq. (9) and (11), the $y_c$ - value satisfies the relation $y_c \leq n_0^{(c)}$ and confirms the validity of above mentioned approach. The physical conclusion here is the appearance of a fast increasing of the polarization, and a $\varepsilon$ - peak appearance very near, but lower then the approximated $x^c$ - value due to peculiarities of the cluster state structure which lead to an avalanche-like order parameter increasing. This circumstance can be manifested as concentration threshold appearance.

The second important peculiarity is connected with the existence of a temperature region within the ferroelectric phase, where $\Delta = (yA)/T$ is fulfilled. Under this resonance condition, the effective tunneling occurs between the excited dipole and non-dipole states. These estimates based on the Gaussian-type vibration wave function approach, with vibration frequencies corresponding to our numerical results (see Fig. 3), show that the corresponding tunnel matrix element can be $\Gamma \cong 4 \cdot 10^9$ Hz. Under these resonance conditions, the dipole relaxation rate due to the resonance tunneling with dissipation can be presented in the form [17]

$$\frac{1}{\tau_{hop}} = (2\Gamma)^2 \tau_o, \qquad (12)$$

where $\tau_o$ is the relaxation time of the non-diagonal components of the corresponding density matrix (in-well relaxation time). The bilinear interaction between the dipole-type degrees of freedom of such relaxator and polarization of the soft TO-mode leads to a central peak phenomenon in the fluctuation spectra of the order parameter. The expressions for the form factor at the top of the central peak and halfwidth of the central peak can be obtained analogously to Ref [18].

The last but not least, peculiarity of the clusters under discussion is the existence of rather soft quasilocal modes in the excited vibration cluster state with on-center behaviour of the central Nb ion and compression of the six other Nb-ions. That is why the thermal occupation of such soft-mode states can cause the cluster-induced displacive-type cooperative dynamics due to the indirect interaction between slow variables of such soft quasilocal cluster modes via more fast lattice TO-mode vibration within a dynamical percolation [1] scenario. The latter can lead to the second component induced ferroelectric ordering of the displacive-type which co-exists with the order-disorder cooperative behaviour discussed above.

## V  CONCLUSIONS

The main result of our consideration is the soundness of the cluster model of induced ferroelectric transitions in perovskite solid solutions.

In addition to the above mentioned off-center effect within the cluster, and appearance of low frequency excitations, the important confirmation of the cluster model is obtained based on the INDO ground state energy calculations. The cluster structure-induced energy lowering, due to the outward extension of the six Nb ions and off-center displacement of the central Nb ion, obtained in the present work ($\sim 0.12$ eV), leads to the conclusion of stability of such self-ordered high symmetry clusters taking into account peculiarities of the crystal growing procedure (slow temperature lowering from $\sim 1000$ K).

It is important to stress that the existence of the above discussed avalanche -like behaviour of cluster-induced polarisation leads to a well-defined peculiarity in the dielectric susceptibility. That is why we can connect such a type of peculiarity with the $\varepsilon$ - peculiarity detected on the experiment [16] and caused by a second component-induced ferroelectric phase transition.

It should be underlined that the cluster model [1] of second component induced ferroelectric-type phase transitions received a strong support in the frame of the present work due to verification of model based on semi-empirical calculations. Moreover, the uncommon cluster state structure discovered here leads to the possibility of characteristic IR-absorption by the cluster states and to definite temperature dependences peculiarities for order parameter and dielectric response which can be studied experimentally.

## ACKNOWLEDGEMENTS

V.S.Vikhnin acknowledges financial support in part by Grant NATO HTECH LG-960540, Grant DFG436RUS113/341/1(R) and RFBR. R.I.Eglitis gratefully acknowledge financial support of the Deutsche Forschungsgemeinschaft (Graduate College) and by the Volkswagen-Stiftung. The authors are grateful to A.V.Postnikov, S.Kapphan, V.V.Lemanov and E.A.Kotomin for their constant interest to the reported problem.

## REFERENCES

1. V.S.Vikhnin, P.A.Markovin, V.V.Lemanov, W.Klemann, IMF-9 report, August 1997, Seoul, Korea; and JKPS, 1997, in press.
2. J. A. Pople and D. L. Beveridge, *Approximate Molecular Orbital Theory*, McGraw-Hill, New York, 1970.
3. A. Shluger, Theoret. Chim. Acta **66**, 355 (1985).
4. E. Stefanovich, E. Shidlovskaya, A. Shluger, and M. Zakharov, Phys. Stat. Sol.(b) **160**(2), 529 (1990).
5. A. Shluger and E. Stefanovich, Phys. Rev. B **42**, 9664 (1990).
6. R. I. Eglitis, A. V. Postnikov, and G. Borstel, Phys. Rev. B **54**, 2421 (1996).
7. A. Stashans, E. A. Kotomin, and J.-L. Calais, Phys. Rev. B **49**, 14854 (1994).
8. E. A. Kotomin, A. Stashans, L. N. Kantorovich, A. Livshitz, A. Popov, I. A. Tale, and J.-L. Calais, Phys. Rev. B **51**, 8770 (1995).
9. A. Stashans, S. Lunell, R. Bergstrom, A. Hagfeldt, and S.-E.Lundqvist, Phys. Rev. B **53**, 159 (1996).
10. E. V. Stefanovich, and A. L. Shluger, J. Phys. Condens. Matter **6**, 4255 (1994).
11. A. Stashans and M. Kitamura, Solid State Commun. **99**, 583 (1996).
12. R. I. Eglitis, A. V. Postnikov, and G. Borstel, Phys. Rev. B **55**, 12976 (1997).
13. R. I. Eglitis, N. E. Christensen, E. A. Kotomin, A. V. Postnikov, and G. Borstel, Phys. Rev. B **56**, 8599 (1997).
14. V.S.Vikhnin, Yu.B.Borkovskaya, Sov. Phys. Solid State **20**, 2082 (1978); V.S.Vikhnin, Sov. Phys. Solid State, **26** 906 (1984).
15. B.Vugmeister, M.Glinchuk, Rev. Mod. Phys., **62**, 993 (1990).
16. U.Höchli, K.Knorr, A.Loidl, Adv. in Physics **39**, 405 (1990).
17. V.S.Vikhnin, Sov. Phys. Solid State, **20**, 771 (1978).
18. G.A.Smolenski et all., *Physics of Ferroelectric Phenomena*, Nauka, Leningrad (1985).

# DAFS Study of Local Structure of Ordered Nanodomains in $PbMg_{1/3}Nb_{2/3}O_3$

A. I. Frenkel,* D. M. Fanning,[†] I. K. Robinson,*,[†] D. L. Adler,[‡] and J. O. Cross[§]

*Materials Research Laboratory,[1]
University of Illinois at Urbana - Champaign, Urbana, Illinois 61801
[†]Loomis Laboratory, Physics Department
University of Illinois at Urbana - Champaign, Urbana, Illinois 61801
[‡]KLA Instruments Corp., San Jose, California
[§]Naval Research Laboratory,[2] Washington, D.C. 20375.

**Abstract.**
We have performed a Diffraction Anomalous Fine Structure (DAFS) study of a single crystal of the relaxor ferroelectric $PbMg_{1/3}Nb_{2/3}O_3$. DAFS measurements were performed at the Nb $K$ edge on a half-order Bragg reflection originating from the 1:1 ordered nanodomains. DAFS data analysis provided us with the local structure around Nb atoms in the ordered nanodomains. The Nb-O distance in the ordered region was determined to be 0.052(10) Å shorter than half the lattice parameter of the bulk crystal, in good agreement with one model of the ferroelectric phase transition broadening [JETP **84**, 994 (1997)].

## INTRODUCTION

Compositional and structural fluctuations of the "relaxor" ferroelectric $PbMg_{1/3}Nb_{2/3}O_3$, or PMN, have been the focus of ongoing investigation since the first report of its synthesis in 1958 [1]. In addition to strong disorder of the Pb atomic positions at the A site of this $ABO_3$ cubic perovskite structure, the mixture of $Mg^{2+}$ and $Nb^{5+}$ cations at the B site have made precise structure determination complicated. Transmission electron microscopy [2] and polarized Raman spectroscopy [3] experiments have shown the existence of small, chemically ordered nanodomains. Recently, the structure and size of the ordered nanodomains in single crystal PMN have been refined by Fanning and Robinson (FR) [4]. They carried out

---

[1)] *Mailing address*: Building 510 E, Brookhaven National Laboratory, Upton, NY 11973. Electronic address: frenkel@bnl.gov. Supported by the DOE grant DEFG02-96ER45439.
[2)] Partially supported by a National Research Council and by the office of Naval Research.

CP436, *First-Principles Calculations for Ferroelectrics*
edited by Ronald E. Cohen
© 1998 The American Institute of Physics 1-56396-730-8/98/$15.00

synchrotron x-ray diffraction measurements of superstructure reflections originating from the ordered regions and measured the nanodomains to be approximately 50 Å in size.

Ordered nanodomains play an important role in several models of the diffuse paraelectric-ferroelectric phase transition in relaxor ferroelectrics [5–7]. The importance of the charge on the B-site for ordering through electrostatic forces in the $A(B'B'')O_3$ ordered regions has been emphasized by Setter and Cross [6] who used $PbSc_{0.5}Ta_{0.5}O_3$ as an example. Comparison between $Mg^{2+}$ and $Nb^{5+}$ ionic radii suggests that Nb-O bonds should be shorter than Mg-O bonds in these ordered domains. The Nb-O bond length is an important parameter which characterizes the charge transfer between $MgO_6$ and $NbO_6$ octahedra, since the shorter Nb-O bonds should be more covalent than the longer Mg-O bonds [8].

Bokov [9] demonstrated that the permittivity broadening parameter $\sigma$ of the ferroelectric phase transition in relaxor ferroelectrics is proportional to the $b/a$ ratio, where $b$ is the displacement of oxygens from their basal planes defined by Pb atoms, and $a$ is the lattice parameter. The coefficient of proportionality is calculated from the model to be 2409 K for PMN [9]. Using ionic radii to calculate $b/a$ in PMN (0.0102), Bokov obtained that the corresponding broadening $\sigma$ is 25 K, in a relatively good agreement with the experimentally measured 30 K broadening of the transition range in the PMN [10].

To check these and other models of the ferroelectricity in PMN and related perovskites more experimental structural data is needed. As of now, direct measurements of the Nb-O bond lengths in the ordered regions have not been available.

Several attempts have been undertaken to obtain the Nb-O bond length in the ordered regions. FR used the Nb-O distance as a fitting parameter in the refinement of their X-ray diffraction data. Their best fit required the shortening of Nb-O bonds by 0.044(3) Å relative to half of the lattice parameter $a = 4.04$ Å of PMN.

An Extended X-ray Absorption Fine-Structure (EXAFS) experiment has been reported at Nb $K$ edge [8]. EXAFS is a local structure probe, but in the case of PMN the data analysis is complicated by the coexistence of two types of local environments around Nb atoms. Assuming that the local structure in the Nb-rich host is rhombohedrally distorted, Prouzet, et al., [8] however, were able to demonstrate the indirect effect of the short Nb-O bond by an observed increase, from 3 to 4, in the average number per atom of short Nb-O bonds in the PMN.

In the present paper, we report a direct measurement of the local structure around the Nb atom in the ordered nanodomains of a single crystal of $PbMg_{1/3}Nb_{2/3}O_3$ by Diffraction Anomalous Fine Structure (DAFS) [11,12]. DAFS is a relatively new structural technique which combines the short-range structural sensitivity of EXAFS spectroscopy with structural specificity of X-ray diffraction. When measured at the superstructure peak, the DAFS signal contains only the contributions from the ordered nanodomains.

# SAMPLE PREPARATION AND DAFS EXPERIMENT

A crystal of pure PMN of roughly 2 mm per side with well formed ⟨100⟩ faces on three sides was used. Details on the crystal growth can be found elsewhere [4]. X-ray diffraction data were measured using a custom-designed 4-circle Kappa diffractometer at the National Synchrotron Light Source at Brookhaven National Laboratory, beamline X16-C. The orientation matrix of the crystal was determined by measuring several bulk Bragg reflections. DAFS measurements of the (0.5, 2.5, 2.5) Bragg peak were taken from 200 eV below to 600 eV above the Nb $K$ edge energy (18986 eV). The program SUPER was used for data collection, after suitable modification for energy-dispersive measurements. A total of 20 scans were taken which were later averaged for better signal/noise ratio. Fluorescence background was measured by repeating the DAFS scan after moving the 2Θ arm of the diffractometer to an off-peak position.

# REDUCING THE DAFS SIGNAL FROM THE BRAGG INTENSITY

The complex DAFS fine-structure function $\chi(E) = \chi'(E) + i\chi''(E)$, which ultimately contains the information about the local structure around the resonant atom, is related to the experimentally measured Bragg peak intensity *via* the anomalous scattering amplitude:

$$\Delta f(E) = f'(E) + if''(E) = f'_a(E) + if''_a(E) + f''_c[\chi'(E) + i\chi''(E)], \qquad (1)$$

where $E$ is the x-ray photon energy, $f'_a$ and $f''_a$ are the atomic parts of $f'$ and $f''$, respectively, and $f''_c(k)$ is the part of $f''_a(k)$ due solely to the excited core electron [13].

The fine-structure background removal and normalization procedure is formally described by

$$\chi''(k) = \frac{f''(k) - f''_a(k)}{f''_c(k)}, \qquad (2)$$

where $k = \sqrt{2m(E-E_0)}/\hbar$ is the photoelectron wave number, $E_0$ is the x-ray photon energy corresponding to the resonant atom absorption edge.

Equation (2) is the diffraction analog of the standard background removal and normalization procedure for EXAFS. In EXAFS experiments, the fine-structure $\chi(E)$ function is simply related to the experimentally measured total absorption cross-section $\sigma(E)$ through the mass absorption coefficient $\mu(E) \propto \sigma(E)$ with

$$\chi(k) = \frac{\mu(k) - \mu_0(k)}{\mu_0(k)}, \qquad (3)$$

where $\mu_0(k)$ is a smooth background function. The methods of determining $\mu_0(k)$ and of analyzing the structural content of $\chi(k)$ are well developed and have been described in detail in several places [14–16]. For DAFS collected in the $\sigma$-scattering geometry, with the x-ray polarization vector normal to the scattering plane, the DAFS $\chi''(k)$ and EXAFS $\chi(k)$ are identical. To obtain $\chi(k)$, EXAFS background removal can be applied to $\mu(E) \propto f''(E)/E$, and the fine-structure analysis follows the same procedures.

The DAFS $\chi''(k)$ is not simply related to the experimentally measured Bragg peak intensity $I(\mathbf{Q}, E)$. In the kinematic approximation, the intensity is proportional to the squared amplitude of the crystallographic structure factor $I(\mathbf{Q}, E) \propto |F(\mathbf{Q}, E)|^2$, with

$$F(\mathbf{Q}, E) = \sum_j [f_0(Q) + \Delta f(E)] e^{i\mathbf{Q}\cdot\mathbf{R}_j} e^{-M_j}, \qquad (4)$$

where the sum $j$ is over every atom in the unit cell, with position $\mathbf{R}_j$ and Debye-Waller factor $M_j$. Isolation of $\chi(k)$ from $I(\mathbf{Q}, E)$ can be accomplished in two ways: 1) spline subtraction and normalization, which leaves a $\mathbf{Q}$-dependent phase residue and amplitude correction in the photoelectron scattering paths [17,18], and 2) iterative Kramers-Krönig decomposition [12,18] of the intensity, which uses the dispersion relations between $f'(E)$ and $f''(E)$ and requires the resonant atom subset of the unit cell to satisfy the symmetry conditions described below. The iterative Kramers-Krönig method was used to isolate $\chi''(E)$ for the analysis presented here.

Since the (0.5, 2.5, 2.5) superstructure peak corresponds to the 1:1 B':B" ordering in the nanodomains where the unit cell size effectively doubles ($a = 8.08$ Å), we can re-define the peak as (1, 5, 5) in terms of a new $Fm3m$ unit cell with a composition Pb(8)B'(4)B"(4)O(24) and four formulae in the unit cell. The structure factor corresponding to the photon wavevector $\mathbf{Q} = \frac{2\pi}{a}(1, 5, 5)$ in the new unit cell is equal to $F(E) = 4i[f_{B'}(E) - f_{B''}(E)]$, where $f(E)$ is the scattering amplitude of a single atom (B' or B"). It should be noted that the bulk integer-order Bragg peaks in this experiment were well separated from the superstructure peaks, thus ensuring that the DAFS signal contained contributions from the ordered nanodomains only. In the DAFS analysis for this problem, we make the simplifying assumption that B' = Mg, B" = Nb and, therefore, the local environment is the same for all Nb atoms in the superstructure unit cell. Thus the anomalous amplitude $\Delta f(E)$ and, therefore, DAFS $\chi(E)$ measured on the superstructure peaks can be brought outside the partial structure factor sum over the resonant sites, i.e.,

$$\sum_n \Delta f(E) e^{i\mathbf{Q}\cdot\mathbf{R}_n} e^{-M_n} = \Delta f(E) \sum_n e^{i\mathbf{Q}\cdot\mathbf{R}_n} e^{-M_n} = \Delta f(E) \alpha(\mathbf{Q}), \qquad (5)$$

where the sum $n$ is over the Nb atoms only. The $Fm3m$ symmetry condition eliminates cross-terms between $f'(E)$ and $f''(E)$ in the intensity, and allows $I(\mathbf{Q}, E)$ to be solved as a simple quadratic equation in $f'(E)$ or $f''(E)$. The remaining part of the structure factor

$$F_0(\mathbf{Q}, E) \approx |F_0(\mathbf{Q})| e^{i\Phi_0(\mathbf{Q})} \tag{6}$$

includes the Thomson scattering from all of the atoms in the unit cell, and the anomalous scattering terms from all of the off-resonance electrons. The iterative Kramers-Krönig method can be used only when the coefficient $\alpha(\mathbf{Q})$ is either pure real or pure imaginary. For the (1, 5, 5) superstructure reflection, $\alpha(\mathbf{Q})$ is pure imaginary, and the intensity can be written in a convenient form for computer modeling:

$$I(\mathbf{Q}, E) = I_0 \left[ (\cos \Phi_0 - \beta(\mathbf{Q}) f''(E))^2 + (\sin \Phi_0 + \beta(\mathbf{Q}) f'(E))^2 \right] + I_{\text{offset}}, \tag{7}$$

where $\beta(\mathbf{Q}) = \alpha/|F_0|^2$, and $I_0$, $\Phi_0$, $\beta$ and $I_{\text{offset}}$ can be used as adjustable fitting parameters in a non-linear least-squares fit to the measured intensity. Since the average structure factor is known, however, $\Phi_0$ and $\beta$ were held fixed and only $I_0$ and a linear offset (introduced to correct for neglecting smooth energy dependence in Eq. (6)) $I_{\text{offset}} = a_1 + a_2 E$ were allowed to vary.

The data were first corrected for the Lorentz-polarization factor and for self-absorption by the sample. The energy dependent Lorentz-polarization correction for $\sigma$ scattering is $(E^3 \sin 2\theta)^{-1}$ at the Bragg angle $2\theta$. The self-absorption correction $1/2\mu(E)$ contains absorption fine-structure not only from the domains contributing to the Bragg peak, but from all of the resonant atoms in the diffracting volume [19]. Fluorescence EXAFS measured simultaneously with the DAFS was used for the self-absorption correction.

The iterative Kramers-Krönig procedure was applied to fit Eq. (7) to the experimental DAFS data and to extract $f'$ and $f''$ using computer programs developed at the University of Washington [18].

Bare-atom $\Delta f_a(E)$ calculated using the method of Cromer-Liberman [20] were used for the initial guess to $\Delta f(E)$. The normalized DAFS data $I(E) = (I_{\text{signal}} - I_{\text{bkg}})/I_0$, where $I_{\text{signal}}$ is the total Bragg intensity, $I_{\text{bkg}}$ is the fluorescence background, and $I_0$ is the incident beam monitor signal, are shown in Fig. 1. The resulting $f'(E)$ and $f''(E)$ are shown in Fig. 2. The calculated absorption cross-section $\sigma(E)$ is shown in Fig. 3.

## DATA ANALYSIS AND RESULTS

The anomalous scattering amplitude $f''(E)$ obtained by the method described above was converted to the mass absorption coefficient $\mu(E)$ and analyzed using the UWXAFS [21] data analysis package. In accordance with Eq. (3), smooth atomic background (Fig. 3) was removed from $\mu(E)$ using computer program AUTOBK [16], and EXAFS signal $\chi(k)$ was obtained (Fig. 4 (a)). Due to the significant amount of noise in the $\chi(k)$ data above 8 Å$^{-1}$, we analyzed the Nb-O (1NN) distances only.

The theoretical EXAFS equation for the single scattering contributions can be expressed in the form [14]:

$$\chi(k) = \frac{S_0^2 N}{kr^2} f(k) e^{-2k^2\sigma^2} \sin(2kr + \delta(k)) e^{-2r/\lambda(k)}, \tag{8}$$

where $S_0^2$ is the passive electron amplitude reduction factor, $N$ is the coordination number of the Nb-O bonds (6), $r$ is half the total scattering path length, $\sigma^2$ is the mean square relative deviation of $r$, $f(k)$ and $\delta(k)$ are the effective scattering amplitude and phase shift, respectively, and $\lambda(k)$ is the photoelectron mean free path.

$f(k)$, $\delta(k)$ and $\lambda(k)$ for Nb-O bonds were calculated with an *ab initio* FEFF6 [22] code using the cubic perovskite structure of the $Fm3m$ space group with $a = 8.08$ Å as a model. Small deviations of the 1NN distances from this model structure by less than 0.1 Å should not affect $f(k)$, $\delta(k)$, and $\lambda(k)$ by more than 5%, which is

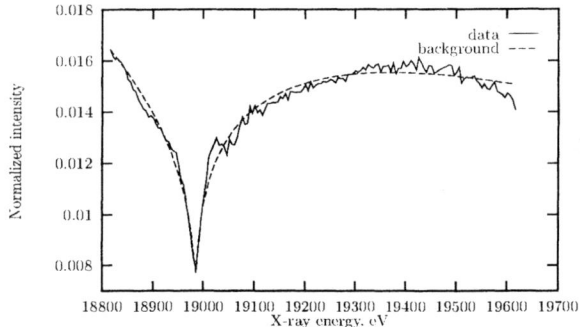

**FIGURE 1.** Normalized DAFS signal (solid) and a smooth background (dash) obtained with Eq. (7).

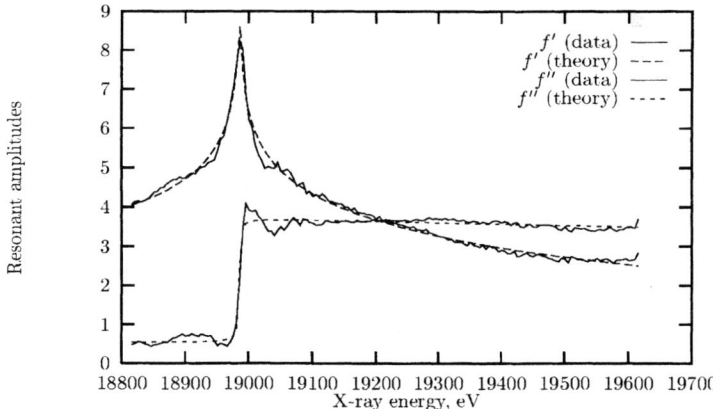

**FIGURE 2.** The results for the $f'(E)$ and $f''(E)$ (solid) as obtained by fit with the Kramers-Krönig iterative procedure. Theoretical bare atom response functions $f'_a(E)$ and $f''_a(E)$ (dash), used as a starting approximation, are shown for comparison.

comparable to the systematic error in FEFF6 theory.

The non-linear least squares fit was performed in $r$ space by Fourier transforming the data and theory within the $\delta k$ range between 2 and 7 Å. The data and theory were $k$-weighted before the transforms. The fitting range in $r$ space $\delta r$ was between 1.1 and 2.2 Å giving 5 relevant independent data points $N_{\text{idp}}$, according to the definition [23]: $N_{\text{idp}} = 2\delta k \delta r / \pi + 2$. The number $P$ of fitting parameters, therefore, should not exceed 5.

We varied 3 parameters in Eq. (8) to account for possible deviations of the local structure from the model. These variables were: correction to the photoelectron energy origin, $\Delta E_0$, and the two leading cumulants of the Nb-O effective pair distribution function, i.e., correction to the model distances ($\Delta r$) and a mean square disorder in the bond lengths ($\sigma^2$). To decrease the number of variables, we fixed $S_0^2$ to be equal 0.9 as found previously in the Nb $K$ edge EXAFS analysis of KNbO$_3$ [24]. An anharmonic correction to the phase of $\chi(k)$, the third cumulant $\sigma^{(3)}$, was not used in our fit procedure. Although this factor is important for the correct distance determination, all our attempts to add it to the fit resulted in unreasonably large values for $\Delta r$ and $\Delta E_0$, because of their correlation with $\sigma^{(3)}$. Previous works demonstrated that the effect of anharmonicity for B-O bonds in ABO$_3$ oxygen perovskites is negligible since they are characterized by the relatively rigid oxygen octahedra, therefore we believe that neglecting the anharmonic correction is a good approximation.

The Nb-O distance in the ordered region was determined to be 1.968(10) Å, i.e., 0.052(10) Å shorter than half the lattice parameter (4.04 Å) of the PMN. This result is in good agreement with the previous EXAFS [8] and x-ray diffraction [4] results.

The mean square disorder $\sigma^2$ in the Nb-O bonds length was determined to be 0.0087(3) Å$^2$. Prouzet, et al., [8] obtained an average $\sigma^2$ of 0.0036 Å$^2$ by analysis of the bulk EXAFS data. Their result, however, included the contributions of all Nb-O bonds and not just those associated with the 1:1 ordered nanodomains.

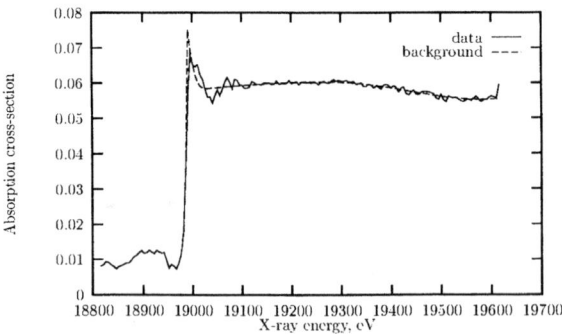

**FIGURE 3.** Normalized absorption cross-section $\sigma(E)$ (solid) and a smooth background function (dash) obtained with AUTOBK.

# DISCUSSION

Our DAFS analysis resulted in the measurement of the Nb-O bond length and its mean square disorder in the ordered region of PMN. We found that this bond is significantly (by 0.052(10) Å) shorter than the average B-O distance in bulk PMN. Our result is an independent measurement in favor of the model suggested by Fanning and Robinson [4] based on their X-ray diffraction analysis of superstructure reflections of PMN. In their model, FR proposed chemical 1:1 ordering of B' and B" atoms, together with concomitant displacements of oxygen atoms towards Nb in a (100) direction. Prouzet, et al. [8], arrived at similar conclusions by analyzing their EXAFS data obtained for the bulk PMN.

Although the DAFS method combines the capabilities of diffraction and EXAFS in a single technique, it has several enhanced sensitivities compared to the separate techniques. First, it can provide EXAFS-like information for the specific subset of

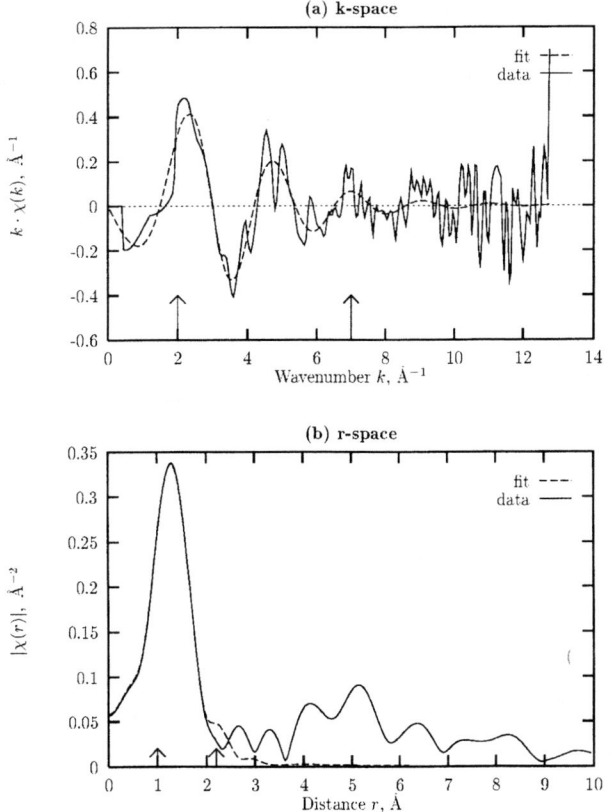

**FIGURE 4.** Fit of the FEFF6 theory (dash) to the data (solid) in (a) $k$-space and (b) $r$-space. Weighting factor $k$ was applied to both theory and data.

atoms selected by the diffraction condition [25]. This makes DAFS clearly advantageous over EXAFS for analysis of PMN. In conventional EXAFS measurement of PMN, the resonant atom resides in both phases, ordered nanodomains and Nb-rich host domains. Therefore, sophisticated, model-dependent analysis is necessary to extract the contribution of the nanodomain structure to the total signal. Using DAFS, we filtered the nanodomain structure out of the total structure factor by aiming the detector at the superstructure reflections, originated from the ordered nanodomains only.

Second, DAFS technique has the same advantage as EXAFS over the diffraction analysis. Since the two former methods contain the phase information of the interference between the outgoing and backscattered photoelectron, their Fourier transforms contain direct information on the local structure in the system. In diffraction techniques, only the intensity of the diffracted wave is measured, and additional modeling is required to determine structural information.

Third, DAFS inherits from EXAFS technique a better spatial resolution over diffraction method. In both EXAFS and diffraction, the spatial resolution is inversely proportional to the upper end of the $q$ range of the data, $1/q_{max}$. In the case of PMN, due to the small size of coherent scattering region (50 Å) both $q$ and $k$ ranges in the XRD and DAFS experiments, respectively, were limited by small intensity of the superstructure Bragg reflections and the low signal/noise ratio. In both cases, $k_{max}$ was approximately $7 \text{ Å}^{-1}$. For DAFS, however, this translates to a spatial resolution by a factor of two better than that of the diffraction measurement since the momentum transfer in DAFS (and EXAFS) experiments is $q = 2k$. Even with this enhanced resolution, however, due to the small nanodomain size and, therefore, limited amount of information in the DAFS data, we were unable to unambiguously resolve subtle details of the Nb local environment beyond the first Nb-O shell. We were unable to determine whether the equilibrium positions of Nb atoms are in the center of symmetry of the oxygen octahedron, or displaced in one of 8 possible $\langle 111 \rangle$ directions.

Our results for the Nb-O distance shortening from its bulk value by 0.052(10) Å, and the corresponding ratio of this displacement to the bulk lattice parameter, $b/a = 0.013(2)$, allow us to calculate a broadening parameter $\sigma$ for the ferroelectric phase transition. Following the model proposed by Bokov [9], we obtain for PMN: $\sigma = 2904 \times 0.013 = 31 \pm 6$ K. This result is in very good agreement with the experimental broadening of 30 K [10].

Recent experiments with La-doped PMN crystals showed an increase in the size of the nanodomains up to 1000 Å [4]. This presents more opportunities for further investigation of the local structure in the ordered regions. For example, analysis of Nb-Pb bonds would help solve the long lasting controversy between the models of disorder of Pb atoms. Multiple-scattering analysis of Nb-O-Mg distances promises more information about charge transfer between $NbO_6$ and $MgO_6$ octahedra.

# CONCLUSIONS

Shortening by 0.052(10) Å of Nb-O bonds in $NbO_6$ octahedra in the ordered nanodomains of $PbMg_{1/3}Nb_{2/3}O_3$ relative to the average Nb-O distance has been established by DAFS. This result is in good agreement with previous results obtained by XRD and EXAFS methods, and with the experimentally observed broadening of the ferroelectric phase transition.

# REFERENCES

1. Smolenskii G. A. and Agranovskaya A. I., *Sov. Phys. – Tech. Phys.* **3**, 1380 (1958).
2. Randall C., Bhalla A., Shrout T., and Cross L. E., *J. Mater. Res.* **5**, 829 (1990).
3. Siny I. and Boulesteix C., *Ferroelectrics* **96**, 119 (1989).
4. Fanning D. M. and Robinson I. K., to be published.
5. Galasso F. S., *Structure, Properties and Preparation of Perovskite-Type Compounds*, Paris: Pergamon Press, 1969.
6. Setter N. and Cross L. E., *J. Mat. Science* **15**, 2478 (1980).
7. Randall C. A., Barber D. J., Whatmore R. W., and Groves P., *J. Mat. Science* **21**, 4456 (1986).
8. Prouzet E., Husson E., de Mathan N., and Morell A., *J. Phys.: Condens. Matter* **5**, 4889 (1993).
9. Bokov A. A., *JETP* **84**, 994 (1997).
10. Smolenskij G. A., Bokov V. A., Isupov V. A., et al., *Physics of Ferroelectric Phenomena* [in Russian], Leningrad: Nauka, 1985.
11. Stragier H., Cross J. O., Rehr J. J., Sorensen L. B., Bouldin C. E., and Woicik J. C., *Phys. Rev. Lett.*, **69**, 3064 (1992).
12. Pickering I. J., Sansone M., Marsch J., and George G. N., *J. Am. Chem. Soc.*, **115**, 6203 (1993).
13. Cross J. O., Newville M., Rehr J. J., Sorensen L. B., Watson G., Bouldin C. E., Bell M. I., Gouder T., and Lander G. H., unpublished.
14. Stern E. A. and Heald S. M., *Handbook on Synchrotron Radiation*, edited by E. E. Koch; New York: North-Holland, 1983, Ch. 10.
15. Koningsberger D. C. and Prins R., editors, *X-Ray Absorption: Principles, Applications, Techniques of EXAFS, SEXAFS, and XANES*, Vol. 92 of *Chemical Analysis*; New York: John Wiley & Sons, 1988.
16. Newville M., Livinš P., Yacoby Y., Rehr J. J., and Stern E. A., *Phys. Rev. B* **47**, 14126 (1993).
17. Arčon I. and Hribar M., *X-Ray Absorption Fine Structure*, edited by S. S. Hasnain; London: Ellis Horwood, 1991, pp. 726-728.
18. Cross J. O., Ph.D. Thesis, University of Washington, 1996.
19. Cullity B. D., *Elements of X-Ray Diffraction*, 2nd edition, Addison-Wesley, 1978, Ch. 4.
20. Cromer D. T. and Liberman D., *J. Chem. Phys.*, **53**, 1891 (1970).

21. Stern E. A., Newville M., Ravel B., Yacoby Y., and Haskel D., *Physica B*, **208 & 209**, 117 (1995).
22. Zabinsky S. I., Rehr J. J., Ankoudinov A., Albers R. C., and Eller M. J., *Phys. Rev. B* **52**, 2995 (1995).
23. Stern E. A., *Phys. Rev. B* **48**, 9825 (1993).
24. Frenkel A. I., Wang F. M., Kelly S., Ingalls R., Haskel D., and Stern E. A., *Phys. Rev. B* **56**, 10869 (1997).
25. Sorensen L. B., Cross J. O., Newville M., Ravel B., Rehr J. J., Stragier H., Bouldin C. E., and Woicik J. C., *Resonant Anomalous X-Ray Scattering*, edited by Materlik G., Sparks C. J., and Fischer K.; Amsterdam, Elsevier Science, 1994, p. 389.
26. You H. and Zhang Q. M., *Phys. Rev. Lett.*, **79**, 3951 (1997).

# Hartree–Fock Studies of the Ferroelectric Perovskites

L. Fu,[1] E. Yaschenko,[1] L. Resca,[1] and R. Resta[1,2]

[1] *Department of Physics, The Catholic University of America, Washington, D.C. 20064*
[2] *INFM–Dipartimento di Fisica Teorica, Università di Trieste, Strada Costiera 11, I-34014 Trieste, Italy*

**Abstract.** Within an *ab–initio* HF scheme, we use both Berry–phase calculations and supercell calculations in order to compute the dynamical charges for lattice dynamics and the electronic dielectric constant for $KNbO_3$ and $BaTiO_3$. Comparison with experimental data indicates that HF provides a description of the electronic properties of this material whose accuracy is of the same order as the LDA one. There are however significant differences between the two sets of results, whose origin is scrutinized. Motivated by the study of surface and domain–boundary properties, we also present some results for $BaTiO_3$ slabs, including both genuinely isolated and periodically repeated slabs with different terminations. The capability of dealing with a genuinely isolated slab is a virtue of the localized–basis implementation adopted here. We demonstrate, amongst other things, the nontrivial dynamical–charge neutrality of $BaTiO_3$ [001] surfaces.

## INTRODUCTION

Density–functional theory (DFT) has been very successful in predicting several physical properties of the ferroelectric perovskites, most implementations being performed within the local–density approximation (LDA). Here we assess the performance of the alternative Hartree–Fock (HF) scheme for these materials, which have an intermediate ionic/covalent character. In general, DFT is particularly convenient and accurate for metals and semiconductors [1], while the HF scheme is very reliable for ionic crystals, and insulators in general [2,3]. In this paper, we address at the HF level some fundamental dielectric properties, such as the dynamical ionic charges and the dielectric constant, in two paradigmatic ferroelectric perovskites: $KNbO_3$ and $BaTiO_3$.

The very first HF study of a ferroelectric perovskite, at the *ab–initio* level, has been published very recently [4]. The case study was $KNbO_3$, and several properties of the electronic ground state were investigated, including the broken–symmetry instability of the tetragonal structure of this material. In this same work the Berry–phase theory of macroscopic polarization [5,6] is implemented within the

HF scheme, using first-principles ingredients. Perhaps the most interesting result that has emerged is that the calculated HF spontaneous polarization of KNbO$_3$ in its tetragonal phase is 0.34 C/m$^2$, quite close to the experimental value [7], which is 0.37 C/m$^2$, as well as previous LDA results [8–10], which are between 0.33 and 0.40 C/m$^2$.

Although the computed value of the spontaneous polarization is almost identical within LDA and HF, we will show that there are significant differences when the dynamical charges of the ions are separately investigated. These polarizabilities are dominated by covalency effects, which plausibly are described in different ways at the HF and LDA levels.

We will then show that, by combining results from both Berry–phase calculations and supercell calculations, one has access to the theoretical value of the electronic dielectric constant $\varepsilon_\infty$ [11]. It is well known that LDA overestimates the value of $\varepsilon_\infty$ in any material, whereas it has been only guessed that HF underestimates it, since no calculation was previously available, for any material. Our HF calculated $\varepsilon_\infty$ of KNbO$_3$, BaTiO$_3$ and MgO are indeed smaller than experimental values.

Finally, we will show some results for BaTiO$_3$ slabs, including both isolated and periodic slabs with different terminations. Since most first–principles schemes are implemented within plane–wave based computational methods, they are very effective for infinite periodic systems but cannot treat the isolated slabs. Even the periodic slabs with moderate vacuum separations prove to be a heavy computational burden for plane–wave methods, and only a few such calculations have been performed for ferroelectric perovskites [12,13]. The capability of dealing with a genuinely isolated slab is a virtue of the localized–basis implementation adopted here [2,14]. We will discuss bulk dielectric properties, surface charge, surface energies, work function, and surface dynamical charges. We will show how the computed values of these quantities change as functions of the number of atomic layers and the thickness of the vacuum layers separating the periodic slabs. Owing to a recently discovered theorem, the dynamical charge of a given ion at a polar surface *cannot* have the same value as in the bulk [15].

# BULK DYNAMICAL CHARGES AND DIELECTRIC CONSTANT

The transverse (Born) charge tensor is defined via the macroscopic polarization $\Delta \mathbf{P}$ linearly induced by a rigid displacement $\mathbf{u}_s$ of the $s$ sublattice, while the field is kept at zero value [16]. Namely,

$$\Delta P_\alpha = \frac{1}{\Omega} \sum_\beta Z^{*(\mathrm{T})}_{\alpha\beta} u_\beta, \qquad (1)$$

where $\Omega$ is the cell volume, and atomic units are used throughout. The longitudinal (Callen) charge tensor $Z^{*(\mathrm{L})}$ is similarly defined, but with the sublattice

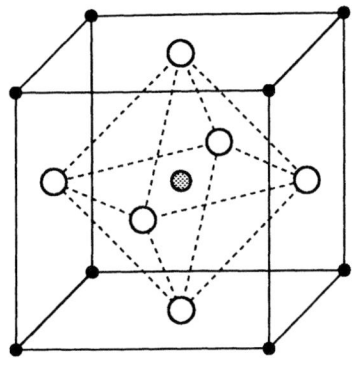

Figure 1. Cubic perovskite structure, with general formula ABO$_3$, where in our cases A is either K or Ba (solid circles) and B is either Nb or Ti (shaded circle). The oxygens (empty circles) form octahedral cages, with B at their centers, and arranged in a simple cubic pattern. We will consider either displacements in the vertical direction, or slabs with horizontal surfaces. In both cases "O$_I$" refers to top and bottom oxygen ions and "O$_{II}$" refers to the four oxygen ions in the basal plane of the octahedron. We have therefore AO$_I$ and BO$_{II}$ planes.

displacement performed in a depolarization field $\Delta \mathbf{E} = -4\pi \Delta \mathbf{P}$. The relation between the two dynamical charges is

$$Z^{*(T)} = \epsilon_\infty Z^{*(L)}, \qquad (2)$$

where $\epsilon_\infty$ is the dielectric constant in the relevant direction. In general, the dynamical charge can be very different from the nominal static charge of an ion. The differences are particularly dramatic when the material has a mixed ionic–covalent character, such as the materials chosen in the present work [17,18].

We calculate the $Z^{*(T)}$ as a Berry–phase [5,6], while we calculate the $Z^{*(L)}$ using a supercell technique. The latter approach was introduced by Martin and Kunc several years ago to compute dynamic charges in semiconductors [19], and is still very much in use today in the framework of LDA pseudopotential calculations with plane–wave basis sets. The same supercell technique in the framework of HF calculations with a localized basis set was implemented for the first time in Ref. [11]. Since the aim of the present work is to assess the HF values versus LDA ones, we only present here results for the cubic structure. For KNbO$_3$ we use the lattice constant 4.016 Å. This is the same value used in the LDA study of Ref. [10], which is based on the low–temperature experimental data [7]. For BaTiO$_3$, we use the lattice constant 4.006 Å. Figure 1 shows the structure of these perovskites.

We use the CRYSTAL95 code, which implements both the HF and LDA self consistent scheme with a linear–combination–of–atomic–orbitals (LCAO) basis set [14]. In our supercell calculations, we displace the sublattice of an atom in the supercell lattice, and compute the planar average of the charge density, potential, and field. We then perform additionally the macroscopic average defined as in Ref. [20], which proves an invaluable tool in the present business. The output of the supercell calculations provide the longitudinal dynamical charges. We use both 3–fold and 4–fold supercells in our supercell calculations, which yield almost identical results. We refer to the original paper for more details [11].

In our Berry–phase calculations, we compute the induced macroscopic polarization $\Delta \mathbf{P}$ in zero field, and thus the corresponding transverse dynamical charges.

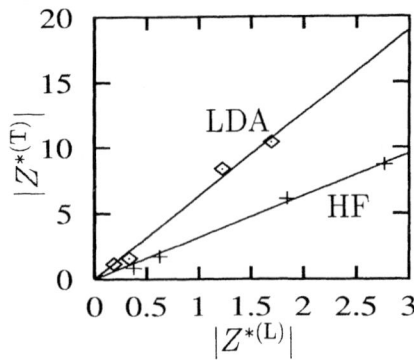

Figure 2. The transverse charge $|Z^{*(T)}|$ versus longitudinal charge $|Z^{*(L)}|$. The slopes of the two solid lines are, respectively, the LDA and HF dielectric constant $\epsilon_\infty$.

In order to compare with the LDA results and examine the performance of our localized basis set, we also run the CRYSTAL95 code in the LDA option for KNbO$_3$. Our results for KNbO$_3$ dynamic charges are reported in Table I.

Table I. Dynamical charges for KNbO$_3$

|  | K | Nb | O$_I$ | O$_{II}$ | sum |
|---|---|---|---|---|---|
| $Z^{*(L)}$, HF, 3–fold supercell | 0.377 | 2.766 | −1.848 | −0.624 | 0.047 |
| $Z^{*(L)}$, LDA, 3–fold supercell | 0.187 | 1.696 | −1.227 | −0.335 | −0.014 |
| $Z^{*(T)}$, HF, Berry–phase | 0.830 | 8.702 | −6.130 | −1.703 | −0.004 |
| $Z^{*(T)}$, LDA, Berry–phase | 1.144 | 10.400 | −8.377 | −1.567 | 0.033 |
| $Z^{*(T)}$, LDA, Ref. [10] | 1.12 | 9.67 | −7.28 | −1.74 | 0.03 |

The last column shows that the acoustic sum rule [16] is satisfactorily obeyed. A few calculations for $Z^{*(T)}$ of KNbO$_3$ at LDA level exist [8–10]: these data are slightly scattered and in some cases refer to slightly different geometries. Here we compare our results with Ref. [10], which uses the same geometry. Our LDA $Z^{*(T)}$ are in reasonable agreement with those of Ref. [10]. The small differences may be due to the fact that our LDA exchange-correlation functional differs somewhat from that of Ref. [10], and our pseudopotentials and basis sets are optimized for HF rather than LDA.

To determine the dielectric constant $\epsilon_\infty$, we plot $|Z^{*(T)}|$ versus $|Z^{*(L)}|$, and use a least–square fitting to determine the slope which yields $\epsilon_\infty$. This is shown in Fig. 2, which indicates that the correlation between our $Z^{*(L)}$ and $Z^{*(T)}$ is reasonably linear. For KNbO$_3$ we obtain the LDA value $\epsilon_\infty = 6.33$, which is close to the LDA value of 6.63 obtained in Ref. [10], and is about 35 % higher than the experimental value 4.69. Similarly, we obtain the HF value $\epsilon_\infty = 3.17$, which is about 32 % smaller than the experimental value.

Table II. Dynamical charges for BaTiO$_3$

|  | Ba | Ti | O$_I$ | O$_{II}$ | sum |
|---|---|---|---|---|---|
| $Z^{*(L)}$, HF, 3–fold supercell | 0.909 | 2.126 | −1.468 | −0.797 | −0.027 |
| $Z^{*(T)}$, HF, Berry–phase | 2.052 | 5.996 | −4.319 | −1.846 | −0.037 |
| $Z^{*(T)}$, LDA, Ref. [21] | 2.74 | 7.29 | −5.75 | −2.13 | 0.02 |

We have performed analogous calculations for BaTiO$_3$, whose results are reported in Table II: the resulting HF dielectric constant is $\epsilon_\infty = 2.76$. This value is about one half of the experimental value 5.40. The LDA value reported in Ref. [21], for a slightly different lattice constant, is 6.73. We have also performed a series of similar calculations for the more simple oxide MgO [22]. We use the cubic lattice constant $a = 4.21$ Å. We obtain HF $Z^{*(L)} = \pm 1.267$ and $Z^{*(T)} = \pm 1.809$, which yields a HF $\epsilon_\infty = 1.428$. This value is about one half of the experimental value 2.94, while the LDA value reported in Ref. [23] is 3.14.

As a general conclusion about the study of bulk dielectric properties, we find that the calculated transverse dynamical charges are about of the same quality within HF and LDA, the trend being that the HF ones are generally closer to the nominal value of the corresponding static charges. This seems to indicate that HF tends to underestimate the covalence mechanism, while LDA tends to overestimate it. The resuts for the dielectric constants confirm this same message.

## SURFACE PROPERTIES VIA SLAB CALCULATIONS

We now switch to the HF investigation of several surface properties of BaTiO$_3$. To this aim we have used both isolated slabs and periodically repeated ones, while at this preliminary stage surface geometries have not been relaxed: all results refer therefore to ideal (truncated bulk) geometries. There are two major differences between the two kinds of calculations. First of all, there are interactions among the periodic slabs, which presumably vanish when the vacuum separating the slabs is sufficiently thick. Using a basis of plane waves (or plane-wave-like, such as LAPW), the increase in the vacuum space drastically increase the number of plane waves necessary for convergence, and therefore the computational workload. Secondly, the isolated slabs are calculated by imposing the boundary condition of vanishing external field, while the periodic slabs are necessarily calculated with the boundary condition of vanishing *average* field (over the supercell). This is a major problem when dealing with supercells built of asymmetrically terminated slabs, even in absence of macroscopic polarization. The reason is that the work functions of the two surfaces are generally different. The supercell is an artificial periodic structure of repeated slabs and vacuum layers, where the electrostatic potential is enforced to be periodical. This periodicity, combined with the vacuum-level difference (due

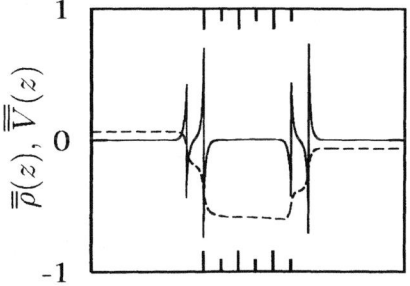

Figure 3. The macroscopic average of charge density ($10^{-1}$ electron/bohr$^3$, solid line) and potential (electron$^2$/bohr, dashed line) for an isolated slab of six atomic layers, where the bulk structure is centrosymmetric. The vertical bars indicate the positions of the atomic layers: short bars for BaO$_\text{I}$ planes, long bars for TiO$_\text{II}$ planes.

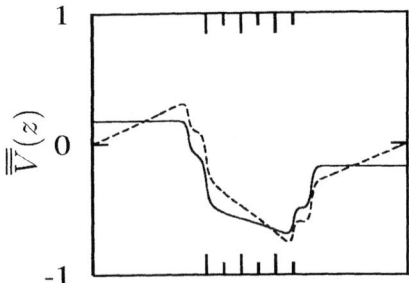

Figure 4. Macroscopic averages of the potential (electron$^2$/bohr) for an isolated slab (solid line) and a periodically repeated slab (dashed line). Both calculations refer to slabs cut from a noncentrosymmetric bulk having the experimental ferroelectric distortions. The meaning of vertical bars is the same as in Fig. 3.

to the different work functions), necessarily originates fictitious electric fields and *static* surface charges. Such unphysical charges can in principle be made arbitrarily small by increasing the thickness of the vacuum layers, but in practice a large enough supercell may be computationally unaffordable.

The problem does not occur when performing calculations for isolated slabs within a localized basis set. As an example we plot in Fig. 3 the macroscopic average of charge density, potential, and field for an isolated slab with asymmetric termination, where however the slab has been cut from a centrosymmetric bulk. There are several very perspicuous features to remark. (1) First of all, despite the asymmetric termination of the slab of Fig. 3, the macroscopic field vanishes both in the bulk region and in the vacuum region. This means that each of the two surfaces is neutral, as required by very general considerations. (2) One observes some structure only in the surface regions. Basically, there is a quadruple charge layer (two double layers) at each surface. We also see that in the middle region of the slab both the charge and field vanish and the potential remains constant. This means that the bulk like behavior is recovered in the middle region of the slab. (3) Although the field in the vacuum regions is zero, the potential goes to two different limits to the left and to the right of the sample. This net change in the potential across the slab is precisely the difference in the work functions of the two nonequivalent surfaces, discussed above. (4) We have additionally calculated all kinds of slabs (asymmetrically terminated as in Fig. 3, doubly Ti-terminated, doubly Ba-terminated): all calculated surface properties look precisely the same, while instead they artificially *do* depend on the choice of the slab in the periodically repeated case.

As a matter of fact, periodically repeated slabs are systematically messed up by the above problems. We show here an example in Fig. 4, where the occurrence of the fictitious field is very perspicuous, despite the large thickness of the vacuum region (6 lattice constants in this calculation). Because of this major problem, very few calculations of a periodically repeated slab (asymmetrically terminated) have ever been performed, for any material. For a discussion of the difficulties, see Refs. [12,15]. By contrast with this, the isolated–slab calculation conveys a very clear message, and the slope of the potential in the bulk region of Fig. 4 is no artifact: the presence of a macroscopic field is a physical feature, related to the spontaneous polarization of the bulk material. This will be discussed in the

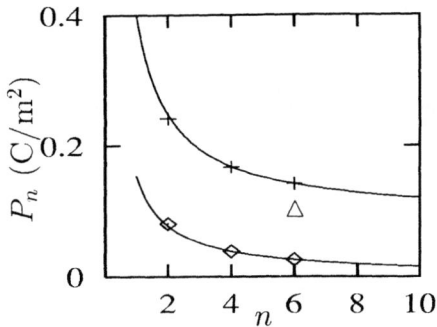

Figure 5. Dipole moment per unit volume as functions of the number of atomic layers, for isolated centrosymmetric (diamonds) and ferroelectric (crosses) slabs. The solid lines are fits of Eq. (4). The triangle is the bulk polarization calculated from the field in the bulk region of the ferroelectric slab.

following.

## Polarization, surface charge, and surface dipoles

Let us consider the total dipole per unit area of a finite slab of $n$ atomic layers: this dipole is easily evaluated from isolated–slab calculations simply looking at the net change in potential across the slab $\Delta V_{\text{slab}}$, and is obviously nonzero even for centrosymmetric slabs (with asymmetric termination), as in Fig. 3. In the case of a ferroelectric slab, an additional contribution to the dipole comes from the spontaneous polarization. In the limit of large $n$, one must recover the appropriate bulk polarization. Since the nominal thickness of the slab is $na/2$, we write the dipole per unit volume as:

$$P_n = \frac{\Delta V_{\text{slab}}}{2\pi na} \simeq \sigma_0 + \frac{2p_0}{n}. \qquad (3)$$

In the latter expression $\sigma_0$ is the charge per unit area which accumulates at the slab surface, and is a bulk property dependent on the spontaneous polarization of the material. Instead $p_0$ is proportional to the work function difference discussed above and is therefore a surface property. Our results are displayed in Fig. 5, which shows the asymptotic behavior of $P_n$: the surface charge $\sigma_0$ is clearly zero in the centrosymmetric case, while we estimate $P_\infty = \sigma_0 = 0.089 \text{C/m}^2$ for the ferroelectric isolated slab.

Besides the above asymptotic extrapolation, an alternative path to evaluating $\sigma_0$ consists in measuring the macroscoic electric field in the middle of the isolated slab. In the centrosymmetric case we get zero field at any thickness, while the value of $\sigma_0$ extracted in this way from the ferroelectric slab calculation, shown in Fig. 5 as a triangle, is $\sigma_0 = 0.095 \text{C/m}^2$, in good agreement with the above mentioned value.

Finally, we wish to compare the above calculated value of $P_\infty$ to the bulk spontaneous polarization of the material, as measured in experiments and as previously calculated within LDA [9,21]. It is important to notice that our $P_\infty$ is by definition the macroscopic polarization in a depolarizing field, while the literature generally refers to the spontaneous polarization in zero field. The latter is related to the former by a factor $\varepsilon_\infty$, much in the same way as the longitudinal and transverse dynamical charges in Eq. (2).

Using our theoretical value of 2.76 for the HF dielectric constant we get a spontaneous polarization of $0.245 C/m^2$, which compares extremely well to the experimental value of $0.263 C/m^2$ [24], and to LDA values in the range 0.286 to $0.363 \ C/m^2$ [9,21].

In order to achieve a more meaningful comparison, we have also run our own Berry phase calculation using identical technical ingredients and geometry as in the slab calculations. We get a value of $0.240 C/m^2$, in very good agreement with the above evaluation of the same quantity from the electric field inside the isolated slab. This shows the internal consistency of the HF properties, calculated at the HF level. Indeed, the small disagreement provides an estimate of the numerical error, within the same given physical approximation. We emphasize that, on the contrary, any periodic slab calculation would be affected by much larger numerical errors.

## Surface energy

We report here a study of the surface energies. We emphasize that—since the surface geometries have not been relaxed—our preliminary calculated values have not to be taken as realistic. What we demonstrate instead is that using a genuinely isolated slab we have a fast convergence to the genuine surface properties, whereas convergence is rather slow in the periodic–slab geometry. Our slabs, made of $n$ atomic layers cut from a centrosymmetric bulk, have symmetric termination for odd $n$, and asymmetric termination for even $n$.

In the present [001] geometry two nonequivalent truncations of the bulk crystal are obviously possible. Whenever this happens—as thoroughly discussed in the literature [25]—the surface energy of each of the nonequivalent surfaces is ill defined and any stability issue must be studied by introducing the appropriate chemical potential. We therefore address the average surface energy (over the two nonequivalent surfaces).

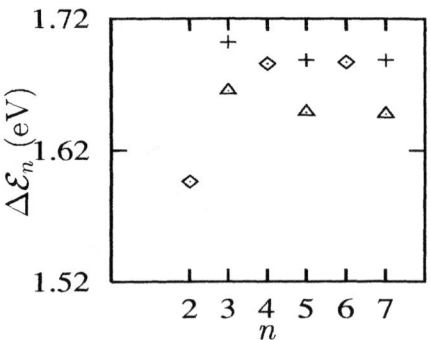

Figure 6. Calculated surface energy as functions of the number of atomic layers. The diamonds and crosses are, respectively, the results from isolated slabs with asymmetric and symmetric terminations. The triangles are the results from periodic slabs with symmetric terminations.

The average surface energy $\Delta\mathcal{E}$ per surface unit cell can be evaluated from slab calculations (with $n$ atomic layers) as the asymptotic value of

$$\Delta\mathcal{E}_n = \begin{cases} [2\mathcal{E}_n - n\mathcal{E}_{\text{bulk}}]/4, & \text{even } n, \\ [\mathcal{E}_n(\text{Ti}) + \mathcal{E}_n(\text{Ba}) - n\mathcal{E}_{\text{bulk}}]/4, & \text{odd } n, \end{cases} \quad (4)$$

where $\mathcal{E}_n(\text{Ti})$ and $\mathcal{E}_n(\text{Ba})$ indicate the energy of a symmetrically terminated slab with Ti or Ba termination, respectively.

From Fig. 6 we see that for isolated slabs, as $n$ increases, the average surface energy converges very fast to the asymptotic value: from above for odd $n$, and from below for even $n$. On the other hand, the results for periodic slabs are sistematically lower than those of isolated slabs. This indicates that interactions among slabs still exist, even with a large vacuum separation: in all calculations reported here, we use a fixed supercell size of 12 interlayer spacings. We mention that the surface energies obtained in previous calculations at LDA level using periodic slabs are 0.921 eV [12]) and 1.237 eV [13] for a relaxed geometry [12]), while 1.358 eV is reported for the unrelaxed geometry [13]. As stated above, our geometry is unrelaxed: the data in Fig. 6 extrapolate to 1.69 eV.

## Surface dynamical charges

We evaluate the surface longitudinal dynamical charges $Z^{*(L)}$ of BaTiO$_3$ from HF calculations performed on symmetrically-terminated slabs of seven atomic layers. The bulk is centrosymmetric, and the reference surface geometry is not relaxed. Upon displacing atoms in the surface region, we get the results in an analogous way as for the bulk longitudinal charges discussed above. The dynamical charges of ions in the first and second outermost surface layer are reported in Tab. III, together with the corresponding bulk values (taken from Tab. II), for the sake of comparison.

Table III. Surface $Z^{*(L)}$ (HF)

|  | Ba | Ti | O$_I$ | O$_{II}$ | Ba+O$_I$ | Ti+2O$_{II}$ |
|---|---|---|---|---|---|---|
| first layer | 1.104 | 2.093 | −1.354 | −0.919 | −0.250 | 0.255 |
| second layer | 0.911 | 2.119 | −1.473 | −0.767 | −0.562 | 0.585 |
| bulk | 0.883 | 2.187 | −1.543 | −0.765 | −0.660 | 0.657 |

We notice that BaO$_I$ planes and TiO$_{II}$ planes are *nominally* charge neutral, in the hypothesis of complete ionicity. But dynamical charges—particularly in a material having mixed ionic–covalent character as the present one—are dramatically different from the static nominal ones. As a matter of fact, the data in the last two columns of Tab. III show that BaO$_I$ planes and TiO$_{II}$ planes are far from being dynamically neutral, both in the bulk and in the surface region.

As observed above, the surface of a centrosymmetric slab is charge–neutral as far as the *static* charge accumulated in the surface region is concerned. This fact implies—owing to a recently discovered theorem [15]—that the surface region must also be *dynamically* neutral. We notice that the dynamical charges in the surface region deviate considerably from their bulk value, although they converge rather fast to it. A corollary of the dynamical–neutrality theorem, as applied to the [001] geometry considered here, states the the sum of the dynamical charges in the surface planes must add up to *one half* of the corresponding bulk value (with the appropriate sign) [15]. This sum rule—which can be regarded as the surface analogue of the popular acoustic sum rule for the bulk [16]—is very accurately fullfilled by our calculations. In fact summing over the two outermost surface layers we get $-0.250 + 0.585 = 0.335$ and $0.255 - 0.562 = -0.307$ for the two nonequivalent surfaces. Both figures are indeed very close (in modulus) to one half of the dynamical charge of a given bulk plane.

## CONCLUSIONS

We have reported several *ab–initio* HF results concerning bulk and surface properties of two important ferroelectric perovskites: $KNbO_3$ and $BaTiO_3$. Though some of our results are preliminary, they indeed assess the accuracy of both the physical approximations and the computational implementation.

About the physical approximations, comparison of our results with the experiment indicates an overall trend: the HF results are about of the same quality as—but on the opposite side of—the LDA results. This trend makes HF an appealing method for "bracketing" a theoretical prediction: it is well exemplified by the case study of the dielectric constant $\varepsilon_\infty$ discussed here for $KNbO_3$, $BaTiO_3$, and even for the simpler oxide MgO.

About the computational implementation within a localized basis set, we have emphasized its merits by showing some case studies of surface properties, where we can study genuinely isolated slabs in vacuo. This is at variance with basically all previous first–principle investigation of surface properties which, for any material, were typically performed within supercell (*i.e.* periodically repeated slab) geometries.

## ACKNOWLEDGMENTS

We are indebted to S. Dall'Olio and R. Dovesi for discussions and help in using CRYSTAL95, as well as for providing us with optimized values of the bulk computational parameters. This work is supported by the Office of Naval Research, through grant N00014-96-1-0689.

# REFERENCES

1. M.L. Cohen, in *Electronic Materials*, edited by J.R. Chelikowsky and A. Franciosi (Springer, Berlin, 1991), p.57.
2. C. Pisani, R. Dovesi, and C. Roetti, *Hartree-Fock Ab-Initio Treatment of Crystalline Systems*, Lecture Notes in Chemistry, Vol. **48** (Springer, Berlin, 1988).
3. *Quantum-Mechanical Ab-initio Calculation of the Properties of Crystalline Materials*, Lecture Notes in Chemistry, Vol **67**, edited by C. Pisani (Springer, Berlin, 1996).
4. S. Dall'Olio, R. Dovesi, and R. Resta, Phys. Rev. B **56**, 10105 (1997).
5. R. Resta, Ferroelectrics **136**, 51 (1992); R.D. King-Smith and D. Vanderbilt, Phys. Rev. B **47**, 1651 (1993); R. Resta, Europhys. Lett. **22**, 133 (1993).
6. R. Resta, Rev. Mod. Phys. **66**, 809 (1994).
7. M.D. Fontana, G. Métrat, J.L. Servoin, and F. Gervais, J. Phys.C **17**, 483 (1984); W. Kleemann, F.J. Schäfer, and M.D. Fontana, Phys. Rev. B **30**, 1148 (1984).
8. R. Resta, M. Posternak, and A. Baldereschi, Phys. Rev. Lett. **70**, 1010 (1993).
9. W. Zhong, R.D. King-Smith, and D. Vanderbilt, Phys. Rev. Lett. **72**, 3618 (1994).
10. C.Z. Wang, R. Yu, and H. Krakauer, Phys. Rev. B **54**, 11161 (1996).
11. L. Fu, E. Yaschenko, L. Resca, and R. Resta, Phys. Rev. B, 15Mar98.
12. R. E, Cohen, J. Phys. Chem. Solids **57**, 1393 (1996); Ferroelectrics **194**, 323 (1997).
13. J. Padilla and D. Vanderbilt, Phys. Rev. B **56**, 1625 (1997).
14. R. Dovesi, V.R. Saunders, C. Roetti, M. Causà, N.M. Harrison, R. Orlando, and E. Aprà, CRYSTAL95 *User's Manual*, University of Turin (Torino, 1996).
15. A. Ruini, S. Baroni, and R. Resta, Phys. Rev. B, 1Mar98 (preprint cond-mat/9701194).
16. R. Pick, M.H. Cohen, and R.M. Martin, Phys. Rev. B **1**, 910 (1970).
17. M. Posternak, R. Resta, and A. Baldereschi, Phys. Rev. B **50**, 8911 (1994).
18. R. Resta, Ferroelectrics **194**, 1 (1997).
19. R.M. Martin and K. Kunc, Phys. Rev. B **24**, 2081 (1981); K. Kunc, in *Electronic Structure, Dynamics and Quantum Structural Properties of Condensed Matter*, edited by J.T. Devreese and P. Van Camp (Plenum, New York, 1985), p. 227.
20. A. Baldereschi, S. Baroni, and R. Resta, Phys. Rev. Lett. **61**, 734 (1988). A complete account can be found in S. Baroni, R. Resta, A. Baldereschi, and M. Peressi, in: *Spectroscopy of semiconductor microstructures*, edited by G. Fasol, A. Fasolino and P. Lugli, NATO ASI Series B, vol 206 (Plenum Publishing, New York, 1989), p 251.
21. P. Ghosez, Ph.D thesis, Universite Catholique de Louvain, 1997.
22. E. Yaschenko, L. Fu, L. Resca, and R. Resta, Phys. Rev. B, in press.
23. O. Schütt, P. Pavone, W. Windl, K. Karch, and D. Strauch, Phys. Rev. B. **50**, 3746 (1994).
24. H.H. Wieder, Phys. Rev. **99**, 1161 (1955).
25. G.-X. Qian, R.M. Martin and D.J. Chadi, Phys. Rev. B **38**, 7649 (1988).

# Effect of nanopolar regions on electrostrictive coefficients of a relaxor ferroelectric

## A. E. Glazounov, J. Zhao, and Q. M. Zhang

*Materials Research Laboratory, The Pennsylvania State University[1], University Park, PA 16802*

**Abstract.** Volumetric electrostrictive coefficient, $Q_h$, of $PbMg_{1/3}Nb_{2/3}O_3$ (PMN) relaxor ferroelectric is investigated using a high-resolution neutron powder diffraction and a conventional strain measurement. Using the results of these experiments we: (1) derive the value of $Q_h$ of prototype cubic lattice of PMN, which is equal to $Q_h = (8.3 \pm 1.0) \times 10^{-2}$ $(m^4/C^2)$; (2) find the temperature dependence of the volume fraction, $\delta_V$, of the polar regions in the material, and estimate that at temperatures around the dielectric permittivity maximum, $T_{max}$, it is equal to $\delta_V \approx 90\,\%$; (3) estimate the magnitude of the contribution of the crystal lattice to the polarization response of PMN at temperatures around and below $T_{max}$. These results are essential for the understanding of the nature of the polarization and strain response of relaxor ferroelectrics.

## INTRODUCTION

In this paper, we investigate electrostrictive coefficients of a "classical" relaxor ferroelectric, complex perovskite $PbMg_{1/3}Nb_{2/3}O_3$ (PMN). The motivation is to understand the relationship between the microstructural feature of PMN, namely the partitioning of the structure into the small regions of local spontaneous polarization with a nanometer scale size, and the electrostrictive response of this material. The key results of this study are summarized in the abstract and provide a new insight into the physics of relaxor ferroelectrics. Particularly, the evidence derived from this study, such as the temperature variation of the volume fraction of the polar regions and the estimate for the relative contribution of the crystal lattice to the total polarization induced in the material, is very important for the understanding of the nature of the polarization response of relaxors. The latter still remains one of the open problems of the contemporary physics of ferroelectrics and challenges both experimentalists and theorists.

---

[1] This work was supported by the Office of Naval Research. The authors also wish to thank Dr.B.Toby and National Institute of Standards and Technology (NIST) for the help in experiment with neutron diffraction.

## Electrostriction in a double-phase mixture

Contemporary models of relaxors consider them as structurally inhomogeneous materials consisting of small polar regions of nanometer scale size which are randomly distributed in a nonpolar matrix. In PMN, the polar regions appear at high temperatures, around $T_d = 600$ K, far above the temperature of the dielectric permittivity maximum, $T_{max}$, which is located around 270 K, and persist down to 0 K [1–3]. They are elongated along the direction of the local spontaneous polarization, $P_s$, (the shape which minimizes the effect of the depolarizing field) which can be oriented in one of eight $\langle 111 \rangle$ pseudocubic directions allowed by the rhombohedral symmetry of the polar phase [3]. The presence of the polar regions results in that the material response (both electric and elastic) to the applied electric field can originate from different mechanisms: **(a)** thermally activated reorientation of vector $\vec{P}_s$ in the polar regions [2,4–7], **(b)** expansion and shrinkage of the regions (similar to the domain wall motion in ordinary ferroelectrics) [8–10], and **(c)** induced phase transformation between nonpolar and polar phases [11]. It is also possible that several mechanisms coexist, and that different mechanisms dominate in different temperature intervals [11]. In addition, one should also take into account a change in the polarization and strain due to the response of the **(d)** crystal lattice, both in polar and nonpolar phases, which is always present and is usually referred to as an intrinsic contribution. Taking into account the above picture of the response of the double-phase mixture "polar regions / nonpolar matrix", the experimentally measured components polarization and strain can be written as:

$$P_i = \sum_\alpha P_i^{(\alpha)}, \qquad S_{ij} = \sum_\alpha S_{ij}^{(\alpha)}, \qquad (1)$$

where $P_i^{(\alpha)}$ and $S_{ij}^{(\alpha)}$ are the contributions to the polarization and strain from different mechanisms, **(a)–(d)**, which are averaged over the crystal. Using Eq.(1) and the definition of the electrostriction, $S_{ij} = Q_{klij} P_k P_l$, the electrostrictive coefficients, $Q_{klij}$, of a relaxor can be written using the matrix notations as:

$$Q_{ij} = \frac{S_j}{P_i^2} = \frac{\sum_\alpha S_j^{(\alpha)}}{\left[\sum_\alpha P_i^{(\alpha)}\right]^2}. \qquad (2)$$

This equation shows that coefficients $Q_{ij}$ derived from strain measurements, which are performed with a reference to the structurally inhomogeneous state, are actually some "apparent" values, $Q_{ij} = Q_{ij}^{app}$, which must depend upon the properties of ensemble of the polar regions, such as their population in the crystal, average size, etc..

*Is it possible to separate contributions to the electrostrictive strain of relaxor from different polarization mechanisms?* One possible solution of this problem has been proposed by Cross [2]. He studied another relaxor, tungsten bronze $(Sr_{0.6}Ba_{0.4})Nb_2O_6$ (SBN), which has a tetragonal structure, and demonstrated that a large qualitative difference observed in the behavior of electrostrictive coefficients $Q_{33}$ (which was measured using electric field applied along the 4-fold axis in the crystal), and $Q_{11}$ (for the field applied perpendicular to it), could be related to the presence of the polar regions, which local spontaneous polarization, $\vec{P}_s$, was along the 4-fold axis. In the present study, we will use another approach in order to separate the contributions to $Q_{ij}$ from different mechanisms.

## *Approach of this study*

The approach of the present study is to choose PMN as a model material and compare its electrostrictive coefficients derived from two experiments. The objective of one experiment is to determine the values $Q_{ij}$ of prototype cubic lattice of PMN. In order to do that, one should eliminate from the material the inhomogeneity related to the presence of the double-phase mixture: "polar regions / nonpolar matrix" [12]. We will do that by measuring the change in the lattice constants related to the structural phase transition into the ferroelectric state induced in PMN during cooling under dc electric field (FC) at temperatures below $T_f = 220$ K, [13,14]. In this method, two measurements must be performed, where the lattice parameters of PMN are measured in the FC and zero-field cooling ZFC regimes. However, it should be stressed that because of the existence of the polar regions which introduce an additional strain in the cubic lattice of PMN [2,15] one cannot use as a reference the lattice parameters measured directly in the ZFC experiment at temperatures below $T_d = 600$ K. Therefore, in order to determine the lattice parameters of the prototype cubic phase at temperatures below 600 K, one should adopt the approach employed in Refs. [1,2]. Namely, at $T < 600$ K the parameters of the prototype cubic phase can be found from the extrapolation of the experimentally measured lattice constants at temperatures above 600 K, which should change linearly according to the normal thermal expansion of the lattice. The comparison of the lattice parameters of rhombohedral ferroelectric state [14,16] with the extrapolated values of the cubic lattice thus will give us the electrostrictive strain related with the structural phase transition.

The second experiment will be a conventional measurement of the electrostrictive strain induced in PMN by small ac fields at temperatures below 600 K, in order to characterize the response of the double-phase mixture: "polar regions / nonpolar matrix".

In the present paper, we will focus on studying the volumetric electrostrictive coefficient, $Q_h = Q_{11} + 2Q_{12}$, of PMN ceramics. We choose $Q_h$ because even though ceramics is the polycrystalline body, the volumetric electrostrictive coefficient remains the same as in the single crystals [17]. Therefore, even though in the experiments we are dealing with ceramics, by measuring $Q_h$ we get the direct information about the properties of PMN single crystals.

# EXPERIMENTAL AND RESULTS

## Electrostrictive coefficient $Q_h$ of prototype cubic phase of PMN

In this experiment, the lattice parameters of PMN were investigated over a wide temperature interval, from 100 K to 1050 K using a high-resolution neutron powder diffraction. The low temperatures were required to study the ferroelectric state induced in PMN during FC below $T_f = 220$ K [13,14], whereas the high temperature interval (600 K – 1050 K) was important to obtain reliable data on the prototype cubic phase to be extrapolated to low temperatures. The choice of neutron diffraction relied on the fact that neutrons have a larger penetration depth compared to conventional X-rays, thus allowing us to study the bulk properties of the material. High-resolution data were required in order to detect the phase transformation with a very small lattice distortion in the rhombohedral ferroelectric phase of PMN from the prototype cubic phase, $(90° - \alpha^{rh}) \leq 0.1°$, [19]. The neutron powder diffraction data were collected using the BT-1 high resolution diffractometer at the NIST Center for Neutron Research reactor, NBSR, and the details of the experiment are described in Ref. [12].

Experiment was first conducted in the ZFC regime where the diffraction patterns of PMN were recorded at several temperatures from 1050 K to 100 K. Afterwards, the dc field $E = 5$ kV/cm was applied to the same sample at 300 K, and the

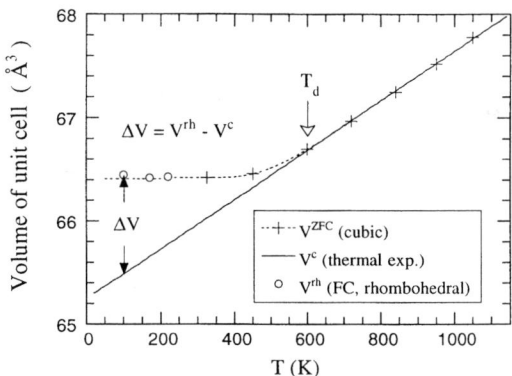

**FIGURE 1.** Temperature dependence of the volume of the unit cell of PMN measured in ZFC (plus signs, and the dotted line drawn through them in order to guide the eye) and FC (open circles) regimes. The linear dependence of the unit cell volume on temperature corresponds to the normal thermal expansion of prototype cubic lattice of PMN (solid line). During the FC regime, PMN is in the ferroelectric state with a rhombohedral structure at $T \leq 220$ K (open circles). $\Delta V$ denotes the volume change in the crystal lattice during the phase transition to the ferroelectric state. $T_d$ is the temperature at which polar regions appear in PMN upon cooling.

sample was cooled down (FC regime). In this regime the diffraction patterns were recorded at several temperatures below $T_f = 220$ K, where the field induced phase transition was expected in PMN for the electric field larger than the threshold field, $E_{th} = 1.7$ kV/cm, [13].

The results of both ZFC and FC measurements were in a good agreement with similar experiments performed by other research groups. For ZFC, the crystal lattice of PMN is cubic in the whole studied temperature interval [3,15], whereas for FC using $E = 5$ kV/cm the crystal structure is rhombohedral at temperatures below 220 K [16,19]. In both cases, unit cell parameters, $a^c$ (for cubic), and $a^{rh}$ and $\alpha^{rh}$ (for rhombohedral), were determined using the structural refinement procedure [12].

Figure 1 presents the temperature dependence of the volume of the unit cell, $V$, of PMN for both ZFC and FC experiments, and also illustrates the approach which we used to calculate the volume strain of the crystal lattice related with the phase transition. In the plot, open circles show the volume $V^{rh}$ of the unit cell of rhombohedral crystal lattice corresponding to the ferroelectric state induced in PMN at $T \leq 220$ K by the dc field $E = 5$ kV/cm. Pluses and dotted line connecting them correspond to the ZFC regime, $V^{ZFC}(T)$, with the cubic structure. Taking into account that no polar regions exist in the material above $T_d = 600$ K, the linear variation of $V^{ZFC}$ in the high-temperature interval is the normal thermal expansion of the prototype cubic lattice of PMN, $[V^c(T) - V^c(T_d)] \propto \beta_V \cdot (T - T_d)$, where $\beta_V$ is the volumetric thermal expansion coefficient and $V^c$ is the volume of the unit cell of the prototype cubic lattice. Below 600 K, $V^{ZFC}(T)$ deviates from the linear dependence due to the appearance of polar regions in PMN, which introduce an additional strain in host cubic lattice equal to $Q_h \langle P_s^2 \rangle$, where $\sqrt{\langle P_s^2 \rangle}$ is the root mean squared local spontaneous polarization [1,2].

In order to find $Q_h$ from the data in Fig. 1, the volumetric strain related to the structural phase transformation was calculated as $\Delta V / V^c$, where $\Delta V = V^{rh} - V^c$, Fig. 1. Below 220 K, the volume $V^c$ of the prototype cubic lattice was found by the extrapolation of the high-temperature linear dependence down to low temperatures, as shown in Fig. 1 with the solid line. Afterwards, the volumetric electrostrictive coefficient was determined from the equation: $Q_h = 1/P_s^2 \cdot [V^{rh} - V^c]/V^c$, where $P_s$ is the macroscopic spontaneous polarization, which was evaluated from saturated polarization hysteresis loops measured on PMN single crystal samples along $\langle 111 \rangle$ direction using the ac field of 30 kV/cm amplitude [12].

**TABLE 1.** Volumetric electrostrictive coefficient $Q_h$ of prototype cubic lattice of PMN relaxor calculated from the results of the structural study of PMN using high-resolution neutron powder diffraction.

| $T$ (K) | 100 | 170 | 220 |
|---|---|---|---|
| $Q_h$ ($10^{-2}$ m$^4$/C$^2$) | $8.2 \pm 1.0$ | $8.1 \pm 1.0$ | $8.5 \pm 1.0$ |

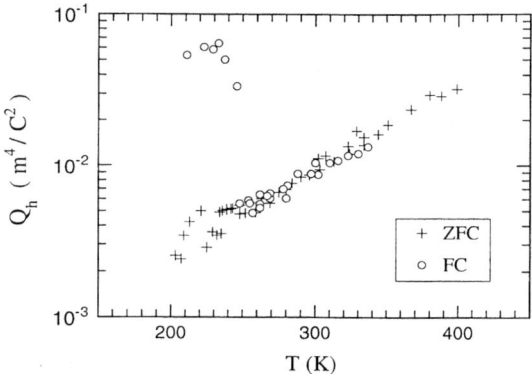

**FIGURE 2.** Temperature dependence of volumetric electrostrictive coefficient, $Q_h$, of PMN measured in ZFC (open circles) and FC under the dc bias $E_b = 3$ kV/cm (plus signs) regimes. The step-like change in $Q_h$ at low temperatures coincides with the structural phase transition from mixed to normal ferroelectric state, which is induced in PMN by large dc bias. Therefore large values of $Q_h$ below the step correspond to the electrostrictive strain of the crystal lattice only. The experimental uncertainty of the evaluation of $Q_h$ is about $0.2 \times 10^{-2}$ (m$^4$/C$^2$).

The $Q_h$ data are summarized in Table 1. From that table, it can be seen that within the error of experiment, the electrostrictive coefficient remains temperature independent in the studied temperature interval of more than 100 K. This result is in agreement with the conventional wisdom that the electrostrictive coefficients coupling electric and elastic properties of the crystal lattice should be "proper" material constants, i.e., should be temperature independent [2].

## *Electrostrictive coefficient $Q_h$ of mixed state of PMN*

The electrostrictive coefficient, $Q_h$, of PMN in its mixed state, "polar regions / nonpolar matrix", was derived from the following experiment [11]. Small ac field of amplitude $E_m = 0.2$ kV/cm and frequency 10 Hz was applied to the sample, and the induced polarization, $P_1$, and longitudinal, $S_1$, and transverse, $S_2$, strains were measured simultaneously. The $P_1$ was measured using a Sawyer – Tower circuit, in which the output signal was detected with a SR 830 lock-in amplifier, and $S_1$ and $S_2$ were measured using the strain gauge technique, where the signal output from the strain gauge was measured with two other SR 830 lock-in amplifiers. Then, the volumetric electrostrictive coefficient was found from: $Q_h = \overline{Q}_{11} + 2\overline{Q}_{12}$, where $\overline{Q}_{11} = S_1/P_1^2$ and $\overline{Q}_{12} = S_2/P_1^2$ are the longitudinal and transverse electrostrictive coefficients of ceramics, respectively. These measurements were performed within the temperature interval from 200 K to 400 K, where the high temperature limit was caused by the increase in the electric conduction of the material at high temperatures, which made the strain measurements very difficult. In addition, the same coefficient was measured in the FC regime, where the small ac probing field, $E_m = 0.2$ kV/cm, was superimposed on large dc bias, $E_b = 3$ kV/cm.

Figure 2 summarizes the results of these experiments by plotting $Q_h$ as a function of temperature. Several features can be noticed in this plot. First, the electrostrictive coefficient of the mixed phase of PMN shows strong temperature dependence, it changes approximately by 10 times within the studied temperature interval. Second, at temperatures around $T_{max} \approx 270$ K, the value of $Q_h$ corresponding to the mixed state is about 10 times smaller than the value derived from the neutron diffraction study, cf. Table 1. Third, when the large dc field is applied to the sample, $Q_h$ undergoes the step-like change at temperatures around 230 K. The position of this step coincides with the temperature of structural phase transition from mixed to normal ferroelectric state, which is induced in PMN by the electric field larger than the threshold field, $E_{th}$, [13]. Thus, the dc bias "removes" the inhomogeneity related with "polar regions / nonpolar matrix", and $Q_h$ measured at temperatures below the step corresponds to the electrostrictive strain of the crystal lattice only. This conclusion is supported by the fact that under the dc bias the electrostrictive coefficient is temperature-independent at low-temperatures, Fig.2, and has a value $Q_h = (6.0 \pm 0.2) \times 10^{-2}$ (m$^4$/C$^2$) which is very close to the average value $Q_h = (8.3 \pm 1.0) \times 10^{-2}$ (m$^4$/C$^2$) derived from the neutron diffraction study of PMN, Table 1.

## DISCUSSION

The plot in Fig.3 combines all the data obtained in the present work. It clearly demonstrates that the temperature dependence of the electrostrictive coefficient $Q_h$, which was derived from strain measurements performed with a reference to the structurally inhomogeneous state (plus signs in Fig.3), most likely is related with the change in the properties of the ensemble of the polar regions in the temperature interval below 600 K. Above $T_d = 600$ K, no polar regions exist in the material, thus electrostrictive strain is only due to the distortion of the prototype cubic lattice of PMN, which is characterized by the electrostrictive coefficient $Q_h^{lat}$. The neutron diffraction data (closed circles in Fig.3) suggest that $Q_h^{lat}$ is temperature independent and is equal to $Q_h^{lat} = (8.3 \pm 1.0) \times 10^{-2}$ (m$^4$/C$^2$), Table 1. When the polar regions appear in the material below 600 K, they, first, introduce an additional volumetric strain in the host cubic lattice, which is equal to $Q_h^{lat} \langle P_s^2 \rangle$ [2,15], and, second, bring "new" mechanisms of the polarization and strain response to the electric field, such as polar vector reorientation, (a), or expansion and shrinkage of the polar regions, (b). Therefore, the electrostrictive coefficient derived from the strain measurement will be now just an "apparent" parameter, $Q_h^{app}$, which must depend upon the properties of ensemble of the polar regions, such as their population in the crystal, average size, etc..

The relationship between the electrostrictive coefficients $Q_h^{lat}$ and $Q_h^{app}$ can be

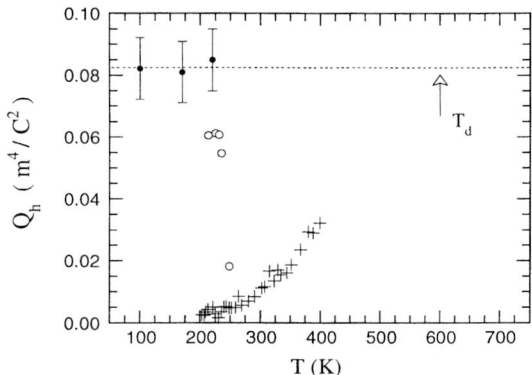

**FIGURE 3.** The plot combines all the data obtained in the present work, and shows the volumetric electrostrictive coefficient of prototype cubic lattice of PMN, $Q_h^{lat}$, (closed circles), and $Q_h^{app}$ which describes the electrostrictive strain in the double-phase mixture, "polar regions / nonpolar matrix", (plus signs). When the structural inhomogeneity is removed by large dc bias field, which induces the normal ferroelectric state in PMN at low temperatures, the measured values of $Q_h^{app}$ tend toward $Q_h^{lat}$, as shown with open circles in the plot. Therefore, the temperature dependence of the electrostrictive coefficient which is derived from strain measurements performed with a reference to the structurally inhomogeneous state is related with the change in the properties of ensemble of the polar regions, which appear in the material at temperatures below $T_d \approx 600$ K (shown with an arrow).

obtained from the comparison of the three states of PMN: *(i)* prototype cubic, *(ii)* mixed, consisting of "polar regions / nonpolar matrix", and *(iii)* completely polarized state. The latter can be obtained in the limit of very high electric field, $E_m$, applied to the sample, so that the whole volume of the crystal will have a homogeneous polarization, $P_s$, along [111] crystal axis (it is reasonable to expect that the symmetry of this state is the same as of the polar regions, i.e., rhombohedral), [20]. Therefore, one can write:

$$Q_h^{app} = \frac{S(E_m)}{P_s^2} = \frac{Q_h^{lat} P_s^2 - Q_h^{lat} \langle P_s^2 \rangle}{P_s^2}, \quad (3)$$

where $Q_h^{lat} \langle P_s^2 \rangle$ is the strain introduced into the cubic lattice due to the appearance of the polar regions in the mixed state, $Q_h^{lat} P_s^2$ is the electrostrictive strain of completely polarized state, and both strains are defined with respect to the prototype cubic lattice. Equation (3) directly leads to the following expression:

$$Q_h^{app} = Q_h^{lat} \cdot \left[ 1 - \frac{\langle P_s^2 \rangle}{P_s^2} \right], \quad (4)$$

which relates both electrostrictive coefficients shown in Fig.3.

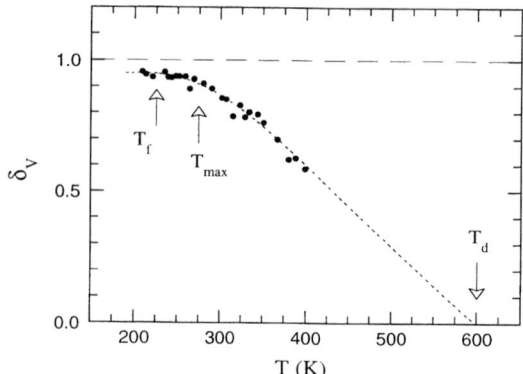

**FIGURE 4.** Temperature dependence of the volume fraction of the polar regions, $\delta_V$, in PMN. Closed circles correspond to the experimental data and the dotted line shows a possible extrapolation of $\delta_V(T)$ toward high temperatures, where the polar regions appear at $T_d \approx 600$ K. Also shown are the temperature of the dielectric permittivity maximum, $T_{max}$, corresponding to the radio-frequency interval, Hz–MHz, and the freezing temperature, $T_f = 220$ K.

In Eq.(4), $\sqrt{\langle P_s^2 \rangle}$ is the root mean squared local spontaneous polarization of PMN. Taking into account that at a given temperature, the absolute value of the local spontaneous polarization of the polar regions most likely is equal to the macroscopic polarization, $P_s$, in the homogeneously polarized state at $E_m$, one can write: $\langle P_s^2 \rangle = \delta_V \cdot P_s^2$, where $\delta_V$ is the volume fraction of the polar regions in the crystal. Therefore, the expression in the brackets in Eq.(4) is equal to: $[1 - \delta_V]$.

Based on the above argument, from the data shown in Fig.3, we can find the volume fraction of the polar regions in PMN using: $\delta_V = 1 - \left[ Q_h^{app}/Q_h^{lat} \right]$. The obtained temperature dependence of $\delta_V$ is plotted in Fig.4. As one can see, within the temperature interval investigated, $\delta_V$ first increases upon cooling from 400 K to 250 K, and then saturates around the freezing temperature $T_f = 220$ K [21]. It is this increase in the volume fraction of the polar regions in the material which causes the decrease in the electrostrictive coefficient, $Q_h^{app}$, of PMN at temperatures below $T_d$, Fig.3. Also, it is worth noting that the data shown in Fig.4 brings new evidence about the absolute values of the volume fraction of the polar regions in PMN. So far, the only estimate which was used in the literature, especially in the works devoted to the theoretical description and modeling of the polarization response of relaxors, was that reported by Mathan et al. [3]. From the profile analysis of the X-ray diffraction data for PMN recorded at 5 K, Mathan et al. obtained that $\delta_V$ is approximately equal to 20 %. Clearly, this estimate is very different from the data obtained in the present study. As one can see from the plot in Fig.4, below the freezing temperature the volume fraction of the polar regions is close to unity, and at temperatures around $T_{max}$ it is equal to $\delta_V \approx 90\%$.

Furthermore, the above picture of the temperature evolution of $\delta_V$ can be correlated with the results of the quasielastic neutron scattering study of PMN by Vakhrushev et al. [22]. That study showed that the correlation length of the polarization fluctuations monotonically increased upon cooling, and at temperatures around $T_f = 220$ K it reached the saturation value equal to 20 nm. The saturation of the correlation length actually implied the stabilization of the average size of the polar regions. Similar behavior can be seen in Fig.4 for $\delta_V(T)$, where it first increases upon cooling and then saturates at temperatures around $T_f = 220$ K.

If only small electric field is applied to the sample, the total polarization and strain response of the material originates from different mechanisms, which were discussed in the *Introduction*. For the volumetric strain, we can divide these mechanisms into those which contribute to the volume change of the crystal [crystal lattice response, (d), and phase switching, (c)], and those which do not involve the volume change [domain wall motion, (b), or reorientation of the polar regions, (a)]. It is apparent that if these two groups of mechanisms coexist, one can write the following expression, [11]:

$$Q_h^{app} \cdot [P_v + P_d]^2 = Q_h^{lat} \cdot P_v^2, \qquad (5)$$

where $P_v$ is the polarization coming from mechanisms involving the volume change, and $P_d$ is the induced polarization which does not produce the volume strain. From Eq.(5), one can obtain that the relative contribution from the crystal lattice, $P_v$ to the total induced polarization, $P_v + P_d$, is given by $\sqrt{Q_h^{app}/Q_h^{lat}}$. Using this expression and the data shown in Fig.3, one can estimate that at temperatures around $T_{max} \approx 270$ K, the contribution from the crystal lattice is about 30 % of the total polarization induced in PMN.

## CONCLUSIONS

In the present work, we investigated the volumetric electrostrictive coefficient, $Q_h$, PMN relaxor using a high-resolution neutron powder diffraction and a conventional strain measurements. Using the results of these experiments we: (1) derive the value of $Q_h$ of prototype cubic lattice of PMN, which is equal to $Q_h = (8.3 \pm 1.0) \times 10^{-2}$ (m$^4$/C$^2$); (2) find the temperature dependence of the volume fraction, $\delta_V$, of the polar regions in the material, and estimate that at temperatures around the dielectric permittivity maximum, $T_{max}$, it is equal to $\delta_V \approx 90\,\%$; (3) estimate the magnitude of the crystal lattice contribution to the polarization response of PMN at temperatures around and below $T_{max}$.

## REFERENCES

1. Burns G. and Dacol F. H., *Solid State Commun.* **48**, 853 (1983).
2. Cross L. E., *Ferroelectrics* **76**, 241 (1987).

3. de Mathan N., et al., *J. Phys.: Condens. Matter* **3**, 8159 (1991).
4. Dorogovtsev S. N. and Yushin N. K., *Ferroelectrics* **112**, 27 (1990).
5. Viehland D. et al., *J. Appl. Phys.* **68**, 2916 (1990).
6. Vugmeister B. E. and Rabitz H., *Ferroelectrics* **201**, 33 (1997).
7. Qian H. and Bursill L. A., *Int. J. Mod. Phys. B* **10**, 2027 (1996).
8. Westphal V., Kleemann W., and Glinchuk M. D., *Phys. Rev. Lett.* **68**, 847 (1992).
9. Glazounov A. E., Tagantsev A. K., and Bell A. J., *Phys. Rev. B* **53**, 11281 (1996).
10. Tagantsev A. K. and Glazounov A. E., *Phys. Rev. B* **57**, 18 (1998).
11. Zhang Q. M. and Zhao J., *Appl. Phys. Lett.* **71**, 1649 (1997).
12. Zhao J., Glazounov A. E., and Zhang Q. M., *Appl. Phys. Lett.*, in press.
13. Sommer R., Yushin N. K., and van der Klink J. J., *Phys. Rev. B* **48**, 13230 (1993).
14. Ye Z.-G. and Schmid H., *Ferroelectrics* **145**, 83 (1993).
15. Bonneau P. et al., *J. Solid State Chem.* **91**, 350 (1991).
16. Arndt H. et al., *Ferroelectrics* **79**, 145 (1988).
17. The electrostrictive coefficients of polycrystalline samples represent the directional average of the single crystal coefficients. Since PMN has a macroscopic cubic symmetry, the average over random orientation of the crystallites in ceramics will yield the following expressions for longitudinal and transverse coefficients [18]: $\overline{Q}_{11} = (3\,Q_{11} + 2\,Q_{12} + Q_{44})/5$ and $\overline{Q}_{12} = (Q_{11} + 4\,Q_{12} - 1/2\,Q_{44})/5$, where $\overline{Q}_{ij}$ represent the directional average values for ceramics and $Q_{ij}$ are the coefficients of the single crystals. From the above expressions one can see that: $\overline{Q}_{11} + 2\,\overline{Q}_{12} = Q_{11} + 2\,Q_{12} = Q_h$.
18. Devonshire A.F., *Philos. Mag.* **42**, 1065 (1951).
19. de Mathan N. et al., *Mater. Res. Bull.* **26**, 1167 (1991).
20. In ferroelectrics, the polarization measured at the tip of the hysteresis loop, at $E_m$, is equal to $P_s + \varepsilon_{lat}\,E_m$, where $P_s$ is the spontaneous polarization and $\varepsilon_{lat}$ is the dielectric permittivity of the crystal lattice. Since $\varepsilon_{lat}\,E_m$ is usually much smaller than $P_s$, we can neglect it.
21. We derive $\delta_V(T)$ from $Q_h^{app}$, which, we think, also saturates at $T \leq 220$ K, because in this temperature interval the difference between the values of the electrostrictive coefficient measured at different temperatures is comparable to the experimental uncertainty, Fig.2.
22. Vakhrushev S. B. et al., *Ferroelectrics* **90**, 173 (1989).

# FERROELECTRICITY AND PROTONIC CONDUCTIVITY IN $Cs_{1-x}(NH_4)_xH_2PO_4$

S. Lanceros-Méndez, S. Meschia, and V. H. Schmidt

*Dept. of Physics, Montana State University, Bozeman, MT 59717, USA*

**Abstract.** Mixed crystals $Cs_{1-x}(NH_4)_xH_2PO_4$ of the ferroelectric $CsH_2PO_4$ (CDP) and the antiferroelectric $(NH_4)H_2PO_4$ were grown with x=0.2 (CADP0.2) in solution. The structural properties of the crystal were analyzed by means of x-ray diffraction. Dielectric measurements at several temperatures and frequencies have been performed along the three crystallographic axes in this sample and also in the fully deuterated CADP0.2 sample (DCADP0.2). Dielectric and NMR experiments in a powdered sample were also performed. The shift of the transition temperature as a function of x and deuteration, and the changes in the properties of the different phases together with the conductivity found in the paraelectric phase will be discussed.

## INTRODUCTION

Crystals that contain mixtures of ferroelectrics (FEs) and antiferroelectrics (AFEs) have interesting properties due to the frustration resulting from the opposing orderings. Many questions remain about the spatial and temporal behavior of dipolar domains and coexistence in such crystals. $CsH_2PO_4$ (CDP) is known to have one dimensional FE order below T=154 K and a superionic transition at T=504 K on heating (1), whereas $NH_4H_2PO_4$ (ADP) is AFE below T=148 K (2). In addition, their paraelectric phases have different structures: CDP is monoclinic with the space group $P2_1/m$ and ADP is tetragonal with the space group $I\bar{4}2d$. In this work we present x-ray, dielectric and NMR results in undeuterated, deuterated and powdered $Cs_{1-x}(NH_4)_xH_2PO_4$ crystals with x=0.2 in solution.

## EXPERIMENTAL METHOD

The crystals were grown by controlled evaporation from aqueous solution of 20% molar concentration of ADP and 80% of CDP. The value of x in the crystal was determined by the Kjeldahl method (3).

In order to perform the x-ray transmission experiments small semi-spherical samples (~1 mm$^3$) were prepared in order to reduce the effect of the extinction coefficients on the transmission spectrums.

For the dielectric measurements samples in the shape of thin slides (<1 mm) were cut perpendicular to the three crystallographic axes, optically polished and silver painted. The temperature (25 K-320 K) and frequency (10 Hz-1 MHz) dependencies of ε' and tan δ (=ε"/ε') were measured with a standard impedance meter (Quad Tech 7600).

For the NMR measurements several crystals were pulverized and desiccated. The variation of the spin-lattice relaxation time as a function of temperature was measured. For each $T_1$ data point the temperature was maintained to within a tenth of a degree over a period of days by means of a Lake Shore DR91C Temperature Controller.

## RESULTS

### X-ray transmission scattering

Crystal quality, a concern for mixed crystals, was excellent. The lattice structure and the space group of CADP0.2 was found to be identical to that of pure CDP at room temperature: monoclinic with the space group P2$_1$/m. The obtained lattice parameters were: a=7.893(67) Å, b=6.383(46) Å, c=4.876(13) Å, and β=107.56(06) deg. There were slightly variations between the parameters of different crystals of the same concentration but, in general, the values were very close to those in the pure compound (4), with the most significant variation being a small increase in length (~0.02 Å) alomg the FE b-axis. The concentration of ammonium present in the crystals was much smaller in the crystal than in the solution from which they grown. Further chemical analysis of the crystals by the Kjeldahl method (3) showed the ammonium concentration to be 0.060±0.002.

### Dielectric measurements

Both of the anomalies in ε'$_b$(T) and ε"$_b$(T) (Figure 1) are pretty similar to those observed in the pure compound. The transition temperature (T$_c$=159.3 K), the shape and the frequency dependence of the anomalies do show small changes due to the amount of ammonium impurities. T$_c$ is shifted ~5 K to higher temperatures with respect to CDP. The fit parameters of the quasi-1D-Ising model (5)-(8) suffer significant variations within the increase of concentration, at the same time the dielectric permittivity along the c- and a-axes clearly show anomalies around the transition temperature into the FE phase (Figure 2). These anomalies do not exist in pure CDP.

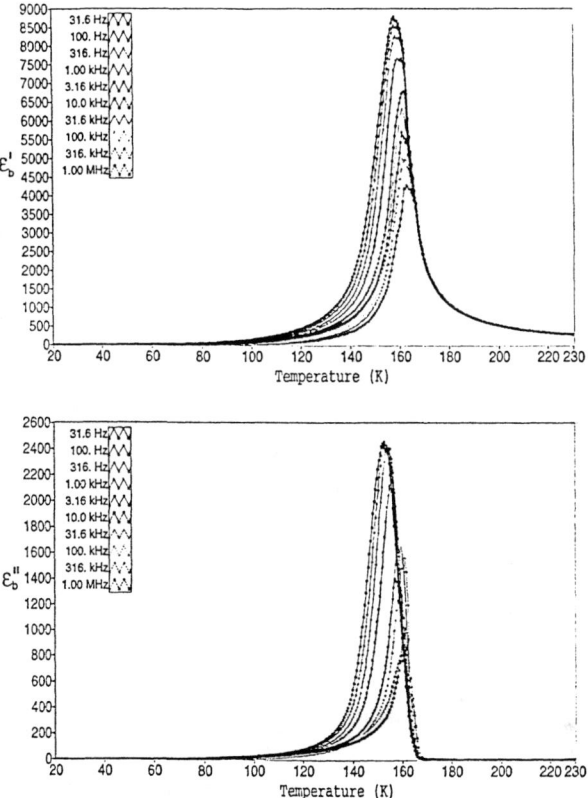

**FIGURE 1.** $\varepsilon'_b(T)$ and $\varepsilon''_b(T)$ for CADP0.2 at several frequencies.

The $\varepsilon'_b(T)$ curves both in the undeuterated and deuterated crystals could be well fitted with the quasi-1D Ising model. The experimental $\varepsilon'_b(T)$-curves were fitted in the temperature regions $T>T_c+(\approx 5$ K) according to the equation

$$\varepsilon'(0) = \varepsilon'_\infty + \frac{C/T}{\exp(-\frac{2J_\parallel}{kT}) - \frac{J_\perp}{kT}},$$

where the $J_\parallel$ is the intrachain and the $J_\perp$ the interchain interaction constant. The values obtained for the fit parameters are summarized in Table 1, and a typical fit is shown in Figure 3.

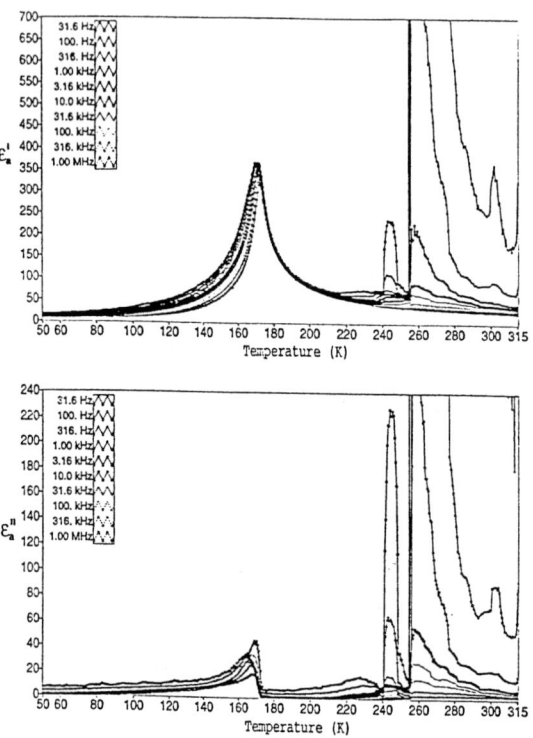

**FIGURE 2.** $\varepsilon'_a(T)$ and $\varepsilon''_a(T)$ for CADP0.2 at several frequencies.

The effect of the increase of the ammonium impurities is a decrease in the interactions along the chains parallel to the FE axis and an increase of the interactions between the chains, perpendicular to that axis (5)-(8). This means an increase in the isotropy of the crystal with respect to pure CDP. This fact was confirmed by the dielectric experiments along the a- and c-axis: in both cases dielectric anomalies were found that do not exist in pure CDP. The anomaly along the a-axis is about 10 times lower than the one along the b-axis (Figure 2); the anomaly along the c-axis is about 100 times lower.

The quasi-one-dimensional Ising correlation length parallel to the FE axis, $x_1(T)$, is given by (10)

$$\xi_1(T) = \left(\frac{kT_c}{2J_1}\right)^{1/2} \left(\tfrac{1}{2}d_1 e^{2\beta J_1}\right) \left|\frac{T_c}{T-T_c}\right|^{1/2},$$

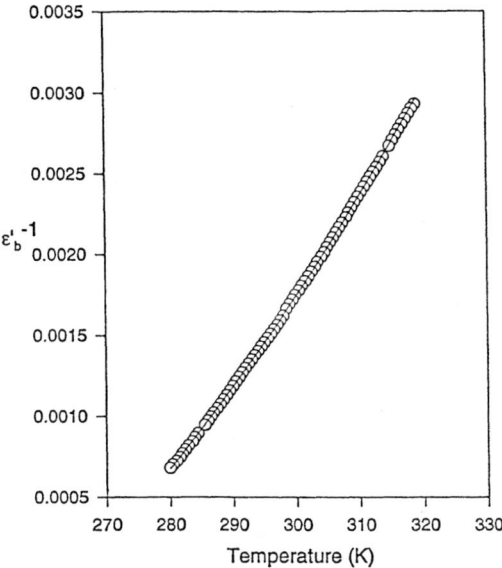

**FIGURE 3.** Fit of the quasi-1D-Ising model for DCADP0.2 at 1 kHz.

where $d_\parallel$ is the distance between spins parallel to the chains and $\beta=1/k_BT$. Assuming CDP bond lengths (11) and employing

$$d_\parallel = (d_{O(2)-H(3)}^2 + d_{O(2)-H(3)'}^2 - 2d_{O(2)-H(3)}\, d_{O(2)-H(3)'}\cos(f))^{1/2},$$

with $d_{O(2)-H(3)}=0.100$ nm, $d_{O(2)-H(3)'}=0.146$ nm and $f=174.4°$, we obtain $d_\parallel=0.246$ nm. For CDP, with $T_c=154$ K and $J_\parallel/k_B=284$ K, we obtain $x_1(T=T_c+1)=31$ nm and $x_1(T=T_c+10)=8.0$ nm. For CADP.20 with $T_c=160$ K and $J_\parallel/k_B=156$ K, we obtain considerably smaller correlation lengths: $x_1(T=T_c+1)=7.7$ nm and $x_1(T=T_c+10)=2.2$ nm.

The prediction of this model concerning the behavior of the transition temperature as a function of $J_\parallel$ and $J_\perp$ is given by (10)

$$T_c \cong 2J_\parallel/\ln(J_\parallel/J_\perp).$$

**TABLE 1.** Fit parameters of the quasi-1D-Ising model for the CADP0.2 and DCADP0.2

| Compound | C | $J_\perp/K$ | $J_\parallel/K$ | $\varepsilon_\infty$ |
|---|---|---|---|---|
| CADP-0.2 | 12860 | 14 | 156 | 48 |
| DCADP-0.2 | 5766 | 17 | 365 | 34 |

Although this equation gives $T_c=188$ K, which is considerably higher than the actual value of $T_c$, 160 K, it accurately predicts the direction of the temperature shift by the addition of ammonium since $T_c(CDP)=128$ K according to this model. The model predicts that $T_c=227$ K for DCADP.20, instead of the actual value of $T_c=276$ K.

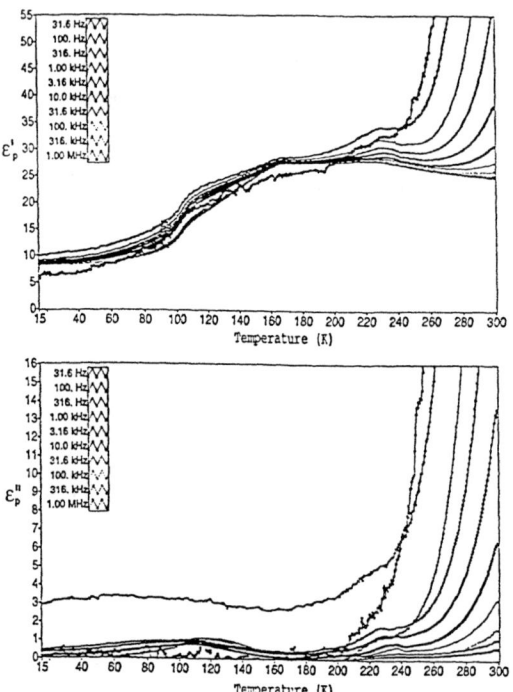

**FIGURE 4.** $\varepsilon'_p(T)$ and $\varepsilon''_p(T)$ for powdered CADP0.2 at several frequencies

A striking feature observed in the dielectric measurements is the apparently thermally activated conductivity within the paraelectric phase along the a- and c-axes. Similar dielectric anomalies occur both in many other mixed crystals isomorphs of KDP containing ammonium and in pure ADP. Comparison with our results from NMR suggests that the large anomaly is related to conductivity arising from the acid protons or deuterons. The large change in transition temperature resulting from deuteration of CDP and CADP0.2 further suggests conduction by hydrogen. Cycling the experiments several times demonstrates that the conductivity effects present a very large (~40 K) and anomalous thermal hysteresis that depends on the thermal history of the crystal: in the heating runs the conductivity starts at lower temperatures than it stops in the cooling runs. The anomalies due to this conductivity were also observed in the deuterated and powdered samples.

The effect of deuteration in CADP0.2 is similar to that in pure CDP: the transition temperature is shifted from 159.3 K to 276 K, as compared a shift from 156 K to 268.3 K in the pure compound (5). The changes in shape, as represented by the values of the fit parameter of the quasi-1D-Ising model (Table 1), the anisotropy and the frequency dispersion are as described for the undeuterated sample. The temperature and shape of the conductivity effects remain almost independent of the amount of deuteration.

Dramatic differences are observed between dielectric responses of crystalline and powdered CADP0.2: the characteristic features of the powdered sample are the disappearance of the high ferroelectric anomaly and the appearance of a small rounded maximum at a temperature 4 K higher, and an abrupt and frequency dependent drop in $\varepsilon'(T)$ accompanied by an anomaly in $\varepsilon''(T)$ (Figure 4). The behavior of the second feature strongly resembles the typical behavior of dipole glasses, which, together with the anomaly around the transition temperature into the ferroelectric phase, prompted some authors to conclude that the related compound CADA exhibited coexistence of ordered (FE or AFE) and glassy phases [9]. The differences between the measurements in crystal or in powder are obvious, but the question remains whether these results are due to grain size, grain composition, boundary and/or some other effects.

## Nuclear Magnetic Resonance

The fact that there were multiple relaxation processes complicated the analysis of the data because we could not assume that the physical process corresponding to the lower $T_1$ at one temperature necessarily remains lower at all temperatures. Instead, in interpolating separate curves through the data, we sought to maximize the continuity of each of the $T_1$ curves. While even minute amounts of paramagnetic impurities can often provide the dominant relaxation mechanism, the relatively long relaxation times and the agreement with theoretical relaxation times suggest that the level of paramagnetic impurities in the sample was small and that the primary mechanism for $^{31}P$ $T_1$ relaxation in our sample was dipolar coupling between hydrogen and phosphorus nuclei in the lattice.

In the case of the ammonium nuclei, reorientation arises from the rotation of the ammonium groups about their lattice positions. In the case of the conduction protons, reorientation arises from hopping between lattice sites. In addition, we will only consider the dipolar contribution to the relaxation time. With these liquid-like assumptions, the $T_1$ relaxation time for nuclei 1 resulting from the interaction between nuclei 1 and 2 with gyromagnetic ratios of $g_1$ and $g_2$ and resonant frequencies of $w_1$ and $w_2$, respectively, is given by (12)

$$T_1 = \frac{160 \pi^2 r^6}{\mu_0 \hbar^2 \gamma_1^2 \gamma_2^2 (\tau / (1+(\omega_1 - \omega_2)^2 \tau^2) + 3 \tau / (1 + \omega_1^2 \tau^2) + 6 \tau / (1 + (\omega_1 + \omega_2)^2 \tau^2))}$$

where $\hbar$ is Planck's constant, $m_0$ is the permeability of the vacuum and $t$ is the relaxation time of the nuclear motion.

The dominant dipolar interaction with phosphorus in CADP occurs with $^1$H because it has a much larger gyromagnetic ratio ($g(^1H)$=26.752156) than any other nucleus, except for the exceedingly rare radioactive $^3$H nucleus, and because the average positions of several hydrogens are closer than those of the other dipolar nuclei present in the crystal, i.e., $^{133}$Cs ($g_{Cs}$=3.533253) and $^{31}$P.

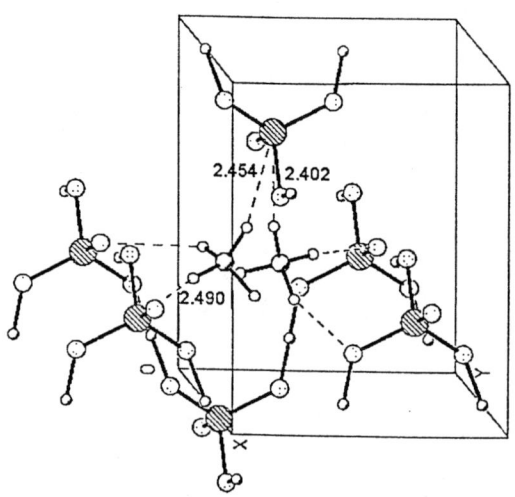

**FIGURE 5.** Model showing minimum ammonium-phosphorus atomic distances (Å).

....At all temperatures, hydrogen is the more mobile nuclear species, implying that the relaxation time $t$ is characteristic of a specific motion of the hydrogen. For a static NMR field $H_0$ giving a resonant frequency of 28 MHz for $^{31}$P, we obtain $T_1$(min)=8.89 $10^{57}$ m$^{-6}$ s $r^6$. Since the diameter of ammonium ($r$=0.148±0.001 nm (13)) is less than that of cesium ($r$=0.167±0.001 nm (14)), for the small concentrations of ammonium incorporated into the lattice, the ammonium ions should experience little hindrance to rotation.

Bond distances were calculated using a software program called Shelxtl by G. M. Sheldrick of Siemens Analytical Instruments of Madison, Wisconsin. The minimum acid H- P bond distance of 0.2603 nm calculated yields a $T_1$ value of 2.77 s at the minimum. Figure 5 shows the result of the Simstat lattice refinement for an ammonium group at a Cs lattice site. The minimum distance between a H atom from an ammonium group and a P atom was calculated to be 0.2402 nm, corresponding to a $T_1$ of 1.71 s.

....Figure 6 shows a plot of the natural logarithm of $T_1$ versus inverse temperature. Assuming that the concentration of paramagnetic impurities was negligible and that other mechanisms were small, we tentatively associate the minimum in Figure 6 at $T=154$ K with ammonium hindered rotation and the minimum at $T=328$ K with hydrogen interbond motion. The values of $T_1$ associated with these minima, $T_1=0.61$ and $T_1=0.90$ s, are about one third of the values of the relaxation times as calculated by the minimum bond distances, $T_1=1.71$ s and $T_1=2.77$ s, suggesting either that the liquid-like assumption is in need of refinement or that contributions to the relaxation time from other P-H couplings are significant.

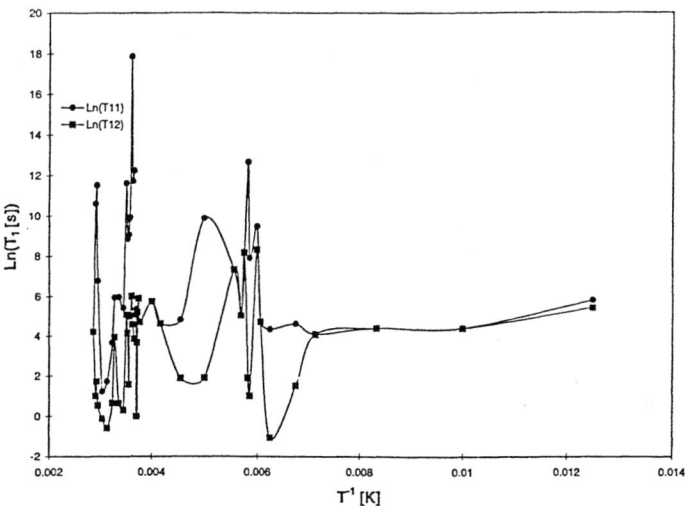

**FIGURE 6.** Natural logarithm of $T_{11}$ and $T_{12}$ versus inverse temperature. $T_{11}$ and $T_{12}$ correspond to the value of the two different processes.

By minimizing $T_1$ as a function of temperature, the frequency of the processes at these minima, $f$, is determined to be 38.9 MHz. Expecting that the effect of the small intrabond proton displacements is too small to detect by $^{31}$P relaxation and assuming an activated process with activation energy, $E_A$, the relaxation times, $t$, of the motions at arbitrary temperatures can be calculated from $t=t_0 exp(E_A/kT)$. Setting the energy of motion equal to the thermal energy at the minimum, we obtain

$$E_A = k_B T ln(kT/hf),$$

where $k_B$ is Boltzmann's constant and $h$ is Planck's constant. This equation yields theoretical values of $E_A$=0.35 eV at $T$=328 K, and $E_A$=0.16 eV at $T$=162 K. These values agree reasonably with the $E_A$ as determined from the slope of the curves of $\ln(T_I)$ versus inverse temperature near the minima: $E_A$=0.46 eV for the acid relaxation at $T$=328 K, and $E_A$=0.19 eV for the ammonium relaxation at $T$=162 K. Since intrabond motion has little effect on $T_I$, the process near $T$=328 K most probably corresponds to an actual mobility between the lattice sites associated with protonic electrical conductivity. Indeed, the activation energy of 0.46 eV is at the lower end of the activation energies found from the dielectric results presented before. The $T_I(T)$ relaxation results for CADP.20 powder are considerably more complicated than those obtained for $^{133}$Cs in CDP, which only exhibits an anomaly at $T_c$ and maybe be characterized by a single relaxation time (15). The second relaxation time must, therefore, be associated with a feature of CADP that is not present in CDP, namely the presence of randomly distributed ammonium ions.

## ACKNOWLEDGEMENTS

Work supported from NSF grant DMR-9520215. S. L.-M. gratefully thanks the support from the Basque Government under grant BFI96.041.

## REFERENCES

[1] Baranov, A. I., Khiznichenko, V. P., and Shuvalov, L.A., *Ferroelectrics*, **100**, 136 (1989).
[2] Eichele K., and Wasylishen, R. E., *J. Phys. Chem.*, **98**, 3108 (1994).
[3.] Simone, A. H., Simone, E. H., and Cresman III, C. P., *J. Sci. Food Ag.* **73**, 39 (1997).
[4.] Uesu, Y., and Kobayashi, J., *Phys. Status Solidi (a)* **34**, 475 (1976).
[5.] Deguchi, K., Okaue, E., and Nakamura, E., *J. Phys. Soc. Jpn* **51**, 3569 (1982).
[6.] Blinc, R., Zeks, B., Levstick, A., Filipic. C., Slak, J., Burgar, M., Zupancic, I., Shuvalov, L. A., and Baranov, A. I., *Phys. Rev. Lett.* **43**, 231 (1979).
[7.] Kanda E., Yoshizawa, M., Yamakani, T., and Fujimura, T., *J. Phys. C* **15**, 6823 (1982).
[8] Imai, K, *J. Phys. Soc. Jpn.* **52**, 3960 (1983).
[9] Song, T. K., Moon, S. E., Noh, K. H., Kwun, S. L., Shin, H. K., and Kim, J. J., *Phys, Rev. B* **50**, 6637 (1994).
10 Scalapino, D. J., and Imry, Y., *Phys. Rev.* **B11**, 5 (1975).
11 Seitz, F., and Turnbull, D, *Solid State Physics -Advances in Research and Applications*, New York: Academic Press, Inc., 1957, p. 155
12 Hennel, J. W., and Klinowski, J., *Fundamentals of Nuclear Magnetic Resonance*, Essex: Longman Scientific and Technical, 1993, p. 65.
13 Pauling, L., *The Nature of the Chemical Bond*, Ithaca: 3$^{rd}$ ed., Cornell University Press, 1960.
14 Weast, R. C., ed., *CRC Handbook of Chemistry and Physics*, Boca Raton: 67$^{th}$ ed., CRC Press, Inc., 1987, p. F160.
15 Schuele, P. J., Ph. D. thesis, Montana State University, Ann Arbor: University Microfilms Intl., 1988.

# First-Principles Study of SrTiO$_3$ in Cubic and Tetragonal Phases

Chris LaSota, Cheng-Zhang Wang, Rici Yu, and Henry Krakauer

*College of William and Mary[1], Department of Physics, Williamsburg, Virginia 23187-8795.*

**Abstract.** Using the first-principles linear response method implemented with an LAPW basis [1], we have investigated the lattice dynamics of cubic SrTiO$_3$ [2] at the LDA equilibrium lattice parameter. Consistent with the antiferrodistortive (AFD) phase transition to the tetragonal structure at 105 K, we found phonon instabilities along the R-M-R line in the Brillouin zone. In addition, a ferroelectric-type (FE) instability was found at the $\Gamma$-point, although no FE structural phase transition is observed. Working in the tetragonal phase, we determine the theoretical equilibrium twist angle for the oxygen octahedra, and show that as the lattice parameter is reduced, the AFD instability is enhanced, while the FE instability is diminished. Furthermore, we find that the tetragonal structure is marginally stable against FE distortions along the c-axis.

Although it possesses the same high-temperature cubic perovskite structure as ferroelectric materials like BaTiO$_3$ and KNbO$_3$, SrTiO$_3$ exhibits qualitatively different temperature-dependent behavior. It transforms to a non-ferroelectric tetragonal phase when the temperature is lowered below T$_c$=105 K. This phase transition is due to a soft-mode instability at the zone-corner (k=$\frac{\pi}{a}$[1, 1, 1]) [3–6], as proposed by Cochran and Zia [7]. The atomic motions in this antiferrodistortive (AFD) mode are characterized by rotations of the Ti-O$_6$ octahedra about a [1 0 0] axis. Below T$_c$, these octahedra are slightly rotated, with the twist angle $\phi$ about the tetragonal c-axis being the order parameter of the transition. The dielectric constant, $\epsilon(T)$, rises in a Curie-Weiss manner until about T=35 K, where it continues to increase, but in a non-singular manner. This deviation was interpreted by Müller and Burkard as the onset of quantum paraelectric behavior in which the zero-point motion of the atoms supress the long-range ferroelectric ordering [8]. At T=3 K and below, $\epsilon(T)$ saturates to values above 10$^4$ [8–11], but a further structural phase transition does not occur. Due to these unusual properties at low temperature, SrTiO$_3$ has been subjected to extensive studies in the past few decades.

In recent years, first-principles density functional theory calculations using the local density approximation (LDA) have achieved remarkable success in determining

---

[1] Supported by the Office of Naval Research Grant No. N00014-97-1-1049.

the physical properties of ferroelectrics, including the equilibrium atomic configurations, lattice vibrational frequencies, and the underlying microscopic mechanisms behind macroscopic behaviors. For instance, the recent first-principles linear response calculations for $KNbO_3$ revealed large planar regions of phonon instabilities in the Brillouin zone [12]. These correspond, in real space, to chains along the [1 0 0] directions of atoms coherently displaced along the chain direction. Subsequent molecular-dynamics simulations of $KNbO_3$ have shown that these chains are dynamic in nature, and that they are preformed even in the high-temperature paraelectric phase [13]. We report here the results of a first-principles study of $SrTiO_3$ in both cubic and tetragonal phases.

To be commensurate with the linear response calculations for $SrTiO_3$ described below, our total energy calculations also used pseudopotentials in the LAPW method [14]. Kerker pseudopotentials [15] and the Wigner interpolation form for the exchange-correlation potential [16] were used. These yielded a cubic-phase LDA theoretical lattice parameter equal to 7.412 Bohr. This is 0.4 and 0.6 percent larger than the experimental values of 7.380 and 7.366 Bohr measured at 298 K and 105 K, respectively. We chose the larger theoretical lattice parameter to perform linear response calculations in the cubic structure.

Theoretical phonon frequencies at the high-symmetry points $\Gamma$ (k=[0 0 0]), X (k=$\frac{\pi}{a}$[1, 0, 0]), R (k=$\frac{\pi}{a}$[1, 1, 1]), and M (k=$\frac{\pi}{a}$[1, 1, 0]), are presented in Table 1, along with corresponding experimental values. The calculated frequencies are generally in good agreement with experiment. In the real material, the modes with imaginary frequencies are stabilized by anharmonic interactions, data from infrared and microwave reflectivity [17,18], neutron scattering [19,20], and Raman scattering experiments [21] show that these modes are soft and their frequencies are temperature-dependent.

We present in Fig.1 the phonon dispersion curves along high-symmetry directions of the Brillouin zone (BZ) [2]. The imaginary frequencies are shown as negative values below the dashed line. We observe a region of instability surrounding the $\Gamma$ point and another region which extends along the entire BZ edges from R-M-R. The unstable mode $R_{25}$ at the R point corresponds to a twisting motion of the $Ti-O_6$ octahedra about a common axis in opposite directions in all adjacent cells. At the M point, the unstable mode $M_3$ is nearly identical to that of $R_{25}$ with one exception. The rotation of the octahedra is in the *same* sense in neighboring cells along the vertical c-axis, but remains opposite in the horizontally adjacent cells. Since there exists a continuous instability extending from R-M-R, linear combinations of phonon modes along this cylindrical tube-like region will also be unstable. Thus, an unstable thin disk or planar region could form in real space wherein the octahedra are rotated in opposite senses in neighboring cells. The instability near $\Gamma$ has characteristics of the familiar ferroelectric-type (FE) TO phonon mode, where the Ti atoms and the oxygen octahedra move in opposite directions, with the Sr atoms essentially at rest. This region of FE instability is relatively large, but does not extend to the zone boundaries as in the $KNbO_3$ calculations [12]. Such a FE mode may be stabilized by pressure. As shown in

**TABLE 1.** A comparison of linear response phonon frequencies (in cm$^{-1}$) with experimental values at high-symmetry points. Subscripts in mode character labeling are as given by Cowely in Ref.[19].

|  | Longitudinal modes | | | | Transverse modes | | | | |
|---|---|---|---|---|---|---|---|---|---|
| Γ-point | $\Gamma_{15}$ | $\Gamma_{25}$ | $\Gamma_{15}$ | $\Gamma_{15}$ | $\Gamma_{15}$ | $\Gamma_{15}$ | $\Gamma_{25}$ | $\Gamma_{15}$ | |
| Linear-response | 146 | 219 | 439 | 751 | 99$i$ | 151 | 219 | 522 | |
| 90 K$^a$ | 170 | 265 | | | 42 | 170 | 265 | | |
| 120 K$^b$ | | | | | 50 | | | | |
| 296 K$^c$ | 175 | | 460 | 830 | 88 | 178 | | 543 | |
| 296 K$^d$ | | | 483 | 817 | 100 | | | 550 | |
| 297 K$^e$ | 169 | 265 | 457 | 823 | 92 | 169 | 265 | 547 | |
| R-point | R$_{25}$ | R$_{15}$ | R$_{15}$ | R$_{25'}$ | R$_{25}$ | R$_{15}$ | R$_{15}$ | R$_{25'}$ | |
| Linear-response | 86$i$ | 122 | 417 | 426 | 86$i$ | 122 | 417 | 426 | |
| 120 K$^b$ | | | | | 20 | 129 | | | |
| 297 K$^e$ | 52 | 145 | 450 | 473 | 52 | 145 | 450 | 447 | |
| M-point | M$_{5'}$ | M$_{5'}$ | M$_1$ | | M$_3$ | M$_{2'}$ | M$_{3'}$ | M$_{5'}$ | M$_5$ |
| Linear-response | 101 | 251 | 444 | | 32$i$ | 94 | 98 | 101 | 311 |
| 120 K$^b$ | | | | | 65 | | | 117 | |
| 297 K$^e$ | 120 | 280 | 468 | | 88 | 111 | 111 | | 332 |
| X-point | X$_{2'}$ | X$_3$ | X$_1$ | | X$_5$ | X$_{5'}$ | X$_5$ | X$_{5'}$ | |
| Linear-response | 157 | 266 | 269 | | 90 | 94 | 154 | 304 | |
| 90 K$^a$ | | 270 | 276 | | 112 | 112 | 188 | 326 | |
| 297 K$^e$ | 115 | 273 | 290 | | 118 | 118 | 196 | 328 | |

$^a$Neutron scattering results [19].
$^b$Neutron scattering results [6].
$^c$Computed in [19] from infrared reflectivity data reported in [17].
$^d$Computed in [19] from infrared reflectivity data reported in [18].
$^e$Neutron scattering results [20].

Fig. 2, we find that the total energy of the cubic cell is already marginally stable against the FE mode distortions when the lattice constant is contracted by 0.6%.

After the tetragonal phase transition occurs at 105 K, $c/a$ increases monotonically

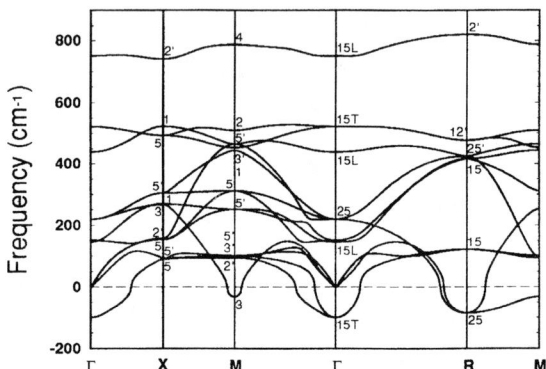

**FIGURE 1.** Theoretical phonon dispersion curves for cubic $SrTiO_3$ at the LDA equilibrium lattice parameter $a$=7.412 Bohr. Negative values appearing below the dashed line correspond to imaginary frequencies of unstable modes. The labeling of the modes at high symmetry points is in notation given by Cowley in Ref.[19].

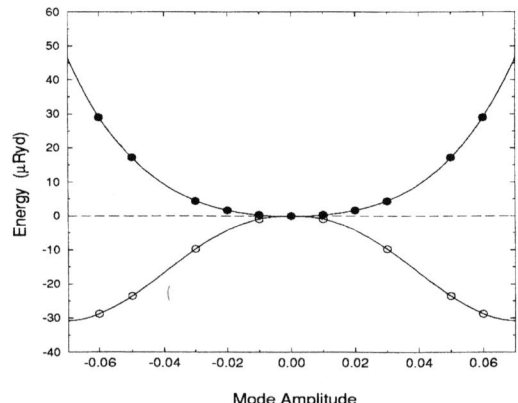

**FIGURE 2.** Total energy calculations at two volumes for FE distortions in cubic structure, taking the undistorted structural energies as references. The data are fitted to fourth-order polynomials. The lower curve is for the larger LDA equilibrium volume and the upper one is for the calculations at the volume measured at 105 K.

from one to 1.0009 near T~0 K [22], and the twist angle $\phi$ increases from 0.3° to 2.1° [3]. The tetragonal cell volume decreases from 799.33 to 798.25 Bohr$^3$ [22,23], changing less than 0.14 percent. We performed total energy calculations in the tetragonal phase for the phonon modes at the R and $\Gamma$ points, which are unstable in the cubic structure. A fixed value of $c/a$=1.0005 was used in all of the following computations. For the zone-corner $R_{25}$ mode, the calculations were done at two volumes, 814.4 Bohr$^3$ ($V_{LDA}$) and 803.9 Bohr$^3$ ($V_{298K}$). $V_{LDA}$ is twice the theoretical LDA volume of 5-atom cubic cell, and $V_{298K}$ is twice the volume of the cubic cell at 298 K. We have plotted in Fig.3 the total energy change with respect to $\phi$=0 of the 10-atom tetragonal cell over a range of $\phi$ values. As expected from the linear response calculations, the $\phi$=0 structure is unstable. The solid lines are fourth-order polynomial fits to the calculations. For $V_{LDA}$ and $V_{298K}$, the energies are minimized at $\phi_{min} = 6.6°$ and 6.9°, respectively. It is evident that decreasing the volume enhances the AFD instability, while increasing the well-depth and $\phi_{min}$. This is in agreement with the experimental observations of Unoki and Sakudo below 105 K [3]. Monte Carlo simulations by Zhong and Vanderbilt [24], using an effective Hamiltonian constructed from first-principles, also reproduced this behavior. However, our values for $\phi_{min}$ are too large, and overestimate the T=0 K value by a factor of three. This discrepancy is also observed in similar LAPW calculations reported by Schwarz [25]. To investigate the possible sources of this discrepancy, we applied the all-electron LAPW+LO method [26], in which the pseudopotential approximation is not used, and the orthogonality is enforced for the semicore and valence orbitals by extending the LAPW basis functions to include local orbitals. A single set of calculations were done at $V_{298K}$, changing the double-well depth from 2.9 mRy to 1.1 mRy. At the same time, $\phi_{min}$ changes

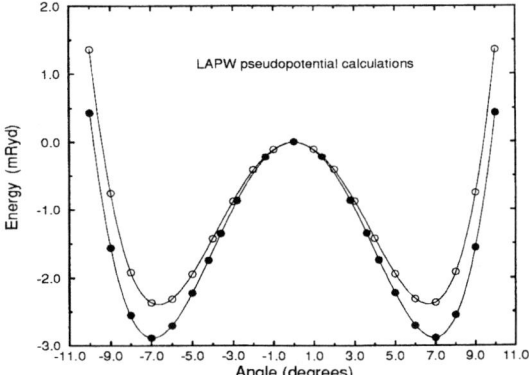

**FIGURE 3.** Total energy as a function of the twist angle of the Ti-O$_6$ octahedra in the 10-atom tetragonal cell at $V_{LDA}$ =814.4 Bohr$^3$ and $V_{298K}$ =803.9 Bohr$^3$.

from 6.9° to 5.4°, but still remains too large compared to experiment. We therefore ascribe this quantitative discrepancy to the limitations of the LDA.

Even though no further phase transition is observed at low temperatures, the sharp rise in the dielectric constant suggests the possibility of a FE instability which may be suppressed by the zero-point motion of the atoms. Such an instability may be revealed by computing the full zone-center dynamical matrix of the tetragonal structure. Considering the large computational burden involved, we instead focus on investigating the influence of the tetragonal transition on the FE instability of the $\Gamma_{15}$ mode. We imposed the calculated $\Gamma_{15}$ FE mode distortion *along the c-axis* in the tetragonal cell, setting the twist angle of the Ti-$O_6$ octahedra at 1.4°, the volume at 799.3 Bohr$^3$, which was measured at 105 K, and $c/a$ at 1.0005 as in the AFD calculations. The total energy as a function of $\Gamma_{15}$ amplitude is shown in Fig.4. We find that the tetragonal structure is marginally stable against the imposed distortions. However we have not yet investigated the structural stability for FE distortions in the plane perpendicular to the c-axis.

To conclude, we have reported the results of linear reponse calculations for $SrTiO_3$ in the cubic phase. Mapping the phonon dispersion in the full Brillouin zone, we found two regions of instability of ferroelectric (FE) and antiferrodistortive (AFD) characters. We minimized the total energy in tetragonal phase with respect to the Ti-$O_6$ octahedral twist angle, which is too large compared to experiment. We also observed that as the volume of the cell is reduced, the AFD instability is enhanced, while the FE instability is diminished.

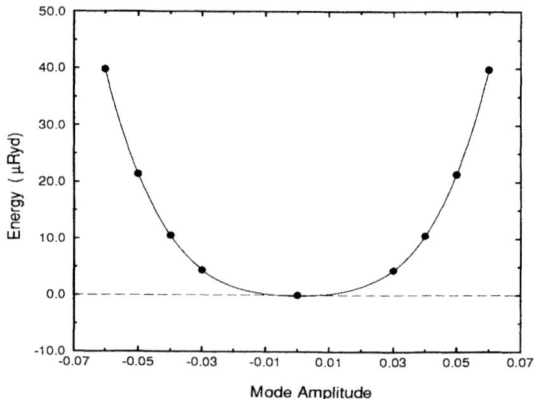

**FIGURE 4.** Total energy calculations by the LAPW+LO method for ferroelectric-type distortions along the c-axis in the tetragonal phase. The volume used is 799.3 Bohr$^3$, the value measured at 105 K.

# REFERENCES

1. Yu, R., and Krakauer, H., *Phys. Rev. B* **49**, 4467-4477 (1994).
2. LaSota, C., Wang, C.-Z., Yu, R., and Krakauer, H., *Ferroelectrics* **194**, 109-118 (1997).
3. Unoki, H., and Sakudo, T., *J. Phys. Soc. Japan* **23**, 546-552 (1967).
4. Müller, K. A., Berlinger, W., and Waldner, F., *Phys. Rev. Lett.* **21**, 814-817 (1968).
5. Fleury, P. A., Scott, J. F., and Worlock, J. M., *Phys. Rev. Lett.* **21**, 16-19 (1968).
6. Shirane, G., and Yamada, Y., *Phys. Rev.* **177**, 858-863 (1969).
7. Cochran, W., and Zia, A., *Phys. Stat. Sol.* **25**, 273-283 (1968).
8. Müller, K. A., and Burkard, H., *Phys. Rev. B* **19**, 3593-3602 (1979).
9. Weaver, H. E., *J. Phys. Chem. Solids* **11**, 274-277 (1959).
10. Sakudo, T., and Unoki, H., *Phys. Rev. Lett.* **26**, 851-853 (1971).
11. Viana, R., Lunkenheimer, P., Hemberger, J., Böhmer, R., and Loidl, A., *Phys. Rev. B* **50**, 601-604 (1994).
12. Yu, R., and Krakauer, H., *Phys. Rev. Lett.* **74**, 4067-4070 (1995).
13. Krakauer, H., Yu, R., Wang, C.-Z., and LaSota, C. *Ferroelectrics* (in press); Krakauer, H., Yu, R., Wang, C.-Z., Rabe, K. M., Waghmare, U. V., available in electronic form at Los Alamos National Laboratory physics e-print archive at http://xxx.lanl.gov/abs/cond-mat/9710088
14. Andersen, O. K., *Phys. Rev. B* **12**, 3060-3083 (1975); Wimmer, E., Krakauer, H., Weinert, M., and Freeman, A. J., *Phys. Rev. B* **24**, 864-875 (1981); Singh, D. J., *Planewaves, Pseudopotentials and the LAPW Method*, Massachusetts: Kluwer Academic Publishers, 1994.
15. Kerker, G. P., *J. Phys. C.* **13**, L189-L194 (1980).
16. Wigner, E., *Phys. Rev.* **46**, 1002-1011 (1934).
17. Spitzer, W. G., Miller, R. C., Kleinman, D. A., and Howarth, L. E., *Phys. Rev.* **126**, 1710-1721 (1962).
18. Barker, A. S., and Tinkham, M., *Phys. Rev.* **125**, 1527-1530 (1962).
19. Cowely, R. A., *Phys. Rev.* **134**, A981-A997 (1964).
20. Stirling, W. G., *J. Phys. C: Solid State Phys.* **5**, 2711-2730 (1972); Stirling, W. G., and Currat, R., *J. Phys. C: Solid State Phys.* **9**, L519-L522 (1976).
21. Worlock, J. M., and Fleury, P. A., *Phys. Rev. Lett.* **19**, 1176-1179 (1967); Fleury, P. A., and Worlock, J. M., *Phys. Rev.* **174**, 613-623 (1968).
22. Heidemann, A., and Wettengel, H., *Z. Phys.* **258**, 429-438 (1973).
23. Liu, M., Finlayson, T., and Smith, T., *Phys. Rev. B* **55**, 3480-3484 (1997).
24. Zhong, W., and Vanderbilt, D., *Phys. Rev. Lett.* **74**, 2587-2590 (1995).
25. Schwarz, K., *Phase Transitions* **52**, 109-122 (1994).
26. Singh, D. J., *Phys. Rev. B* **43**, 6388-6392 (1991).

# Maximally-localized Wannier functions in perovskites: Cubic $BaTiO_3$

## Nicola Marzari and David Vanderbilt

*Department of Physics and Astronomy, Rutgers University, Piscataway, NJ 08854-8019*

**Abstract.** The electronic ground state of a periodic crystalline solid is usually described in terms of extended Bloch orbitals; localized Wannier functions can alternatively be used. These two representations are connected by families of unitary transformations, carrying a large degree of arbitrariness. We have developed a localization algorithm that allows one to iteratively transform the extended Bloch orbitals of a first-principles calculation into a unique set of *maximally localized* Wannier functions. We apply this formalism here to the case of cubic $BaTiO_3$. The purpose is twofold. First, a localized-orbital picture allows a meaningful band-by-band decomposition of the whole Bloch band complex. In perovskites, these Wannier functions are centered on the atomic sites and display clearly a $s$, $p$, $d$, or hybrid character. Second, since the centers of the Wannier functions map the polarization field onto localized point charges, the ground state dielectric properties become readily available. We study the Born effective charges of the paraelectric phase of $BaTiO_3$. We are able to identify not only the contributions that come from a given group of bands, but also the individual contributions from the "atomic" Wannier functions that comprise each of these groups.

## INTRODUCTION

The electronic ground state of a periodic solid, in the independent particle approximation, is naturally labelled according to the prescriptions of Bloch's theorem: single-particle orbitals are assigned a quantum number **k** for the crystal momentum, together with a band index $n$. Although this choice is widely used in electronic structure calculations, alternative representations are available. The Wannier representation [1], essentially a real-space picture of localized orbitals, assigns as quantum numbers the lattice vector **R** of the cell where the orbital is localized, together with the band index $n$. Wannier functions can be a powerful tool in the study of the electronic and dielectric properties of materials: they are the solid-state equivalent of "localized molecular orbitals" [2], and thus provide an insightful picture of the nature of chemical bonding, otherwise missing from the Bloch picture of extended orbitals. In addition, the modern theory of polarization [3] directly relates the centers of the Wannier functions to the macroscopic polarization of a crystalline insulator.

CP436, *First-Principles Calculations for Ferroelectrics*
edited by Ronald E. Cohen
© 1998 The American Institute of Physics 1-56396-730-8/98/$15.00

Wannier functions are strongly non-unique. This is a consequence of the phase indeterminacy $e^{i\phi_n(\mathbf{k})}$ that Bloch orbitals $\psi_{n\mathbf{k}}$ have at every wavevector $\mathbf{k}$. This indeterminacy is actually more general than just the phase factors: Bloch orbitals belonging to a composite group of bands (i.e. bands that are connected between themselves by degeneracies, but separated from others by energy gaps) can undergo arbitrary unitary transformations $U^{(\mathbf{k})}$ between themselves at every $\mathbf{k}$. We have recently developed a procedure [4] that can iteratively refine these otherwise arbitrary degrees of freedom, so that they lead to Wannier functions that are well defined and that are localized around their centers (in particular, they minimize the second moment around the centers). Such a procedure can be applied either to a whole band complex of Bloch orbitals, or just to some isolated subgroups.

As a natural first application of this technique, we present here results for the case of $BaTiO_3$ in the cubic phase. Perovskite ferroelectrics, of which $BaTiO_3$ is a paradigmatic example, owe their very rich phenomenology to the subtle competition of several degrees of freedom, balancing the long-range dipole-dipole interaction with short-range Pauli repulsion. One of the striking features is the display of anomalously large Born effective charges [5]. Their origin is understood in a simple tight-binding picture [6]: the change in the bond length (Ti-O in this case) corresponds to a dynamic charge transfer that is stronger when the bonding is borderline between ionic and covalent. Localized Wannier functions can thus be used to investigate the nature of this bonding, and to monitor the changes that follow a ferroelectric distortion. Additionally, the displacement of each Wannier center relates directly to the effective charge contribution of its orbital, and can be used to identify the nominal and the anomalous contributions to the polarization induced by an atomic displacement.

## METHOD

Electronic structure calculations are carried out using periodic boundary conditions. This is the most natural choice to study perfect crystals and to minimize finite size-effects in the study of several non-periodic systems (e.g. surfaces, or impurities). The one-particle effective Hamiltonian $\hat{H}$ then commutes with the lattice-translation operator $\hat{T}_{\mathbf{R}}$, allowing one to choose as common eigenstates the Bloch orbitals $|\psi_{n\mathbf{k}}\rangle$,

$$[\hat{H}, \hat{T}_{\mathbf{R}}] = 0 \Rightarrow \psi_{n\mathbf{k}}(\mathbf{r}) = e^{i\phi_n(\mathbf{k})} u_{n\mathbf{k}}(\mathbf{r}) e^{i\mathbf{k}\cdot\mathbf{r}} , \qquad (1)$$

where $u_{n\mathbf{k}}(\mathbf{r})$ has the periodicity of the Hamiltonian. There is an arbitrary phase $\phi_n(\mathbf{k})$, periodic in reciprocal space, that is not assigned by the Schrödinger equation and that we have written out explicitly. We obtain a (non-unique) Wannier representation using any unitary transformation of the form $\langle n\mathbf{k} | \mathbf{R}n \rangle = e^{i\varphi_n(\mathbf{k}) - i\mathbf{k}\cdot\mathbf{R}}$:

$$|\mathbf{R}n\rangle = \frac{V}{(2\pi)^3} \int_{BZ} |\psi_{n\mathbf{k}}\rangle e^{i\varphi_n(\mathbf{k}) - i\mathbf{k}\cdot\mathbf{R}} d\mathbf{k} . \qquad (2)$$

Here $V$ is the real-space primitive cell volume. It is easily shown that the $|\mathbf{R}n\rangle$ form an orthonormal set, and that two Wannier functions $|\mathbf{R}n\rangle$ and $|\mathbf{R}'n\rangle$ transform into each other with a translation of a lattice vector $\mathbf{R} - \mathbf{R}'$ [7]. The arbitrariness that is present in $\varphi_n(\mathbf{k})$ [or $\phi_n(\mathbf{k})$] propagates to the resulting Wannier functions, making the Wannier representation non-unique. Since the electronic energy functional in an insulator is also invariant with respect to a unitary transformation of its $n$ occupied Bloch orbitals, there is additional freedom associated with the choice of a full unitary matrix (and not just a diagonal one) transforming the orbitals between themselves at every wavevector $\mathbf{k}$. Thus, the most general operation that transforms the Bloch orbitals into Wannier functions is given by

$$|\mathbf{R}n\rangle = \frac{V}{(2\pi)^3} \int_{BZ} \sum_m U_{mn}^{(\mathbf{k})} |\psi_{m\mathbf{k}}\rangle e^{-i\mathbf{k}\cdot\mathbf{R}} d\mathbf{k} \ . \quad (3)$$

The Wannier functions $w_n(\mathbf{r} - \mathbf{R}) = |\mathbf{R}n\rangle$, for non-pathological choices of phases, are "localized": for a $\mathbf{R}_i$ far away from $\mathbf{R}$, $w_n(\mathbf{R}_i - \mathbf{R})$ is a combination of terms like $\int_{BZ} u_{m\mathbf{k}}(0) e^{i\mathbf{k}\cdot(\mathbf{R}_i-\mathbf{R})} d\mathbf{k}$, which are small due to the rapidly varying character of the exponential factor [7].

## Maximally-localized Wannier functions

Several heuristic approaches have been developed that construct reasonable sets of Wannier functions, reducing the arbitrariness in the $U_{mn}^{(\mathbf{k})}$ with symmetry considerations and analiticity requirements [8], or explicitly employing projection techniques on the occupied subspace spanned by the Bloch orbitals [9]. At variance with those approaches, we introduce a well-defined *localization criterion*, choosing the functional

$$\Omega = \sum_n \left[ \langle r^2 \rangle_n - \bar{\mathbf{r}}_n^2 \right] \quad (4)$$

as the measure of the spread of the Wannier functions. The sum runs over the $n$ functions $|\mathbf{0}n\rangle$; $\langle r^2\rangle_n$ and $\bar{\mathbf{r}}_n = \langle \mathbf{r}\rangle_n$ are the expectation values $\langle \mathbf{0}n|r^2|\mathbf{0}n\rangle$ and $\langle \mathbf{0}n|\mathbf{r}|\mathbf{0}n\rangle$. Given a set of Bloch orbitals $|\psi_{m\mathbf{k}}\rangle$, the goal is to find the choice of $U_{mn}^{(\mathbf{k})}$ in (3) that minimizes the values of the localization functional (4). We are able to express the gradient $G = \frac{d\Omega}{dW}$ of the localization functional with respect to an infinitesimal unitary rotation of our set of Bloch orbitals

$$|u_{n\mathbf{k}}\rangle \rightarrow |u_{n\mathbf{k}}\rangle + \sum_m dW_{mn}^{(\mathbf{k})} |u_{m\mathbf{k}}\rangle \ , \quad (5)$$

where $dW$ an infinitesimal antiunitary matrix $dW^\dagger = -dW$ such that

$$U_{mn}^{(\mathbf{k})} = \delta_{mn} + dW_{mn}^{(\mathbf{k})} \ . \quad (6)$$

This provides an equation of motion for the evolution of the $U_{mn}^{(\mathbf{k})}$, and of the $|\mathbf{R}n\rangle$ derived in (3), towards the minimum of $\Omega$; small finite steps in the direction opposite to the gradient decrease the value of $\Omega$, until a minimum is reached.

# 1 Real-space representation

There are several interesting consequences stemming from the choice of (4) as the localization functional, that we briefly summarize here. Adding and subtracting the off-diagonal components $\tilde{\Omega} = \sum_n \sum_{\mathbf{R}m \neq \mathbf{0}n} |\langle \mathbf{R}m|\mathbf{r}|\mathbf{0}n\rangle|^2$, we obtain the decomposition $\Omega = \Omega_{\mathrm{I}} + \Omega_{\mathrm{D}} + \Omega_{\mathrm{OD}}$, where $\Omega_{\mathrm{I}}$, $\Omega_{\mathrm{D}}$ and $\Omega_{\mathrm{OD}}$ are respectively

$$\Omega_{\mathrm{I}} = \sum_n \left[ \langle r^2 \rangle_n - \sum_{\mathbf{R}m} |\langle \mathbf{R}m|\mathbf{r}|\mathbf{0}n\rangle|^2 \right],$$

$$\Omega_{\mathrm{D}} = \sum_n \sum_{\mathbf{R} \neq 0} |\langle \mathbf{R}n|\mathbf{r}|\mathbf{0}n\rangle|^2,$$

$$\Omega_{\mathrm{OD}} = \sum_{m \neq n} \sum_{\mathbf{R}} |\langle \mathbf{R}m|\mathbf{r}|\mathbf{0}n\rangle|^2.$$

It can be shown that all terms are *positive-definite* (in particular $\Omega_{\mathrm{I}}$, see Ref. [4]); more importantly, $\Omega_{\mathrm{I}}$ is also *gauge-invariant*, i.e., it is invariant under any arbitrary unitary transformation (3) of the Bloch orbitals. The minimization procedure thus corresponds to the minimization of $\tilde{\Omega} = \Omega_{\mathrm{D}} + \Omega_{\mathrm{OD}}$. At the minimum, the elements $|\langle \mathbf{R}m|\mathbf{r}|\mathbf{0}n\rangle|^2$ are as small as possible, realizing the best compromise in the simultaneous diagonalization, within the space of the Bloch bands considered, of the three position operators $x$, $y$ and $z$ (which do not in general commute when projected within this space).

# 2 Reciprocal-space representation

As shown by Blount [7], matrix elements of the position operator between Wannier functions take the form

$$\langle \mathbf{R}n|\mathbf{r}|\mathbf{0}m\rangle = i \frac{V}{(2\pi)^3} \int d\mathbf{k}\, e^{i\mathbf{k}\cdot\mathbf{R}} \langle u_{n\mathbf{k}}|\nabla_{\mathbf{k}}|u_{m\mathbf{k}}\rangle \tag{7}$$

and

$$\langle \mathbf{R}n|r^2|\mathbf{0}m\rangle = -\frac{V}{(2\pi)^3} \int d\mathbf{k}\, e^{i\mathbf{k}\cdot\mathbf{R}} \langle u_{n\mathbf{k}}|\nabla_{\mathbf{k}}^2|u_{m\mathbf{k}}\rangle. \tag{8}$$

These expressions provide the needed connection with our underlying Bloch formalism, since they allow us to express the localization functional $\Omega$ in terms of the matrix elements of $\nabla_{\mathbf{k}}$ and $\nabla_{\mathbf{k}}^2$. We thus determine the Bloch orbitals $|u_{m\mathbf{k}}\rangle$ on a regular mesh of **k**-points, and use finite differences to evaluate the above derivatives. For any given **k**-point in a regular cubic mesh (sc, fcc, bcc), we have a star **b** of $Z$ **k**-points that are first-neighbors; their weights in the evaluation of derivatives are $w_b = 3/Zb^2$. We define $M_{mn}^{(\mathbf{k},\mathbf{b})} = \langle u_{m\mathbf{k}}|u_{n,\mathbf{k}+\mathbf{b}}\rangle$ as the matrix elements between Bloch orbitals at neighboring **k**-points. The $M_{mn}^{(\mathbf{k},\mathbf{b})}$ are a central quantity in our formalism, since we can then express all the contributions to the localization functional using the connection made by Blount, together with our finite-difference evaluations of the gradients. After some algebra we obtain [4]

$$\Omega_\mathrm{I} = \frac{1}{N} \sum_{\mathbf{k},\mathbf{b}} w_b \left( N_\mathrm{bands} - \sum_{mn} |M_{mn}^{(\mathbf{k},\mathbf{b})}|^2 \right), \qquad (9)$$

$$\Omega_\mathrm{OD} = \frac{1}{N} \sum_{\mathbf{k},\mathbf{b}} w_b \sum_{m \neq n} |M_{mn}^{(\mathbf{k},\mathbf{b})}|^2, \qquad (10)$$

and

$$\Omega_\mathrm{D} = \frac{1}{N} \sum_{\mathbf{k},\mathbf{b}} w_b \sum_n \left( -\mathrm{Im}\, \ln M_{nn}^{(\mathbf{k},\mathbf{b})} - \overline{\mathrm{Im}\, \ln M_{nn}^{(\mathbf{k},\mathbf{b})}} \right)^2. \qquad (11)$$

From these, we can calculate the change in the localization functional in response to an infinitesimal unitary transformation of the Bloch orbitals, as a function of the $M_{mn}^{(\mathbf{k},\mathbf{b})}$; once these steepest-descents are available, it is straightforward to construct a procedure that updates the $U_{mn}^{(\mathbf{k})}$ towards the minimum of the functional.

## RESULTS: THE CASE OF BATIO$_3$

We study here the cubic phase of BaTiO$_3$, using a plane-wave total-energy pseudopotential approach with the local-density approximation to the exchange-correlation functional. We use norm-conserving pseudopotentials in the Kleinman-Bylander representation, with $q_c$ kinetic-energy tuning [10] for the oxygen atom and a Troullier-Martins procedure [11] for the titanium, to bring the cutoff convergence down to 900 eV. The $3s$ and $3p$ levels of titanium have been included in the valence. The Brillouin zone is sampled with a $4 \times 4 \times 4$ Monkhorst-Pack mesh; the lattice parameter used is 3.98Å.

## Wannier functions of cubic BaTiO$_3$

The minimization of the total energy provides the Kohn-Sham Bloch orbitals on a regular mesh of **k**-points, that are then used as a starting point for the construction of the Wannier functions. The subsequent minimization of the localization functional determines the $U_{mn}^{(\mathbf{k})}$ that correspond to the maximally-localized Wannier functions. In BaTiO$_3$ there are several groups of bands that are separated by gaps. In order of increasing energy, we have the band groups corresponding to the Ti $3s$ (1), Ti $3p$ (3), Ba $5s$ (1), O $2s$ (3), Ba $5p$ (3) and O $2p$ (9) levels (in parenthesis are the number of bands in each group). We initially consider each group of

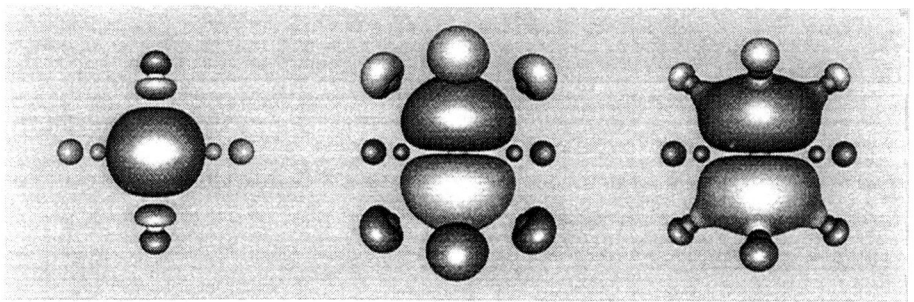

**FIGURE 1.** Left panel: oxygen-centered Wannier function from the O $2s$ 3-band group (the O atom is surrounded by four Ba atoms on the sides, and two Ti atoms on top and bottom). Center and right panels: barium-centered Wannier function from the Ba $5p$ 3-band group, and barium-centered Wannier function from the Ba $5p$ and the O $2s$ 6-band group (the Ba atom is surrounded by 12 oxygens).

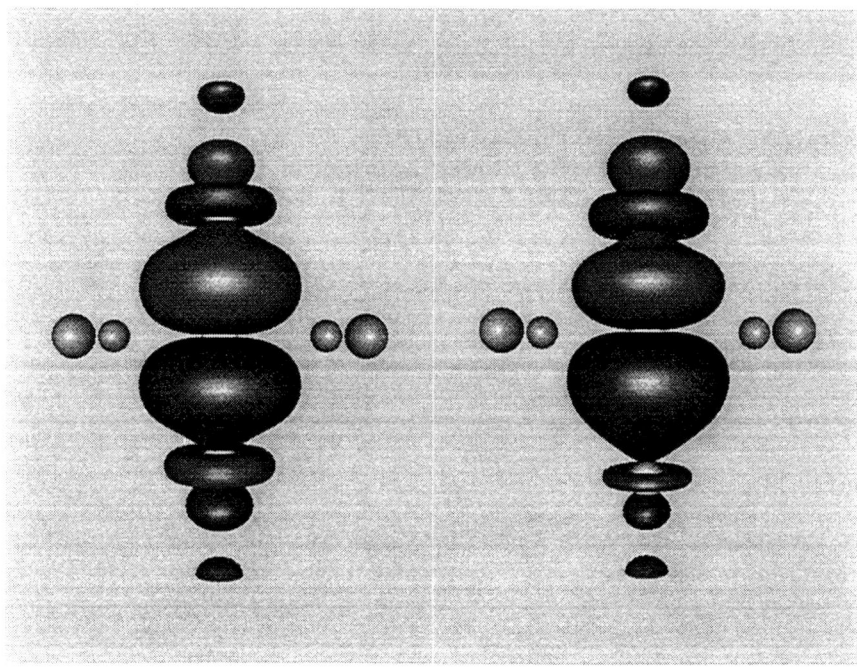

**FIGURE 2.** Oxygen-centered $\sigma$ Wannier function from the localization of the O $2p$ 9-band group. The orbital is oriented along the Ti-O-Ti bond; the four Ba atoms neighboring the central oxygen and the two Ti atoms on top and bottom are also shown. Left panel: ideal atomic positions. Right panel: same, but with the titanium atoms displaced downwards.

bands separately, and perform the minimization on the 6 subspaces of dimensions $1 \times 1$, $3 \times 3$, $1 \times 1$, $3 \times 3$, $3 \times 3$, and $9 \times 9$.

The Wannier functions determined from the the Ti $3s$ or the Ti $3p$ groups strongly resemble atomic orbitals, slightly deformed by the underlying crystal potential, and are not shown here. We show instead in the left panel of Fig. 1 one of the three oxygen-centered Wannier functions that are derived from the oxygen $2s$ bands. In each unit cell there are three such functions, sitting on each of the oxygens. The orbital shows its atomic $s$ character; there are some contributions on the Ti, with $p$ and/or $d$ character (the titanium is slightly embedded in a $d_{z^2}$ orbital). Interesting results emerge from the localization of the Ba $5p$ bands (Fig. 1, center and right panels). These bands correspond to three orbitals in each unit cell, all centered on the barium and oriented along the three cristallographic directions ($p_x$, $p_y$ and $p_z$). It can easily be seen (center panel) that, in addition to the distinctive atomic $p$ orbital on the barium, there are significant $sp$-like contributions sitting on 8 of the 12 neighboring oxygens. This supports the suggestion that barium in this compound has some covalent character [12], and it is consistent with the anomalous effective-charge contributions that come from this group of Wannier functions (see next subsection). It is interesting to note that if we decide to treat the Ba $5p$ bands together with the O $2s$ bands, we can (obviously) increase the degree of localization of each orbital. In this latter case (right panel) the $sp$ contributions on the oxygens decrease, being transferred to the $2s$ orbitals localized on the oxygens themselves.

Finally, we examine the 9 oxygen $2p$ bands that result from the hybridization of the O $2p$ electrons with the Ba $6s$ and the Ti $4s$ and $3d$. There are three localized orbitals on each oxygen, oriented along the Ti-O-Ti bonds. We label two of these orbitals as $\pi$ and one as $\sigma$, according to their symmetry along the Ti-O-Ti axis. One of the $\sigma$ orbitals is shown in Fig. 2, first with the atoms in their ideal positions (left panel) and then with the Ti atoms displaced along the Ti-O bond (right panel). The $\sigma$ orbital shows clearly the hybridization between the oxygen $p$ orbital oriented along the Ti-O-Ti direction and the $d_{z^2}$ orbital of the titanium. The $\sigma$ and even more the $\pi$ orbitals have strong anomalous contributions to the effective charges, that can be visualized with the large charge transfer from the oxygen atoms in response to the titanium displacements.

## Band-by-band decomposition of the Born effective charges

The Born dynamical effective charges describe the change in macroscopic polarization that is induced by the displacement of a given ion. As such, they play a fundamental role in determining the dynamical properties of insulating crystals, and are a powerful tool to investigate the dielectric and ferroelectric properties of materials. They also determine the splitting of the infrared-active optical modes; in simpler compounds (e.g. GaAs) they can be unambiguously determined from the experimental phonon dispersions. Perovskite ferroelectrics display anomalous, large effective charges, that can be almost double their nominal ionic value. The

**TABLE 1.** Born-effective charges decomposed into atomic contributions, for BaTiO$_3$. The $Z^*_{el}$ are calculated by displacing a Ba, Ti or a O sublattice (O$_1$ is parallel to the Ti-O direction, O$_2$ perpendicular) by $\delta a_0$, where $\delta = 0.002$. The numbers in parenthesis are those of Ghosez et al. ($a_0$=3.94Å). Horizontal lines group band complexes that have been treated together in the Wannier minimization. Top line in each group is the total for that group. We have added at the bottom the O 2p decomposition obtained in a calculation performed using a Ti atom with the 3s and 3p levels frozen in the core (Ti$^{4+}$).

|  | Ba $(\delta,0,0)$ | | Ti $(\frac{1}{2}+\delta,\frac{1}{2},\frac{1}{2})$ | | O$_1$ $(\frac{1}{2},\frac{1}{2},\delta)$ | | O$_2$ $(\frac{1}{2}+\delta,\frac{1}{2},0)$ | |
|---|---|---|---|---|---|---|---|---|
| Ti 3s (1) | 0.01 | (0.01) | -2.04 | (-2.03) | 0.03 | (0.02) | 0.00 | (0.00) |
| Ti 3p (3) | 0.02 | (0.02) | -6.19 | (-6.22) | 0.21 | (0.21) | -0.02 | (-0.02) |
| $(\frac{1}{2},\frac{1}{2},\frac{1}{2})$ | 0.00 | | -2.21 | | 0.00 | | 0.00 | |
| $(\frac{1}{2},\frac{1}{2},\frac{1}{2})$ | 0.01 | | -1.99 | | 0.21 | | -0.02 | |
| $(\frac{1}{2},\frac{1}{2},\frac{1}{2})$ | 0.01 | | -1.99 | | 0.00 | | 0.00 | |
| Ba 5s (1) | -2.09 | (-2.11) | 0.04 | (0.05) | 0.01 | (0.01) | 0.02 | (0.02) |
| O 2s (3O) | 0.65 | (0.73) | 0.20 | (0.23) | -2.45 | (-2.51) | -2.21 | (-2.23) |
| $(0,\frac{1}{2},\frac{1}{2})$ | -0.11 | | 0.48 | | -0.04 | | -0.01 | |
| $(\frac{1}{2},0,\frac{1}{2})$ | 0.38 | | -0.14 | | -0.04 | | 0.01 | |
| $(\frac{1}{2},\frac{1}{2},0)$ | 0.38 | | -0.14 | | -2.36 | | -2.21 | |
| Ba 5p (3Ba) | -7.20 | (-7.38) | 0.31 | (0.36) | -0.11 | (-0.13) | 0.50 | (0.58) |
| $(0,0,0)$ | -2.46 | | 0.04 | | -0.06 | | 0.21 | |
| $(0,0,0)$ | -2.37 | | 0.14 | | 0.01 | | 0.01 | |
| $(0,0,0)$ | -2.37 | | 0.14 | | -0.06 | | 0.29 | |
| O 2p (9) | 1.26 | (1.50) | 3.01 | (2.86) | -9.57 | (-9.31) | -6.35 | (-6.50) |
| $(0,\frac{1}{2},\frac{1}{2})\sigma$ | -0.01 | | 0.81 | | -0.07 | | 0.03 | |
| $(0,\frac{1}{2},\frac{1}{2})\pi$ | -0.20 | | 1.78 | | -0.17 | | -0.49 | |
| $(0,\frac{1}{2},\frac{1}{2})\pi$ | -0.20 | | 1.78 | | -0.01 | | 0.00 | |
| $(\frac{1}{2},0,\frac{1}{2})\pi$ | 0.22 | | -0.17 | | -0.01 | | 0.01 | |
| $(\frac{1}{2},0,\frac{1}{2})\pi$ | 0.51 | | -0.27 | | -0.17 | | 0.04 | |
| $(\frac{1}{2},0,\frac{1}{2})\sigma$ | 0.10 | | -0.24 | | -0.07 | | 0.02 | |
| $(\frac{1}{2},\frac{1}{2},0)\pi$ | 0.22 | | -0.17 | | -3.10 | | -1.89 | |
| $(\frac{1}{2},\frac{1}{2},0)\sigma$ | 0.10 | | -0.24 | | -2.86 | | -1.81 | |
| $(\frac{1}{2},\frac{1}{2},0)\pi$ | 0.51 | | -0.27 | | -3.10 | | -2.26 | |
| Total $Z^*_{el}$ | -7.35 | | -4.67 | | -11.87 | | -8.06 | |
| Core | 10.00 | | 12.00 | | 6.00 | | 6.00 | |
| Total $Z^*$ | 2.65 | (2.77) | 7.33 | (7.25) | -5.87 | (-5.71) | -2.06 | (-2.15) |
| Ref. [5] | 2.75 | | 7.16 | | -5.69 | | -2.11 | |
| Ti$^{4+}$ | | | | | | | | |
| O 2p (9) | 1.29 | (1.50) | 2.33 | (2.86) | -8.93 | (-9.31) | -6.36 | (-6.50) |
| $(0,\frac{1}{2},\frac{1}{2})\sigma$ | -0.02 | | 0.49 | | -0.13 | | 0.01 | |
| $(0,\frac{1}{2},\frac{1}{2})\pi$ | -0.19 | | 1.48 | | -0.19 | | -0.41 | |
| $(0,\frac{1}{2},\frac{1}{2})\pi$ | -0.19 | | 1.48 | | -0.02 | | -0.01 | |
| $(\frac{1}{2},0,\frac{1}{2})\pi$ | 0.22 | | -0.16 | | -0.02 | | 0.01 | |
| $(\frac{1}{2},0,\frac{1}{2})\pi$ | 0.51 | | -0.25 | | -0.19 | | 0.04 | |
| $(\frac{1}{2},0,\frac{1}{2})\sigma$ | 0.11 | | -0.15 | | -0.13 | | 0.03 | |
| $(\frac{1}{2},\frac{1}{2},0)\pi$ | 0.22 | | -0.16 | | -2.88 | | -1.89 | |
| $(\frac{1}{2},\frac{1}{2},0)\sigma$ | 0.11 | | -0.15 | | -2.50 | | -1.87 | |
| $(\frac{1}{2},\frac{1}{2},0)\pi$ | 0.51 | | -0.25 | | -2.88 | | -2.28 | |

origin of this effect lies in the large dynamical charge transfer that takes place when moving away from the high-symmetry cubic phase (i.e., going from more ionic to more covalent bonding). Orbital hybridization is necessary for this transfer to take place; for this reason, our localized-orbitals approach provides an insightful tool in examining these effects. In the language of the modern thory of polarization [3], the anomalous contribution is determined by the relative displacement of the Wannier centers with respect to the ion that is being moved. If the bonding were purely ionic, electrons (and thus Wannier centers) would be firmly localized on each anion, and move rigidly with it. This is not the case in perovskite oxides. The anomalous contribution is often traced [13] to substantial hybridization between the oxygen $p$ orbitals and the $d$ orbitals of the atom in the B site (Ti, in this case). The picture can be somewhat more complex, with other group of bands playing a role in the anomalous dielectric behavior.

We present in Table 1 a full decomposition of the effective charges in $BaTiO_3$ coming from the different groups of bands; and, inside each group, coming from the individual Wannier functions identified by the localization procedure. We compare the results for the groups of bands with those obtained by Ghosez et al. (Ref. [12]). We find very good agreement, given the difference in the pseudopotentials used and our choice of lattice parameter. These results underline the conclusion that the decomposition into band groups is consistently defined in the linear-response formalism only when the calculations are performed in the so-called *diagonal gauge* [12]. The effective-charge tensor reduces to a scalar for barium and titanium, while it is diagonal with two inequivalent components (those parallel and perpendicular to the Ti-O bonds) for the oxygens. Several facts stand out from an inspection of the table. The anomalous effective charges originate not only from the oxygen $2p$ bands; there are also sizeable contributions originating from the barium $5p$ orbitals and even from the oxygen $2s$. More notably, there is a wide range of *compensating effects* between groups of bands and between different orbitals inside each group. The partial cancellation of such large orbital polarizabilities hints again at the complexity of these materials, which exhibit such a wide range of equilibrium properties as a consequence of the existence of many of these competing effects. Titanium shows the strongest deviations from a naive ionic picture. The $\sigma$ and $\pi$ oxygen orbitals carry a *positive* electronic $Z^*$ contribution, equal respectively to 0.81 and 1.78. It should be noted that it is always the O $2p$ $\pi$ orbitals that carry the largest anomalous charge. The O $2p$ contributions to the $O_1$ effective charges (i.e. in the direction of the Ti-O bond) are also anomalous, up to $-1.10$ for each $\pi$ orbital (in addition to the nominal $-2.00$ for each orbital).

Finally, we present at the bottom of Table 1 the oxygen $2p$ decomposition performed in a calculation where the $3s$ and $3p$ orbitals of the titanium have been removed from the pseudopotential (i.e., removed from the valence and frozen in the core). It is interesting to note that most contributions are completely unchanged; differences arise only for the anomalous contributions for the Ti and the $O_1$ displacements (whose anomality, incidentally, is reduced if this more approximate formalism is employed).

## CONCLUSIONS

We have summarized here our formalism for obtaining maximally-localized Wannier functions from the Bloch orbitals of an ab-initio electronic structure calculation. This formalism can be very helpful in understanding the chemical and dielectric properties of materials. Perovskite ferroelectrics are a particularly promising class of systems to be studied, since the nature of the bonding and hybridization can have a striking influence on the dielectric properties and on the development of ferroelectricity. At variance with other approaches, our method allows for a decomposition of electronic properties (e.g., the effective charges) into meaningful atomic contributions. In the case of $BaTiO_3$, it elucidates in particular the origins of the large anomalous contributions to the effective charges.

## ACKNOWLEDGMENTS

This work was supported by ONR Grant N00014-97-1-0048, and NSF Grants DMR-96-13648 and ASC-96-25885. We would like to thank Ph. Ghosez and X. Gonze for providing an early copy of their work on the effective-charge decompositions.

## REFERENCES

1. G. H. Wannier, Phys. Rev. **52**, 191 (1937).
2. S. F. Boys, in *Quantum Theory of Atoms, Molecules, and the Solid State*, P.-O. Löwdin, ed. (Academic Press, New York, 1966), p. 253.
3. R. D. King-Smith and D. Vanderbilt, Phys. Rev. B **47**, 1651 (1993); R. Resta, Rev. Mod. Phys. **66**, 899 (1994).
4. N. Marzari and D. Vanderbilt, Phys. Rev. B **56**, 12847 (1997).
5. W. Zhong, R. D. King-Smith and D. Vanderbilt, Phys. Rev. Lett. **72**, 3618 (1994).
6. W. A. Harrison, *Electronic Structure and the Properties of Solids*, (Dover, New York, 1980).
7. E.I. Blount, Solid State Physics **13**, 305 (1962).
8. B. Sporkmann and H. Bross, J. Phys.: Condens. Matter **9**, 5593 (1997); H. Teichler, Phys. Stat. Sol. (b) **43**, 307 (1971); H. Bross, Z. Physik **243**, 311 (1971); B. Sporkmann and H. Bross, Phys. Rev. B **49**, 10869 (1994).
9. S. Satpathy and Z. Pawlowska, Phys. Stat. Sol. (b) **145**, 555 (1988); U. Stephan and D. A. Drabold, to be published.
10. M.-H. Lee, J.-S. Lin, M. C. Payne, V. Heine, V. Milman, and S. Crampin, to be published.
11. N. Troullier and J. L. Martins, Phys. Rev. B **43**, 1993 (1991).
12. Ph. Ghosez, X. Gonze, Ph. Lambin, and J.-P. Michenaud, Phys. Rev. B **51**, 6765 (1995); Ph. Ghosez and X. Gonze, to be published.
13. M. Posternak, R. Resta, and A. Baldereschi, Phys. Rev. B **50**, R8911 (1994).

# Stress-Induced Phase Transition in Pb(Zr$_{1/2}$Ti$_{1/2}$)O$_3$

Nicholas J. Ramer,* Steven P. Lewis,* E. J. Mele[†,‡] and Andrew M. Rappe[*,‡]

*Department of Chemistry, †Department of Physics and
‡Laboratory for Research on the Structure of Matter
University of Pennsylvania, Philadelphia, PA 19104

**Abstract.** We investigate, within local density functional theory, the structural phase stability of the piezoelectric material Pb(Zr$_{1/2}$Ti$_{1/2}$)O$_3$ (PZT) under finite applied uniaxial stress. Previous theoretical analyses of piezoelectric properties have examined the behavior of the system near the ground-state structure. However, because some piezoelectrics can accept large (> 1%) strains reversibly [J. Appl. Phys. **82**, 1804 (1997)], these crystals are clearly operating beyond their lowest-order behavior, and insight can be gained from studying piezoelectricity at finite strain. We have studied the structural properties of the (111)-PZT crystal over a range of positive and negative uniaxial strains. For each strain state, we identity two metastable structures: a tetragonal phase and a low-symmetry rhombohedral-like phase. For each phase, we determine the evolution of the magnitude and direction of polarization in the crystal as a function of applied stress. We also analyze the energetic and structural profile of the stress-induced tetragonal-to-rhombohedral phase transition. In particular, we compute the transition stress, and we identify the accompanying atomic motions in terms of larger structural motifs such as cation motion within oxygen cages and oxygen-cage tilting.

## INTRODUCTION

PbZr$_x$Ti$_{1-x}$O$_3$ (PZT) is an important material for use in actuator applications. The structural phase diagram of PZT and its dependence on composition and temperature have been studied experimentally [1–6] for the past forty years and more recently, theoretically [7,8] in much detail. Most PZT ceramics employed in modern solid-state devices are synthesized with a lead zirconate to lead titanate ratio that is close to the tetragonal-rhombohedral morphotropic phase boundary (approximately 50/50 batch composition). Near this boundary, PZT ceramics have high electro-mechanical coupling and low coercive fields.

Structural phase transitions in ferroelectric materials induced by hydrostatic pressure have been extensively studied experimentally [9]. For example, 95/5 PZT has been found to depolarize completely under hydrostatic pressure around 290 MPa [10]. In addition it has been found that 56/44 PZT will undergo partial depolarization for hydrostatic pressures of 480 MPa [10]. Most theoretical studies of PZT have been limited to phenomenological models parametrized to experimentally determined composition-temperature-pressure data points [7]. These models are based on a simplified Landau-Devonshire theory [11,12] for the free energy of ferroelectric materials. Recently, a first-principles study of the hydrostatic-pressure-induced phase transition in zero-temperature $PbTiO_3$ was conducted using full-potential muffin-tin orbitals within the local density approximation [13]. It was found that at a critical pressure of 16.4 GPa, $PbTiO_3$ would undergo a tetragonal-to-cubic phase transition. The computed compressibilities and their dependence on pressure were in good agreement with available experimental data.

Because many modern piezoelectric device applications involve longitudinal or uniaxial stresses, experiments have investigated the response of piezoelectric materials under uniaxial stress [14,15]. Recently, experimental techniques have been developed and implemented to determine accurately the piezoelectric coefficients and dielectric constants of commercially-available compositions of PZT [16]. For PZT compositions near the morphotropic phase boundary, the tetragonal-rhombohedral phase transition induced by uniaxial compressive stress has recently been suggested as a possible cause of polarization reorientation in pulsed high-voltage electric fields [17].

With the advent of piezoelectric materials that can accept large strains, a microscopic understanding of these materials at finite strain has become more important. In this work, we study the effect of applying a uniaxial stress to a 50/50 PZT crystal. (We will refer to this composition as PZT for the remainder of this paper.) By examining structural phases of PZT on both sides of the morphotropic phase boundary, we elucidate the evolution of structural features (and therefore the polarization) as a function of applied stress.

## METHODOLOGY

The first-principles calculations presented in this paper are performed within density functional theory [18] in the local density approximation (LDA). The single electron wave functions are expanded in a plane-wave basis using a cutoff energy of 50 Ry.

To describe the electron-nuclear interaction, optimized pseudopotentials [19] in fully separable nonlocal form [20] are used. An additional feature of our nonlocal pseudopotentials [21] is their improved transferability over a wide range of electronic configurations, obtained by judiciously exploiting the flexibility contained in the separation of the local and nonlocal parts of the pseudopotential. By designing the form of the local potential so that the pseudo-eigenvalues and all-electron eigenvalues agree at an additional charge state, it is possible to improve the transferability of the potential across the charge states lying between the original reference

**TABLE 1.** Computed equilibrium lattice constants for tetragonal and rhombohedral (111)–Pb(Zr$_{1/2}$Ti$_{1/2}$)O$_3$ (PZT). Experimental lattice constants are given for randomly ordered PZT ceramics close to the 50-50 batch composition.

|  | Present | Experiment[a] |  | Present | Experiment[b] |
|---|---|---|---|---|---|
|  | **Tetragonal (111)–PZT** |  |  | **Rhombohedral (111)–PZT** |  |
| $c$(Å) | 8.199 | 8.279 | $c$(Å) | 8.052 | 8.163 |
| $c/a$ | 1.042 | 1.027 | $\angle$(°) | 89.64 | 89.79 |

[a] Ceramic **3** reference [24].
[b] Ceramic **5** reference [24].

state and this second state.

In order to obtain the high accuracy needed in examining ferroelectric phenomena, semi-core shells are in the generation of the pseudopotentials. We include as valence states the $3s$ and $3p$ for Ti and the $4s$ and $4p$ for Zr. The $5d$ shell is included for Pb. Furthermore, scalar relativistic effects are included in the generation of the Pb pseudopotential [22]. Brillouin zone integrations were approximated accurately as sums on a $4 \times 4 \times 4$ Monkhorst-Pack $k$-point mesh [23].

We have applied uniaxial stress along the (100) direction to two structurally distinct phases of a (111)-PZT superlattice. In the tetragonal phase, the polarization direction is (100), parallel to the applied stress. In the rhombohedral-like phase, the polarization in the unrelaxed crystal is oriented along (111), which is the stacking direction for planes of like cations. Upon structural relaxation, the polarization in the rhombohedral-like phase gains a (100) component in response to the applied stress.

Complete relaxation of both atomic positions and in-plane lattice constant were performed for a variety of fixed unit cell heights. We have neglected the shear response to uniaxial stress. We estimate that the contribution of shear to the energetics of the rhombohedral-like phase is less that 0.01 eV/unit cell. The relaxation of the in-plane lattice constants introduces a component of isotropic pressure into the analysis. However, by permitting this structural relaxation, we can more closely simulate the experimental conditions.

# RESULTS

The experimental and theoretical lattice parameters are contained in Table 1. Our results agree well with experiment for both phases. The theoretical lattice constants are smaller than experiment by 1–1.5%, as is usual for calculations done with the LDA [25,26]. Figure 1 shows the equations of state (total energy as function of unit cell height) for the two phases of the (111)-PZT superlattice. We find

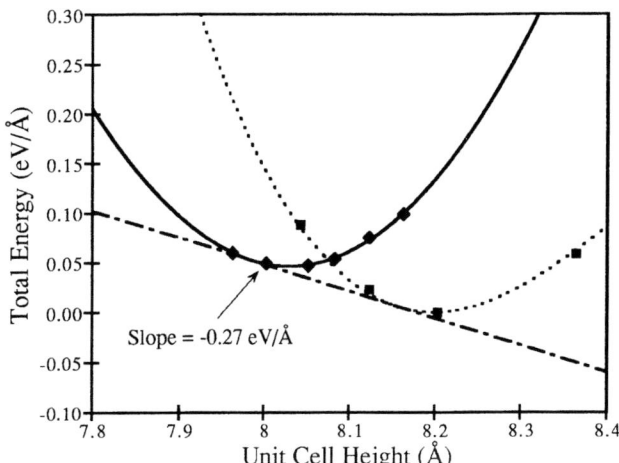

**FIGURE 1.** Equations of state for the tetragonal (dotted-line) and rhombohedral-like (solid-line) phases of (111)–Pb(Zr$_{1/2}$Ti$_{1/2}$)O$_3$. The solid line represents the rhombohedral-like phase and the dotted line shows the tetragonal phase. The heights and energies are for a 40-atom unit cell.

that the tetragonal phase is the ground-state structure. However, under compressive stress, the rhombohedral-like phase becomes more favorable. Additionally, we find that the tetragonal phase shows significant anharmonicity versus strain, while anharmonicity is small for the rhombohedral-like phase. Specifically, we find a 13% change in the transition stress if the cubic fit shown is replaced by a quadratic fit to the data that are near the equilibrium.

In order to examine the interatomic structural motifs that characterize the two phases as a function of stress, we have plotted illustrative distances from fixed features within the unit cell. For brevity, we include only the TiO$_6$ octahedron response. We have omitted the ZrO$_6$ octahedron response because of its similarity to the TiO$_6$ octahedron and PbO$_{12}$ dodecahedron response due to its complexity. In the rhombohedral-like phase, there are four crystallographically unique oxygen atoms in the TiO$_6$: two equatorial and two axial. In the tetragonal phase, there are three crystallographically unique oxygens: one equatorial and two axial.

In Figure 2 we have plotted the distance of the Ti atom and axial oxygens from the equatorial plane of oxygen within the TiO$_6$ octahedron. In the case of the rhombohedral-like phase, the equatorial plane is not constrained by symmetry to lie perpendicular to the (100). Therefore we have computed an average distance from the plane.

The width of the TiO$_6$ can be analyzed by computing the diagonal distance of the equatorial plane of oxygens. We have plotted this distance as a function of unit cell height for both phases in Figure 3.

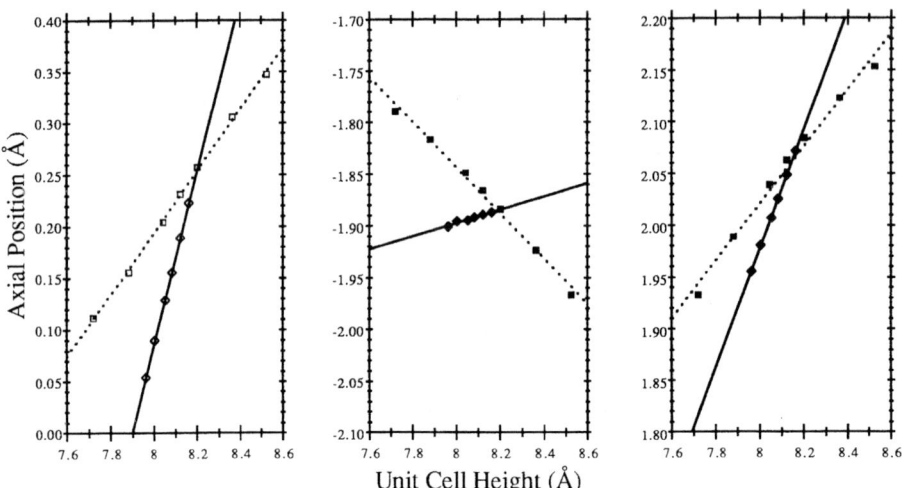

**FIGURE 2.** Positions relative to the equatorial oxygen plane in the $TiO_6$ octahedron of Ti (left) and axial O atoms (center, right) as a function of unit cell height for the tetragonal and rhombohedral-like phases of $(111)$–$Pb(Zr_{1/2}Ti_{1/2})O_3$. The solid lines and diamonds represent the rhombohedral-like phase, and the dotted lines and squares show the tetragonal phase.

## GIBBS CONSTRUCTION

For materials under applied uniaxial force, $F$, we can construct a generalized enthalpy:

$$H(L) = E(L) + FL \qquad (1)$$

Any structure adjusts its unit cell height, $L$, to minimize its enthalpy:

$$\frac{dE}{dL} = -F \qquad (2)$$

At zero temperature, a phase transition occurs when two competing phases have equal enthalpy.

$$E_1 + FL_1 = E_2 + FL_2 \qquad (3)$$

Rearranging (3) yields

$$-F = \frac{(E_2 - E_1)}{(L_2 - L_1)} = \left.\frac{dE_2}{dL}\right|_{L_1} = \left.\frac{dE_1}{dL}\right|_{L_2} \qquad (4)$$

where the last two equalities in equation (4) come from equation (2).

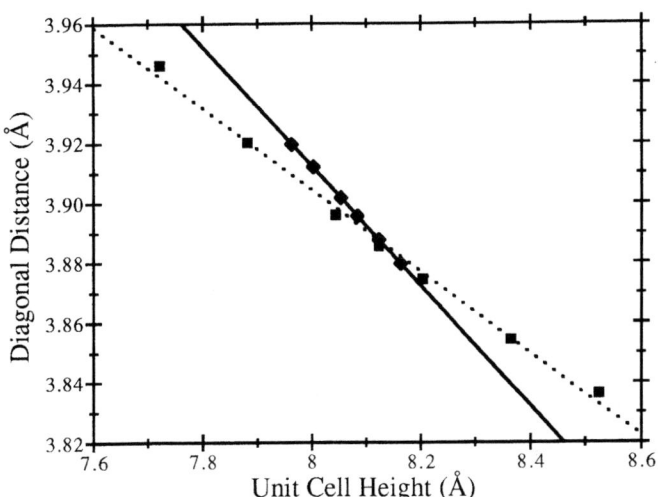

**FIGURE 3.** Equatorial oxygen plane diagonal distance as a function of unit cell height for the tetragonal and rhombohedral-like phases of $(111)$–$Pb(Zr_{1/2}Ti_{1/2})O_3$. The solid lines and diamonds represent the rhombohedral-like phase and the dotted lines and squares show the tetragonal phase.

Thus, a first-order phase transition occurs between points on the equations of state of the two phases that are connected by a common tangent line. Then, the transition stress, $\sigma$, is found via $F = \sigma A$, where $A$ is the unit cell area, and $-F$ is the slope of the common tangent.

The line of common tangent for the PZT tetragonal and rhombohedral-like phases has been plotted in Figure 1. The slope of this line is -0.27 eV/Å. This represents the uniaxial force per conventional surface unit cell necessary to cause the phase transition. The sign of the transition force shows that compressive stress will cause a phase transition from the tetragonal phase to the rhombohedral-like. The unit cell height at the transition is 7.987Å for the rhombohedral-like phase and 8.157Å for the tetragonal phase. By examining the in-plane lattice constant relaxations, we can determine the area of the unit cell perpendicular to the direction of uniaxial stress at these transition cell heights. From this we compute the compressive transition stress to be 669 ± 6 MPa.

# DISCUSSION

## Anharmonicity in Tetragonal Phase

As stated above, our finite stress approach has revealed some anharmonicity in the tetragonal phase equation of state. In addition, our analysis of structural features shows a slight nonlinearity in the B-metal/oxygen bond lengths as a function of applied stress. Also, the change in the $TiO_6$ equatorial size with applied stress shows a nonlinearity for the tetragonal phase (see Figure 3). These results further highlight the importance of going beyond lowest order in understanding the behavior of these materials. The origin of this anharmonicity is difficult to ascertain from the present study, and its examination is ongoing. Anharmonicity should also be found in the rhombohedral-like phase for stresses larger than those applied in the present study.

## Trends in Structural Motifs

### Rhombohedral-like Phase

For the uniaxial compression of the rhombohedral-like phase, we find decreasing axial distances from the equatorial plane for the Ti atom and the oxygen atom near it (see Figure 2). We find that as a compressive uniaxial stress is applied, the Ti atom moves toward the equatorial plane. The O in the same half of the octahedron as the Ti atom moves with it while the O on the other side remains nearly stationary. These motions indicate a decrease in magnitude of the (100) component of the local polarization in the $TiO_6$ octahedron as this phase is compressed. We have also completed an analysis of the (011) motions of the axial atoms relative to the center of the equatorial plane. These motions are negligible on the scale of the axial motions parallel to the applied stress. In response to uniaxial stress, the (100) component of polarization decreases while the other components remain intact.

### Tetragonal Phase

The response of the tetragonal phase of (111)-PZT to uniaxial compression is quite different from that of the rhombohedral-like phase. Upon compression we find that all the axial atoms relax towards the equatorial oxygen plane (see Figure 2). In fact, the Ti and both axial oxygens atoms move at nearly the same rate. This response of all three axial atoms is in sharp contrast with the rhombohedral-like phase which showed larger response in the Ti and the O in the same half of the octahedron, but negligible response of the other axial O atom. The magnitude of response (both axial and equatorial) is smaller in the tetragonal phase. This is because to the (100) compression is parallel to the direction of polarization for the tetragonal phase.

## CONCLUSIONS

In this density functional study, we have made the first prediction of a uniaxial stress-induced phase transition from first-principles methods. We have examined the tetragonal and rhombohedral-like phases of the (111)-Pb($Zr_{1/2}Ti_{1/2}$)$O_3$ crystal under (100) uniaxial stress. Characterization of the two metastable structural phases emphasizes the importance of strain coupling to polarization, since the simplest Landau theory of polarization, in which strain is neglected, does not permit more than one metastable structure [27]. We have found that a compressive stress of 669 MPa will cause a tetragonal-to-rhombohedral phase transition in (111)-Pb($Zr_{1/2}Ti_{1/2}$)$O_3$. Our calculations use LDA predictions of lattice parameters throughout, and this may affect the predicted transition stress. Quantitative comparison of the computed transition stress with experiment is difficult due to the absence of experimental studies on this phase transition. Most modern studies [28,29] of uniaxial stress on PZT materials focus on donor-doped (soft) materials that undergo polarization switching at lower stresses than undoped PZT. The work by Schäufele presents the closest comparison to the current study. In their work, neodymium-doped soft PZT ($Pb_{0.97}Nd_{0.02}$($Zr_{1/2}Ti_{1/2}$)$O_3$) is subjected to uniaxial compression. For this material, 50-200 MPa was needed to completely reorient the polarization of the material under various applied electric fields. Since our calculation does not incorporate donor doping or structural disorder, our transition stress must therefore be considered an upper limit to transition stresses of real PZT materials.

It is thought that for many perovskite materials, the *magnitude* of polarization remains nearly constant as stress is applied, and that the major contribution to the piezoelectricity is from *rotation* of the polarization vector. In order to assess the effect of uniaxial stress on (111)-PZT, we have analyzed relevant atomic distances in the $TiO_6$ octahedron in order to isolate distortions that would contribute to certain components of the polarization. For the rhombohedral-like phase, we find a reduction of the (100) component of polarization upon uniaxial compression, but almost no change in the polarization in other directions. This polarization response is inconsistent with a pure rotation model. In the tetragonal phase, we find atomic motions parallel to the direction of uniaxial compression. These motions give rise to a decrease in the magnitude of polarization.

Finite stress analyses done in conjunction with spontaneous polarization studies will aid the understanding of lead zirconate-titanate ceramics and other piezoelectric materials.

## ACKNOWLEDGMENTS

The authors would like to thank Karin Rabe and David Vanderbilt for valuable discussions during this project. This work was supported by the Laboratory for Research on the Structure of Matter and the Research Foundation at the University of Pennsylvania as well as NSF grants DMR 93-13047 and DMR 97-02514 and the Petroleum Research Fund of the American Chemical Society (Grant No. 32007-G5). Computational support was provided by the National Center for Supercomputing Applications and the San Diego Supercomputer Center.

# REFERENCES

1. Shirane, G., Suzuki, K., and Takeda, A., *J. Phys. Soc. Japan* **7**, 12 (1952).
2. Shirane, G., and Suzuki, K., *J. Phys. Soc. Japan* **7**, 333 (1952).
3. Shirane, G., and Takeda, A., *J. Phys. Soc. Japan* **7**, 5 (1952).
4. Clarke, R., and Glazer, A. M., *Ferroelectrics* **12**, 207 (1976).
5. Multani, M. S., Gokarn, S. G., Vijayaraghavan, R., and Palkar, V. R., *Ferroelectrics* **37**, 652 (1981).
6. Zhang, H., Uusimäki, A., Leppävuori, S., and Karjalainen, P., *J. Appl. Phys.* **76**, 4292 (1994).
7. Porat, Y., Imry, Y., Aharony, A., and Bransky, I., *Ferroelectrics* **37**, 591 (1981).
8. Haun, M. J., Furman, E., McKinstry, H. A., and Cross, L. E., *Ferroelectrics* **99**, 27 (1989).
9. Hegenbarth, E., *Ferroelectrics* **22**, 79 (1978).
10. Data provided by Gulton Industries, Inc., cited in Mock, W., and Holt, W. H., *J. Appl. Phys.* **49**, 5846 (1978).
11. Devonshire, A. F., *Phil. Mag.* **40**, 1040 (1949).
12. Devonshire, A. F., *Phil. Mag.* **42**, 1065 (1951).
13. Mryasov, O. N., Novikov, D. L., and Freeman, A. J., *Ferroelectrics* **164**, 279 (1995).
14. Berlincourt, D., and Krueger, H., *J. Appl. Phys.* **30**, 1804 (1959).
15. Cao, H., and Evans, A. G., *J. Am. Ceram. Soc.* **76**, 890 (1993).
16. Zhang, Q. M., Zhao, J., Uchino, K., and Zheng, J., *J. Mater. Res.* **12**, 226 (1997).
17. Sadykov, S. A., Bondarenko, E. G., and Agalarov, A. S., *Tech. Phys.* **38**, 965 (1993).
18. Payne, M. C., Teter, M. P., Allan, D. C., Arias, T. A., and Joannopoulos, J. D., *Rev. Mod. Phys.* **64**, 1045 (1992), and references therein.
19. Rappe, A. M., Rabe, K. M., Kaxiras, E., and Joannopoulos, J. D., *Phys. Rev. B* **41**, 1227 (1990).
20. Kleinman, L., and Bylander, D. M., *Phys. Rev. Lett.* **48**, 1425 (1982).
21. Ramer, N. J., and Rappe, A. M., *Phys. Rev. B*, submitted.
22. Koelling, D. D., and Harmon, B. N., *J. Phys. C* **10**, 3107 (1977).
23. Monkhorst, H. J., and Pack, J. D., *Phys. Rev. B* **13**, 5188 (1976).
24. Jaffe, B., Roth, R. S., and Marzullo, S., *J. Res. Nat. Bur. Stand.* **55**, 239 (1955).
25. Ramer, N. J., Mele, E. J., and Rappe, A. M., *Ferroelectrics*, in press, (1997).
26. Saghi-Szabo, G., and Cohen, R. E., *Ferroelectrics*, in press, (1997).
27. Vanderbilt, D. H., private communication, (1998).
28. Lynch, C. S., *Acta Mater.* **44**, 4137 (1996).
29. Schäufele, A. B., and Härdtl, K. H., *J. Am. Ceram. Soc* **79**, 2637 (1996).

# First-Principles Studies of Local Order in Relaxor Ferroelectrics

## Mark Wensell and Henry Krakauer

*College of William and Mary[1], Department of Physics*
*P.O. Box 8795, Williamsburg, VA 23187-8795*

**Abstract.** A key to optimizing the growth of the new single-crystal relaxor ferroelectrics [1] is resolving basic questions concerning their structural properties and energetics. We report on initial first-principles total energy and force calculations, examining the energetics of local order in PZN type relaxors.

## INTRODUCTION

The optimization of the electromechanical properties of the new single-crystal relaxor ferroelectrics [1] and the challenge of growing large crystals raise several fundamental questions. What is the structure of these materials, and how is it related to their properties? We first briefly review the present understanding of these questions and the capabilities of current theoretical approaches in addressing them.

Although some structural features of relaxor ferroelectrics are understood, their precise microscopic structure is unknown at the present time. (This is, of course, even more true of the new single crystal relaxor ferroelectrics.) The task of determining the atomic geometry is complicated by the long-range disorder inherent in relaxor ferroelectrics and the infinite-range Coulomb interaction between ions, which imposes severe constraints on the structure. It is perhaps not surprising then that despite years of extensive effort there are many fundamental questions about the atomic geometry of relaxors that are still unsettled.

For example, in PMN type relaxors with the stoichiometry $Pb(B^{2+}_{1/3}B^{5+}_{2/3})O_3$ (here $B^{2+}$=Mg, Co, Ni, Zn; $B^{5+}$=Nb, Ta), the B-sites must accommodate a 2:1 distribution of cations with very different positive charges. Experimental evidence indicates that PMN crystals do this by forming 1:1 ordered nanometer scale domains dispersed in a disordered matrix. [4–9] The ordered domains, probed by diffraction and high-resolution TEM studies, are seen to have a 1:1 distribution of two different cation sites (designated $\beta'$ and $\beta''$) arranged on a face-centered NaCl-type

---

[1] Supported by the Office of Naval Research grant N00014-97-1-0049.

lattice with a doubled perovskite repeat along the [111] direction. This much seems to be generally accepted. Other details about the structure are considerably less certain, as discussed below, with several basic questions that are still not settled: 1) Which atoms reside on $\beta'$ and $\beta''$ sites? 2) Are the 1:1 nano-domains intrinsically size-limited by the Coulomb interaction or are they the result of particular crystal growth and annealing procedures? It is evident that the $\beta'$ and $\beta''$ site assignments play a crucial role in answering the second question. For if the nano-domains are not charge-neutral, Coulomb interactions would be expected to limit the growth of these domains.

One answer to these questions is provided by the widely accepted space-charge model, which relates the observed relaxor behavior to postulated nanometer scale concentration inhomogeneities. [2,4,10,11] The space-charge model postulates that the nano-domains are $B^{5+}$- deficient, with the $\beta'$ and $\beta''$ sites occupied exclusively by the $B^{2+}$ and $B^{5+}$ cations respectively, in a NaCl like structure. The ordered regions thereby carry a net negative charge. This charge imbalance is then compensated by a $B^{5+}$-rich disordered matrix. The resulting large self-Coulomb repulsion of the charged regions would be expected to limit the size of the nano-domains, and the random distribution of nano-domain polarizations is an appealing explanation for the observed diffuse and frequency dependent permitivity of the relaxor ferroelectrics. Support for this model includes an observed lack of domain coarsening with long-term annealing in all but the most recent experiments, discussed below. This has contributed to widespread acceptance of the space-charge model, despite the absence of direct evidence for this type of nano-level compositional segregation. [2,9]

Recently, however, Akbas and Davies have reported experiments on $Pb(Mg_{1/3}Ta_{2/3})O_3$ - $PbZrO_3$, in which the size of fully chemically ordered 1:1 domains (evidenced by x-ray diffraction) were increased by two orders of magnitude through annealing conducted at 1325 °C; moreover, fully ordered ceramics comprised of large domains were found to retain relaxor behavior. [2] (Previous annealing was restricted to below 970 °C.) The concentration inhomogeneities of the space-charge model are inconsistent with these recent results, since Coulomb repulsion would limit the size of the $B^{5+}$-deficient 1:1 domains. There are several important implications of this recent experiment. First, the distribution of $B^{2+}$ and $B^{5+}$ concentrations must be homogeneous at the atomistic scale to permit the growth of large 1:1 domains. Second, the retention of relaxor behavior of the fully ordered ceramics suggests that the relaxor behavior is due to random local (at the atomistic scale) polarization fluctuations. Akbas and Davies proposed a previously dismissed charge-balanced *random-site* model for cation ordering as a possible candidate for atomistic scale homogeneity. According to this scheme, the $\beta''$ sites are occupied exclusively by $B^{5+}$, while the $\beta'$ sites are occupied by a 50-50 random mixture of $B^{2+}$ and $B^{5+}$ ions. The structural formula can be represented as $Pb[B^{2+}_{2/3}B^{5+}_{1/3}]_{1/2} [B^{5+}_{1/2}]O_3$. The random-site model is consistent with the domain coarsening observed in this experiment.

It is also consistent with the observations of Teslic *et al.* [3] using pulsed neutron atomic pair-distribution function measurements of PMN, PZ and PZT, which show that the atomic environments, particularly that of Pb, are similar in all these compounds. Teslic *et al.* suggest that a large portion of the ferroelectric polarization is provided by Pb displacements (to accommodate the lone-pair electrons), and these displacements are closely coupled to the rotation of the $BO_6$ octahedra. This picture provides another possible framework for understanding relaxor behavior at the atomic level. Since the coupling of the Pb polarization to the environment and the ease of rotation of the $BO_6$ will have different energy scales, the random distribution of the B cations on the $\beta'$ sites might be expected to lead to relaxor behavior.

The above illustrates the status and shortcomings in the present understanding of the geometry of even the conventional relaxor ferroelectrics. Even less is known about the single crystal relaxors. Yet knowledge of the atomic geometry is certainly a necessary prerequisite for understanding and optimizing their remarkable properties. A complete description of the atomic geometry will have to resolve questions like the space-charge versus random-site models. To do this, local effects such as atomic relaxation (i.e. Pb displacements and $BO_6$ octahedra rotations) away from ideal perovskite structure will also have to be treated. Such short-range order likely plays a crucial role in defining the properties and the relative stability of different structures. If the atomistically homogeneous models of the relaxors are in fact correct, then this is likely to be the case, because in this scenario the nano-domains are not intrinsically limited in size by Coulomb effects, but inhibited in growth by geometrical frustration, similar to spin glasses. [3] Indeed, the role of $PbTiO_3$ admixture in PMN-PT single-crystal relaxors may be to reduce frustration and facilitate single-crystal growth.

A powerful theoretical approach to characterize the short-range order is to use first-principles calculations to determine local atomistic structures and resolve the energetics of the space-charge versus random-site model. In this initial study, we employ supercells containing 15 - 30 atoms to model the short-range order. To efficiently handle the computational burden incurred in such large scale calculations, we employ our first-principles mixed-basis projector method. [12,13]

Looking ahead, we eventually want to model growth processes as well as long-range disorder and domain interactions. A direct first-principles approach, however, would require very large simulation cells, a task that exceeds the capability of any presently available first-principles method. Nevertheless, complementary methods have been developed to extend the reach of the first-principles methods. One such method is based on using first-principles derived effective Hamiltonians. These Hamiltonians act in a reduced sub-space of the full Hilbert space, retaining only the most important degrees of freedom.

Molecular dynamics and Monte Carlo simulations with such effective Hamiltonians have demonstrated their ability to successfully predict phase transitions and temperature-dependent static and dynamic properties for the pure constituents systems like $BaTiO_3$, $KNbO_3$, and $PbTiO_3$. [14,15] Thus, the correct phase sequence of ferroelectric phase transitions was obtained in these materials, showing that the first-principles effective Hamiltonian used in the simulations captures the essential behavior of the microscopic fluctuations driving the transitions. In $KNbO_3$ and $BaTiO_3$ molecular dynamics simulations have revealed the existence of localized preformed dynamic chain-like structures, which are present even in the high-temperature paraelectric phase, well above the cubic-tetragonal phase transition. [15] The molecular dynamics simulations also reproduced the essential features of the diffuse x-ray scattering measurements and the weak temperature dependence of measured diffuse streak patterns. These studies provided a framework for understanding both the displacive and order-disorder characteristics of these phase transitions as both arising from the softening of an *entire branch* of unstable transverse-optic phonon modes.

We hope that the present calculations for PZN can lay the foundation for extending such finite temperature studies to the relaxor ferroelectrics. Extending the effective Hamiltonian approach to alloy systems presents many challenges, and may ultimately be unsuccessful. In that case other approaches, such as effective interatomic potentials or generalized shell models may be better suited to the problem. In any case, any of these complimentary methods will be based on first-principles results of the type presented below.

## METHODOLOGY

Self-consistent local density functional approximation (LDA) calculations were carried out using a first-principles mixed basis projector method. [12] The method uses a Kerker type pseudopotential to remove the chemically inert innermost core-electron states from the Hilbert space. For the 30-atom supercells, there were 134 occupied bands, and the basis set consisted of about 3600 plane waves, corresponding to about a 19 Ry kinetic energy cut-off. This relatively low kinetic energy cut-off is achieved by the inclusion of 148 local basis functions, including Pb $5d$, Zn $3d$, Nb $4s$, $4p$, $4d$, and O $2s$, $2p$ orbitals. The semicore Nb $4s$ and $4p$ states were retained in the variational basis, due to their spatially extended character. The method benefits from Car-Parinello CPU-time scaling, facilitating the treatment of these large supercells. It achieves this using fast Fourier transforms and the fact that the local orbitals are strictly confined within non-overlapping muffin-tin spheres. Thus there are no two- or three-center integrals, and the local orbitals overlap only with the plane waves. The muffin-tin sphere radii were 2.0, 1.9, 1.6, and 1.65 a.u. for Pb, Zn, Nb, and O atoms, respectively. In these initial calculations, we chose the volume to correspond to the experimentally observed volume of $PbTiO_3$, but allowed the internal atomic positions to relax to the lowest energy. Calculations were carried out for 30- and 15-atom supercells, using one and two k-points, respectively. This corresponds, roughly, to a 2x2x2 Monkhorst-Pack mesh.

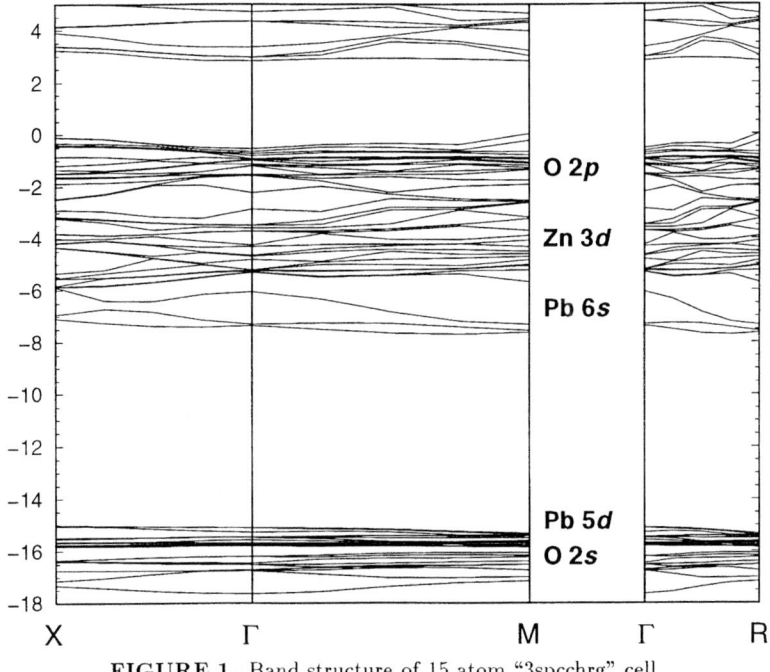

**FIGURE 1.** Band structure of 15 atom "3spcchrg" cell

## RESULTS AND DISCUSSION

We performed calculations on five different structural models, based on [111] B cation-plane stacking. Two are based on [111] planes containing either Zn or Nb. The first is a 30 atom model that has the stacking sequence Zn, Nb, Zn, Nb, Nb, Nb,Zn ... , and we refer to this model as "6spcchrg", containing 6 perovskite units. It features the 1:1 ordering proposed by the space charge model and in addition, crudely simulates a Nb rich region. The second is a 15 atom model, "3spcchrg", with the stacking sequence Zn, Nb, Nb .... We have imposed $R3m$ space group symmetry on both of these models.

In the remaining three models the [111] B-site planes do not all contain pure Zn or Nb. There are two 30 atom models, each of which is based on the stacking sequence $\beta'$ $\beta''$ ..., where $\beta'$ is $[Zn_{2/3}Nb_{1/3}]_{1/2}$ and the $\beta''$ site is pure Nb. These two differ by the in-plane lattice vectors and have different symmetry. One of these, "6tria", has symmetry $Imm2$ and is based on a triangular arrangement of the Zn and Nb atoms. The other, "6lin", has symmetry $Fmm2$ and features adjacent rows of Zn atoms separating rows of Nb atoms. The last structural model is a 15 atom model patterned on 6tria with each [111] B-site plane containing $[Zn_{1/3}Nb_{2/3}]$, which we name "3tria".

We begin by presenting the general features of the electronic structure of these

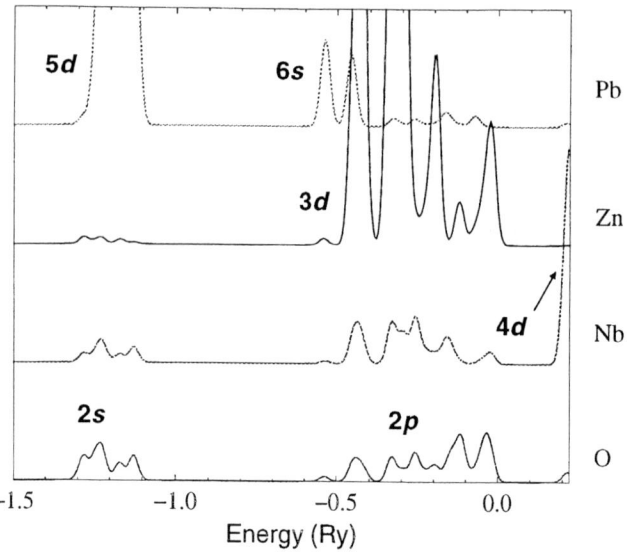

**FIGURE 2.** Density of states of 15 atom 3-space-charge cell. The $s-p-d$ character of Pb, Zn, Nb and O were determined using sphere sizes of 2.0, 1.9, 1.6 and 1.65 bohr, respectively.

models. Figure 1 shows the band structure for 3spcchrg along high-symmetry directions in the Brillouin zone, referenced to the ideal PbTiO$_3$ Brillouin zone, and Figure 2 shows the corresponding density of states projected onto the various atomic species. Strong Zn $3d$ - O $2p$ bonding is evident in these figures. Other characteristic features of PbTiO$_3$ [17], are also evident, such as strong Nb $4d$ - O $2p$ bonding and Pb $6s$ - O $2p$ bonding.

The distribution of nearest neighbor distances is presented in Figure 3 as pair distribution functions, artificially gaussian broadened. The dashed curves represent the ideal perovskite structure. There is large differences in the distribution of nearest neighbor distances in the different structures. The ranges of values are also tabulated in Table I. The characteristic ferroelectric splitting of Nb-O bondlengths is evident in all the structural models, with one bond shorter than ideal and the other longer. The widest variation occurs in the 6lin model. The Zn-O distances are all closely clustered around the ideal value, the widest variation again occuring for the 6lin structure. By contrast with Nb-O, however, the Zn-O distances are mostly equal to or larger than the ideal perovskite B-O distance. By far the largest differences from ideal occur for the Pb-O bond lengths. The widest variation occurs for the 3tria structure. In all cases, the lower end of the range is less than the ideal value, while the upper end is larger than ideal. Interestingly, the smallest variation occurs for the lowest energy 6tria structure. The ranges of values for the B-O distances agree well with the XAFS measurements of Chen et al. [18] for PZN, but less well for the Pb-O bond length (only the shortest measured Pb-O bond was

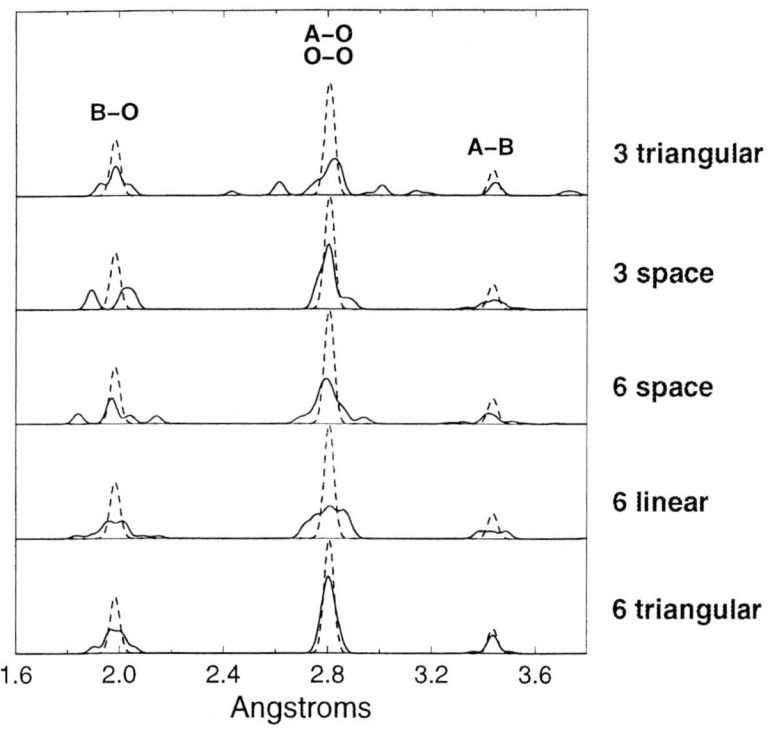

**FIGURE 3.** Pair distribution functions for relaxed and unrelaxed PZN structures

|  | Distances (Å) | | | $E_T$ |
|---|---|---|---|---|
|  | Pb-O | Nb-O | Zn-O | (mRy) |
| 6-triangle | 2.79–2.83 | 1.90–2.06 | 2.01 | 0 |
| 3-triangle | 2.43–3.19 | 1.93–1.98 | 1.98–2.04 | +30 |
| 3-space | 2.76–2.86 | 1.89–2.05 | 2.02 | +39 |
| 6-space | 2.69–2.95 | 1.84–2.14 | 1.98–2.04 | +63 |
| 6-linear | 2.71–2.87 | 1.84–2.15 | 2.01–2.09 | +73 |
| Ideal | 2.81 | 1.98 | 1.98 | – |
| Expr.[†] | 2.38, 2.40* | 1.92–2.12 | 2.03 | – |

[†] I-Wei Chen, *et al.*, 1995
*Egami *et al.*, 1995

**TABLE 1.** Summary of PZN bond-lengths and total energies

reported). The exception is the 3-tria model, whose smallest value is 2.43 Å.

The total energies (per 15 atom PZN formula unit) of each of the five structures are compared in Table I, with all energies referenced to the 6tria model, which had the lowest total energy. The next most stable structures are the 3tria and 3spc models, with the 6lin and 6spc being least stable. This suggests that the random-site model (in the 6tria form) has the lowest energy. However, we must still investigate the effect of increasing the k-point sampling and reducing the imposed symmetry.

## CONCLUSIONS

We have performed first-principles total energy and force studies of models of the solid solution relaxor ferroelectric PZN. Five different models were investigated incorporating the effects of [111] B-site ordering in both the space charge and random-site models. In all the calculations sizeable Z-O bonding occurs, but the Z-O bond distances are all close to the ideal perovskite value. Comparisons of fully relaxed total energies in these initial calculations suggest that the random site model has the lowest energy. The effects of denser k-point sampling and reducing the imposed symmetry have yet to be investigated, however. Nb-O and Zn-O distances are consistent with XAFS measurements but the measured Pb-O bond length is significantly smaller than all but one of the models. Detailed comparisons with neutron pair-distribution measurements would be desirable.

## REFERENCES

1. S.-E. Park and T. R. Shrout, *Journal of Materials Research Innovations*, 1997, in press.
2. M. A. Akbas and P. K. Davies, preprint.
3. S. Teslic, T. Egami, and D. Viehland, *Ferroelectrics* **194**, 271 (1997).
4. L. E. Cross, *Ferroelectrics* **151**, 305 (1994); *ibid.* **76**, 241 (1987).
5. H. B. Krause, J. M. Cowley, and J. Wheatley, *Acta. Cryst.* **A35**, 1015 (1979).
6. A. D. Hilton, C. A. Randall, D. J. Barber, and T. Shrout, *Ferroelectrics* **93**, 379 (1988).
7. H. D. Rosenfeld and T. Egami, *Ferroelectrics* **164**, 133 (1995).
8. D. Viehland and J. -F. Li, *J. Appl. Phys.* **74**, 4121 (1993).
9. For a recent discussion of possible nanometer scale concentration inhomogeneities arising from Coulombic constraints see for example Refs. [1, 7] and A. G. Khachaturyan, *Phys. Rev. B* **48**, 2949 (1993) and references therein.
10. R. Laiho, S. Lushnikov, and I. Siny, *Ferroelectrics* **125**, 493 (1992).
11. D. A. Sagala and S. Nambu, *J. Am. Ceram. Soc.* **75**, 2573 (1992).
12. D. J. Singh, H. Krakauer, C. Haas, and A. Y. Liu, *Phys. Rev. B* **46**, 13065 (1992)
13. D. J. Singh, H. Krakauer, C. Haas, and W. E. Pickett, *Nature* **365**, 39 (1993).

14. K. M. Rabe and U. V. Waghmare, *Phys. Rev. B* **52**, 13236 (1995), and references therein.
15. H. Krakauer, R. Yu, C.-Z. Wang, and C. LaSota, Ferroelectrics, in press, 1997; H. Krakauer, R. Yu, C.-Z. Wang, K. Rabe, and U. Waghmare, submitted; and references therein.
16. Y Liu, D. J. Singh, and H. Krakauer, *Phys. Rev. B* **49**, 17424 (1994).
17. R. E. Cohen and H. Krakauer, *Ferroelectrics* **136**, 65 (1992).
18. I-w. Chen, P. Li and Y. Wang, *J. Phys. Chem Solids* **57**, 1525 (1996).

# The quantum-mechanical position operator and the polarization problem

Raffaele Resta

*INFM – Dipartimento di Fisica Teorica,*
*Università di Trieste, Strada Costiera 11, I-34014 Trieste, Italy*
*and*
*Department of Physics, The Catholic University of America, Washington, D.C. 20064*

**Abstract.** The position operator (defined within Schrödinger representation as usual) becomes meaningless when the usual Born-von Kármán periodic boundary conditions are adopted: this fact is at the root of the polarization problem. I show how to define the position expectation value by means of rather peculiar many-body (multiplicative) operator acting on the wavefunction of the extended system. This definition can be regarded as the generalization of a precursor work, apparently unrelated to the polarization problem. For uncorrelated electrons, the present finding coincides with the so-called "single-point Berry phase" formula, which can hardly be regarded as the approximation of a continuum integral, and is computationally very useful for disordered systems. Simulations which are based on this concept are being performed by several groups.

## INTRODUCTION

In a phenomenological description of dielectric media the concept of macroscopic polarization is the basic one [1]. One formally defines the polarization of a macroscopic sample as its electrical dipole moment, divided by the volume. The problem is that the dipole of a finite system is dominated by surface effects, which seem to make polarization ill defined as a property of the bulk. In condensed matter theory, one typically gets rid of undesirable surface effects by dealing with extended systems within Born-von Kármán periodic boundary conditions (BvK). Unfortunately, this does not solve the polarization problem: the dipole is in fact the expectation value of the quantum-mechanical electronic position, which in its usual form is an ill defined operator within BvK.

Because of the above reasons, even *defining* what bulk dielectric polarization is remained a major challenge for many years. If one assumes the polarization elements as discrete, à la Clausius-Mossotti, then the polarization mechanism can be safely understood: but such oversimplified picture does not apply to a real dielectric, where the electronic distribution is continuous, and often very delocalized.

Typically, textbooks attempt to explain polarization of a bulk periodic solid via the dipole moment of a unit cell, or something of the kind [2,3]. These definitions are incorrect [4]: according to the modern viewpoint, bulk macroscopic polarization is a physical observable *completely independent* from the periodic charge distribution of the polarized dielectric.

The breakthrough was fostered by the 1992 Williamsburg meeting. Macroscopic polarization was *defined* in terms of the wavefunctions, not of the charge. This definition has an unambiguous thermodynamic limit, such that BvK and Bloch states can be used with no harm [5]. In the following months a modern theory of macroscopic polarization in crystalline dielectrics has been completely established [6,7], thanks to a major advance due to R.D. King-Smith and D. Vanderbilt [8], who expressed polarization in terms of a Berry phase [9,10]. An early comprehensive account of the modern theory exists [7]. Other less technical presentations are available as well [11,12]; for an oversimplified outline see Ref. [13].

In its original form, the Berry–phase theory of polarization was based on independent–electron wavefunctions, such as the Kohn–Sham orbitals of density-functional theory [7,14]. Its generalization to correlated many–electron wavefunctions is due to Ortíz and Martin [15]. Both the independent–electron and the correlated–electron versions of the Berry–phase theory of polarization rely on some lattice periodicity, and define macroscopic polarization in the form of a reciprocal-space integral. In particular, the Ortíz–Martin polarization of a correlated many–electron system takes the form of a peculiar "ensemble average", since it is an integral over a set of many–body ground states. This viewpoint is indeed correct and very valuable: the Ortíz–Martin theory has even been implemented to study the polarization of interesting model systems of highly correlated electrons [16–18]. However, the fact that macroscopic polarization could not be even *defined* by means a "pure state" expectation value (over a single many-electron ground state) was a disturbing drawback. This has been overcome thanks to the advance of Ref. [19], where a novel solution to the polarization problem is provided. The compact formula arrived at in Ref. [19], Eq. (14) below, is *apparently* unrelated to the Berry–phase concept; it applies on the same footing to correlated systems and to independent–electron systems, as well as to crystalline and to disordered systems. Lattice periodicity and integration in reciprocal space are no longer needed in order to *define* what polarization is. Polarization can in fact be cast as the expectation value of a rather peculiar operator over the many–electron ground wavefunction (in the thermodynamic limit).

In the present work, I present the main achievement of Ref. [19] under a rather different light than in the original paper. I will show that its main result can be regarded as the many–electron generalization of a precursor work, apparently unrelated to the polarization problem [20]. I will also discuss in detail the independent-electron case, linking in particular the results of Ref. [19] to the so-called single-point Berry phase [21,22], a concept which proves very useful, particularly in the study of disordered systems.

# THE HILBERT SPACE OF CONDENSED MATTER

Almost invariably, in condensed matter theory one adopts BvK for the electronic wavefunctions, when dealing with either crystalline systems or disordered systems. There are several good reasons for such choice. One reason is that one eliminates any surface by construction, thus getting rid of indesirable surface effects. Another reason is that in the crystalline case one can exploit the virtues of Bloch's theorem. I stress that the BvK choice is tantamount to defining the Hilbert space where our solutions of Schrödinger equation live.

For the sake of simplicity, the present work will deal with the one–dimensional case. For a single–particle wavefunction BvK reads $\psi(x+L) = \psi(x)$, where $L$ is the imposed periodicity, chosen to be large with respect to atomic dimensions. Notice that lattice periodicity is *not* assumed, and BvK applies to disordered systems as well. An operator maps any vector of the given Hilbert space into another vector belonging to the same space: the multiplicative position operator $x$ is *not* a legitimate operator when BvK are adopted for the state vectors, since $x\,\psi(x)$ is not a periodic function whenever $\psi(x)$ is such. Of course, any periodic function of $x$ is a legitimate multiplicative operator: this is the case *e.g.* of the nuclear potential acting on the electrons.

In the following, we will need to explicitly introduce many–body wavefunctions as well. BvK then imposes periodicity in each electronic variable separately:

$$\Psi_0(x_1, \ldots, x_i, \ldots, x_N) = \Psi_0(x_1, \ldots, x_i+L, \ldots, x_N). \tag{1}$$

Our interest is indeed in studying a bulk system: $N$ electrons in a segment of length $L$. Eventually the thermodynamic limit is taken: $L \to \infty$, $N \to \infty$, and $N/L = n_0$ constant. We will also assume throughout the ground state nondegenerate, and we deal with insulating systems only: this means that the gap between the ground eigenvalue and the excited ones remains finite for $L \to \infty$.

In order to illustrate where the main problem is, let us start with an Hilbert space *different* from the BvK one, using instead the boundary conditions which are appropriate for the bound states of an isolated atom or molecule. In this case the many–body wavefunction $\Phi_0(x_1, \ldots, x_i, \ldots, x_N)$ goes to zero exponentially whenever $|x_i|$ gets large. Within this Hilbert space there is no problem with the position, which is trivially defined as the multiplicative operator

$$\hat{X} = \sum_{i=1}^{N} x_i \tag{2}$$

within the Schrödinger representation. The ground–state expectation value is then simply:

$$\langle X \rangle = \langle \Phi_0 | \hat{X} | \Phi_0 \rangle = \int dx \, x \, n(x), \tag{3}$$

where $n(x)$ is the one–particle density. The value of $\langle X \rangle$ scales with the system size, and the quantity of interest is indeed the dipole per unit length, which coincides with macroscopic polarization. The expectation value in Eq. (3) is dominated by surface effects, thus making the definition of a bulk quantity problematic. If instead one adopts BvK, Eq. (1), then the operator defined by Eq. (2) is no longer a legitimate one, as explained above.

The major goal is therefore to define the expectation value of the electronic position $\langle X \rangle$ in the BvK Hilbert space, and to prove that our definition provides in the thermodynamic limit the physical macroscopic polarization of the sample: this has been very recently achieved in Ref. [19]. Before illustrating this major advance, we discuss an important precursor work, apparently unrelated with the polarization problem, where nonetheless the expectation value of the position operator plays the key role.

## THE ELECTRON–IN–BROTH FORMULA

Some years ago, A. Selloni *et al.* [20] addressed the properties of electrons dissolved in molten salts at high dilution, in a paper which at the time was commonly nicknamed the "electron in broth". The physical problem was studied by means of a mixed quantum–classical simulation, where a lone electron was adiabatically moving in a molten salt (the "broth") at finite temperature. The simulation cell contained 32 cations, 31 anions, and a single electron. KCl was the original case study, which therefore addressed the liquid state analogue of an F center; other systems were studied afterwards [23]. The motion of the ions was assumed as completely classical, and the Newton equation of motions were integrated by means of standard molecular dynamics (MD) techniques, though the ionic motion was coupled to the quantum degree of freedom of the electron. The electronic ground wavefunction was determined solving the time–dependent Schrödinger equation at each MD time step. Two snapshots of the simulation, reproduced from the original paper, are shown in Fig. 1.

As usual in MD simulations, periodic boundary conditions are adopted for the classical ionic motion. Ideally, the ionic motion occurs in a simulation cell which

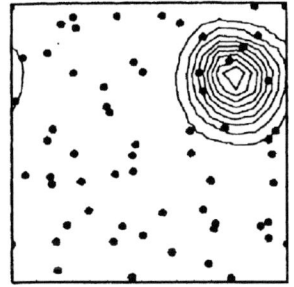 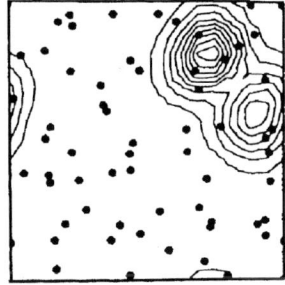

**FIGURE 1.** Contour plot of the electronic density (integrated along the sight line) at two different time steps. The dots are projections of the ionic positions. After Ref. [20].

is surrounded by periodic replicas: inter-cell interactions are accounted for, thus avoiding surface effects. Analogously, the electronic wavefunction is chosen in the work of Selloni et al. to obey BvK over the simulation cell, and therefore features periodic replicas as well. In Fig. 1, the "center" of the electron distribution is close to one of the borders of the simulation cell: the contour plot clearly shows the tail of its replica spilling into the opposite border.

One of the main properties investigated in Ref. [20] was the electronic diffusion, where the thermal ionic motion is the driving agent (within the adiabatic approximation). In order to perform this study, one has to identify first of all where the "center" of the electronic distribution is. Intuitively, the distributions in Fig. 1 appear to have a "center", which however is defined only modulo the replica periodicity, and furthermore *cannot* be evaluated simply as $\int d\mathbf{r}\, \mathbf{r}|\psi(\mathbf{r})|^2$ precisely because of BvK. Selloni et al. solved the problem by means of a very elegant and far–reaching formula. Here I write its one–dimensional analogue in the form:

$$\langle x \rangle = \frac{L}{2\pi} \text{Im} \log \int_0^L dx\, e^{i\frac{2\pi}{L}x} |\psi(x)|^2. \tag{4}$$

This formula provides the center of the electronic distribution, which is defined unambiguously though modulo $L$, as it must be.

## HOW THE FORMULA WORKS

The density of the single–particle wavefunction $n(x) = |\psi(x)|^2$ obeys BvK periodicity and can therefore be written as a Fourier series:

$$n(x) = \sum_{s=-\infty}^{\infty} C_s\, e^{i\frac{2\pi s}{L}x}. \tag{5}$$

The electron–in–broth formula, Eq. (4) becomes:

$$\langle x \rangle = \frac{L}{2\pi} \text{Im} \log \int_0^L dx\, e^{i\frac{2\pi}{L}x} n(x) = \frac{L}{2\pi} \text{Im} \log C_{-1}, \tag{6}$$

a quantity which is well defined (modulo $L$) whenever the $s = -1$ Fourier coefficient is nonvanishing.

In general, any periodic function $n$ can be written—in a nonunique way—as the superposition of a localized function $n_{\text{loc}}$ and of its periodic replicas. We write in the most general case the superposition as

$$n(x) = \sum_{n=-\infty}^{\infty} n_{\text{loc}}(x - x_0 - nL), \tag{7}$$

where $x_0$ is hitherto arbitrary. Then the Fourier coefficients of $n(x)$ are related to the Fourier transform of $n_{\text{loc}}(x)$ as:

$$C_s = \frac{1}{L}\tilde{n}_{\text{loc}}\left(\frac{2\pi s}{L}\right) e^{-i\frac{2\pi s}{L}x_0}. \tag{8}$$

We therefore express $\langle x \rangle$, Eq. (4), as:

$$\langle x \rangle = \frac{L}{2\pi}\text{Im log}\left[e^{i\frac{2\pi}{L}x_0}\tilde{n}_{\text{loc}}\left(-\frac{2\pi}{L}\right)\right] = x_0 + \frac{L}{2\pi}\text{Im log }\tilde{n}_{\text{loc}}\left(-\frac{2\pi}{L}\right). \tag{9}$$

We now specialize to the only interesting case: $n_{\text{loc}}(x)$ is a real positive function, which integrates to one over $(-\infty, \infty)$; furthermore $n_{\text{loc}}(x)$ is centered at $x = 0$ and decays over distances of the order of $\Delta x$. In the special case where $n_{\text{loc}}(x)$ is an even function, its Fourier transform is real and Eq. (9) simplifies to $\langle x \rangle \equiv x_0$ (mod $L$), which is indeed the expected result. In the general case where $n_{\text{loc}}(x)$ is *not* an even function, an expansion of its Fourier transform is needed. The Fourier transform $\tilde{n}_{\text{loc}}(k)$ decays over a reciprocal distance of the order of $1/\Delta x$: therefore supposing $\Delta x \ll L$ (as e.g. is clearly the case in Fig. 1) the expansion yields:

$$\tilde{n}_{\text{loc}}\left(-\frac{2\pi}{L}\right) = \int_{-\infty}^{\infty} dx\, n_{\text{loc}}(x) + i\frac{2\pi}{L}\int_{-\infty}^{\infty} dx\, x n_{\text{loc}}(x) + \mathcal{O}(L^{-2}). \tag{10}$$

Since by hypothesis $n_{\text{loc}}(x)$ integrates to one, Eq (9) to leading order yields:

$$\langle x \rangle \simeq x_0 + \int_{-\infty}^{\infty} dx\, x n_{\text{loc}}(x) \quad (\text{mod } L), \tag{11}$$

again a very meaningful expression for the center of the electronic distribution $n(x)$.

## POLARIZATION AS A MANY–ELECTRON PROBLEM

In the previous two Sections we have discussed how to define the expectation value of the position operator within BvK, when only a single electron is present in a large cell. When dealing instead with macroscopic polarization, we are faced with a system of many electrons, having a finite density. As anticipated above, our main interest is in studying a bulk system: $N$ electrons in a segment of length $L$. Eventually the thermodynamic limit is taken: $L \to \infty$, $N \to \infty$, and $N/L = n_0$ constant. If one arrives at defining a suitable position expectation value $\langle X \rangle$, analogue of Eq. (3) for the much less trivial BvK case, then the electronic contribution to macroscopic polarization (dipole per unit length) is:

$$P_{\text{el}} = \lim_{L \to \infty} \frac{e\langle X \rangle}{L}, \tag{12}$$

where $e$ is the electron charge.

So the main issue is as follows: can the electron–in–broth formula, Eq. (4), be generalized to provide an expectation value $\langle X \rangle$ appropriate to the case of many electrons in the given BvK periodicity? The answer is yes, but a smart generalization is needed. The most straightforward generalization, where one simply identifies $n(x)$ with the electron density of the extended system, does not work at all. There is a very good reason for this.

Suppose the system is crystalline, and furthermore assume independent electrons for the sake of symplicity. According to the modern viewpoint [7] the macroscopic polarization is a bulk observable *completely independent* from the periodic charge distribution of the polarized dielectric. While the density is obtained from the square modulus of the Bloch orbitals, polarization can only be obtained from their *phase*. Indeed, if one attempts to actually evaluating Eq. (4) upon replacement of $|\psi(x)|^2$ with the bulk density $n(x)$ of the extended system a meaningless result is reached. In fact, if the lattice constant of the crystalline solid is $a$, the BvK periodicity is then taken over a multiple of $a$: $L = Ma$. Since $n(x)$ is periodic with period $a$, its Fourier coefficients $C_s$ in Eq. (5) vanish unless $s$ is a multiple of $M$: therefore $C_{-1} = 0$, and the electron-in-broth formula, Eq. (6), is ill defined.

In order to arrive at the correct many-electron generalization of Eq. (4), one first introduces the multiplicative many-body operator $e^{i\frac{2\pi}{L}\hat{X}}$, where $\hat{X}$ is the simple operator defined in Eq. (2). In terms of this, one then defines:

$$\langle X \rangle = \frac{L}{2\pi} \text{Im log } \langle \Psi_0 | e^{i\frac{2\pi}{L}\hat{X}} | \Psi_0 \rangle. \quad (13)$$

The similarity with Eq. (4) is self evident; as in the one-body case, the position expectation value is defined only modulo $L$. The main ingredient of our generalized electron-in-broth formula is a many-body expectation value: an $N$-electron integral where a genuine *many-body operator* appears. In general, one defines an operator to be one-body whenever it is the *sum* of $N$ identical operators, acting on each electronic coordinate separately: for instance, the $\hat{X}$ operator itself is such. In order to express the expectation value of a one-body operator the full many-body wavefunction is not needed: knowledge of the reduced one-body reduced density matrix $\rho$ is enough: this is e.g. the case in Eq. (3). I stress that, instead, the expectation value of $e^{i\frac{2\pi}{L}\hat{X}}$ over a correlated wavefunction *cannot* be expressed in terms of $\rho$, and knowledge of the $N$-electron wavefunction is explicitly needed. The uncorrelated case where $|\Psi_0\rangle$ is a single determinant will be discussed in the next Section.

The electronic polarization, Eq. (12), becomes then:

$$P_{\text{el}} = \lim_{L \to \infty} \frac{e}{2\pi} \text{Im log } \langle \Psi_0 | e^{i\frac{2\pi}{L}\hat{X}} | \Psi_0 \rangle. \quad (14)$$

As anticipated in the Introduction, this compact and general expression for the macroscopic polarization applies on the same footing to correlated systems and to independent-electron systems, as well as to crystalline and to disordered systems. Notice that in any case $L \to \infty$ is an highly nontrivial limit, since the exponential operator in Eq. (14) goes formally to the identity, but the size of the system and the number of electrons in the wavefunction increase with $L$. Therefore some *ad-hoc* procedure is needed to demonstrate Eq. (14): this is reported in detail in Ref. [19], and will not be repeated here.

# INDEPENDENT PARTICLES AND SINGLE–POINT BERRY PHASE

We focus here on an uncorrelated system of independent electrons, whose $N$-electron wavefunction $|\Psi_0\rangle$ is a Slater determinant. In this special case $|\Psi_0\rangle$ is uniquely determined by the one–body reduced density matrix $\rho$ (which is the projector over the set of the occupied spinorbitals): therefore the expectation value $\langle X \rangle$, Eq. (13), is uniquely determined by $\rho$. The expression for $\langle X \rangle$—as well as the related one for the electronic polarization—is however most simply expressed directly in terms of the orbitals, as I am going to show. Such an expression, which goes under the name of "single–point Berry phase", was first proposed in a series of lecture notes [21]; it has been further scrutinized in some detail in a recent paper by Yaschenko et al. [22].

Suppose $N$ is even, and $|\Psi_0\rangle$ is a singlet. The Slater determinant has thus the form:

$$|\Psi_0\rangle = \frac{1}{\sqrt{N!}} |\varphi_1 \overline{\varphi}_1 \varphi_2 \overline{\varphi}_2 \ldots \varphi_{N/2} \overline{\varphi}_{N/2}|, \qquad (15)$$

where $\varphi_i$ are the single-particle orbitals. It is then expedient to define

$$|\tilde{\Psi}_0\rangle = e^{i\frac{2\pi}{L}\hat{X}} |\Psi_0\rangle : \qquad (16)$$

even $|\tilde{\Psi}_0\rangle$ is indeed a Slater determinant, where each orbital $\varphi_i(x)$ of $|\Psi_0\rangle$ is multiplied by the plane wave $e^{i\frac{2\pi}{L}x}$. According to a well known theorem, the overlap amongst two determinants is equal to the determinant of the overlap matrix amongst the orbitals. We therefore define the matrix (of size $N/2 \times N/2$):

$$S_{ij} = \int_0^L dx\, \varphi_i^*(x) e^{i\frac{2\pi}{L}x} \varphi_j(x), \qquad (17)$$

in terms of which we easily get

$$\langle X \rangle = \frac{L}{2\pi} \text{Im} \log \langle \Psi_0 | \tilde{\Psi}_0 \rangle = \frac{L}{\pi} \text{Im} \log \det S, \qquad (18)$$

where the factor of 2 accounts for spin.

Notice that nowhere we have assumed crystalline symmetry: the $\varphi_i$ orbitals will therefore have the most general form, while obviously obeying BvK at the sample boundary. Eq. (18) applies on the same ground to crystalline and to disordered systems: in the crystalline case it can be shown to be equivalent to a discretization of the usual continuum Berry phase [7], where crystalline Bloch orbitals are used. A simple test case is studied by Yaschenko et al. [22], where the numerical equivalence is actually demonstrated.

The single–point Berry phase has been recently implemented quite successfully to calculate infrared spectra of disordered systems [24,25]. In a typical Car–Parrinello simulation [26], the thermodynamic limit is approximated using BvK over a large simulation cell, which contains several tens of atoms. The time evolution of the ions is followed along discrete time steps, while the electrons follow adiabatically. In order to evaluate the infrared spectrum, one needs to evaluate the electronic macroscopic current traversing the simulation cell. This is done on the fly with the use of the single–point Berry phase, where $\langle X \rangle$, Eq. (18), is evaluated at each time step, and its time derivative is approximated with a finite difference [24].

## CONCLUSIONS

The formula originally proposed by Selloni et al. [20] for dealing with a lone electron in a large liquid sample, and named here the "electron–in–broth formula", is quite naturally generalized to the many–body case. It provides then a novel solution to the polarization problem, Eq. (14): such solution, though apparently unrelated to the Berry–phase concept, is indeed a discretized Berry phase in disguise [19].

Since any Berry phase is by definition a phase, it may appear surprising that in Eq. (14) only the square modulus of the wavefunction enters: any phase information is then apparently obliterated. One has to bear in mind, however, that $|\Psi_0\rangle$ is the *many–body* wavefunction of the $N$–electron system. In the independent–electron case, this wavefunction embeds the relevant information about the relative *phases* of the one–electron orbitals, and this is what one needs in order to evaluate the macroscopic polarization; in the correlated–electron case, the wavefunction carries in a more complex way the analogue information.

## ACKNOWLEDGMENTS

Discussions with M. Bernasconi and E. Yaschenko are gratefully acknowledged. Work partly supported by the Office of Naval Research, through grant N00014-96-1-0689.

## REFERENCES

1. L.D. Landau and E.M. Lifshitz, *Electrodynamics of Continuous Media* (Pergamon Press, Oxford, 1984).
2. C. Kittel, *Introduction to Solid State Physics*, 7th. edition (Wiley, New York, 1996).
3. N.W. Ashcroft and N.D. Mermin, *Solid State Physics* (Saunders, Philadelphia, 1976).
4. R.M. Martin, Phys. Rev. B **9**, 1998 (1974).
5. R. Resta, Ferroelectrics **136**, 51 (1992).
6. R. Resta, Europhys. Lett. **22**, 133 (1993).
7. R. Resta, Rev. Mod. Phys. **66**, 899 (1994).
8. R.D. King-Smith and D. Vanderbilt, Phys. Rev. B **47**, 1651 (1993).
9. M.V. Berry, Proc. Roy. Soc. Lond. A **392**, 45 (1984).

10. *Geometric Phases in Physics*, edited by A. Shapere and F. Wilczek (World Scientific, Singapore, 1989).
11. R. Resta, Ferroelectrics, **151**, 49 (1994).
12. R.M. Martin and G. Ortíz, Solid State Commun. **102**, 121 (1997).
13. R. Resta, Europhysics News **28**, 18 (1997).
14. *Theory of the Inhomogeneous Electron Gas*, edited by S. Lundqvist and N.H. March (Plenum, New York, 1983).
15. G. Ortíz and R.M. Martin, Phys. Rev. B **43**, 14202 (1994).
16. R. Resta and S. Sorella, Phys. Rev. Lett. **74**, 4738 (1995).
17. G. Ortíz, P. Ordejón, R.M. Martin, and G. Chiappe, Phys. Rev. B **54**, 13515 (1995).
18. R. Resta, Ferroelectrics **194**, 1 (1997).
19. R. Resta, Phys. Rev. Lett., in press (preprint cond-mat/9709306).
20. A. Selloni, P. Carnevali, R. Car, and M. Parrinello, Phys. Rev. Lett. **59**, 823 (1987).
21. R. Resta, *Berry Phase in Electronic Wavefunctions*, Troisième Cycle Lecture Notes (Ecole Polytechnique Fédérale, Lausanne, Switzerland, 1996). Available online (194K) at the URL:
http://ale2ts.ts.infn.it:6163/~resta/publ/notes_trois.ps.gz.
22. E. Yaschenko, L. Fu, L Resca, and R. Resta, Phys. Rev. B, in press.
23. E.S. Fois, A. Selloni, M. Parrinello, and R. Car, J. Phys. Chem. **92**, 3268 (1988); F. Ancilotto and F. Toigo, Phys. Rev. B **45**, 4015 (1992).
24. M. Bernasconi and M. Parrinello, unpublished; A. Debernardi, M. Bernasconi, M Cardona, and M. Parrinello, Appl. Phys. Lett. **71**, 2692 (1997); P.L. Sivestrelli, M. Bernasconi and M. Parrinello, Chem. Phys. Lett. **277**, 478 (1997).
25. A. Pasquarello and R. Car, Phys. Rev. Lett. **79**, 1766 (1997).
26. R. Car and M. Parrinello, Phys. Rev. Lett. **55**, 2471 (1985).

# Quantum Effects in Ferroelectrics

### S.A. Ktitorov* and L. Jastrabik**

*A.F. Ioffe Physical & Technical Institute, Russian Academy of Sciences,
Polytechnicheskaia str. 26, 194021 St. Petersburg, RUSSIA,
** Institute of Physics, Czech Academy of Sciences,
Na Slovance 2, 180 40, Prague, the CZECH REPUBLIC

**Abstract.** The problem of a crossover between the classical and quantum behavior is studied. A contribution of the dynamic clusters into the thermodynamic and dynamic properties is discussed. The Berry's phase in the lattice dynamics is considered on the base of the Hamiltonian path integral. The novel model describing quantum fluctuations in the Josephson junction with the quantum ferroelectric is suggested and analyzed.

## I  INTRODUCTION

A few years ago K.A. Muller [1] observed that some anomalous phenomena take place in quantum paraelectrics (QP) or virtual ferroelectrics (VF).We will use these terms as synonyms. But, notwithstanding that just an existence of solids is based on quantum physics, it is not easy to find an observable quantum effect connected with phase transitions due to the classical nature of the phase transition phenomena.. In particular, only a few known paraelectrics can be considered as quantum ones at temperatures below 30K. We discuss here some quantum aspects of the phase transition in ferroelectrics and some connected problems.

## II  DIMENSIONAL CROSSOVER IN QUANTUM PARAELECTRICS

From the point of view of the phase transition theory, quantum effects in QP means just the dimensional reduction while we go from a vicinity of the zero temperature point to higher temperatures: when the temperature is getting more than the soft-mode frequency, we can neglect all nonvanishing Matsubara frequencies and the theory is getting three-dimensional in opposite to the low-temperature case, where the Euclidean space including the imaginary time of the statistical physics is four-dimensional. The space dimension increase makes an order parameter more stable because fluctuations are getting weaker. It is well known now how zero-point fluctuations and tunneling inhibit a tendency to the FE phase transition, but there are some points which has to be clarified in the field-theory language [2]. We begin from the path-integral for the partition function

$$Z(\beta) = \int D\Phi \exp[-S(\beta)/\hbar], \qquad (1)$$

with the Euclidean action
$S(\beta) = \int_0^{\beta\hbar} d\tau \int d^3x (\rho/2)[(s\nabla\Phi_a)^2 + (\partial\Phi_a/\partial\tau)^2 + \kappa(\Phi_a)^2 + (g/4)(\Phi_a^2)^2$
where $\beta = T^{-1}$, $\rho$ is the mass density, $s$ is the soft-mode dispersion parameter, $\kappa$ is the unrenormalized inverse susceptibility. Introducing the auxiliary field and carrying out the Hubbard-Stratonovich transformation we arrive at the following effective action

$$S_{eff} = -H^{-1}\int_0^B d\tau \int d^d(\Psi^2)/2 + (N/2)tr\log(-\Delta + sign\kappa + \Psi), \qquad (2)$$

where $H = (\hbar\omega_0)/[(l)^d\Phi_0\omega_0^2)$, $B = \hbar\omega_0\beta$, $\tau = \omega_0 t$, $y = x/l$, $l$ is a length scale.
Within the saddle-point approximation which is asymptotically exact in the limit $N \to \infty$ we get

$$M^2 = \kappa - (N/2)trG(0, 0, M^2), \qquad (3)$$

where $M^2$ is the renormalized inverse susceptibility; the temperature Green's function describing the dimensional crossover can be written as

$$G_\beta(0) = G_\infty(0) + \int_0^\Lambda ds(4\pi s)^{-(d+1)/2}\exp[-M^2 s][\theta_3(0|iB^2/4\pi s) - 1], \qquad (4)$$

where $\theta_3(z|\tau)$ is the Jacoby's theta-function. Taking into account the leading terms we get in the quantum limit $(d = 4)$:

$$M^2 = \kappa - NM^2\log(\Lambda/M^2), \qquad (5)$$

where $\Lambda$ is the lattice cut-off. All non-singular terms were absorbed into the temperature shift. The important feature of (6) is that the function $M = M(\kappa)$ is double-valued at $\kappa \to 0$ so that $M$ goes to a finite limit while $\kappa \to 0$ except the thermodynamically unstable segment. We see, that summing up one-loop graphs non-perturbatively (using 1/N-approximation), we proved impossibility for the phase transition to occur in the quantum regime due to the dynamical "mass" generation effect. In the classical regime we get the single-valued function $M(\kappa)$ that makes the phase transition possible if $\kappa$ goes to zero in this region. On the other hand, this result can be interpreted as forming of dynamic clusters (instantons) so that $<\Phi> = 0$, but $<\Phi^2>$ increases while the temperature decreases on the left from the crossover point.

# III  DYNAMIC AND STATIC CLUSTERS

The intriguing problem of QP is that just in these crystals in rather narrow temperature and composition regions the classical-quantum regime crossover, dipolar glass and strong nonlinearities can be observed.

Consider in the beginning a possible effect of static impurity clusters. Static impurities lowering the local transition temperature can lead to the essential Griffiths singularity [3]. Within the one-loop approximation they can be described in terms of the Green's function:

$$F = N/(2V)tr \ln G_0^{-1}, \tag{6}$$

$$G_0^{-1} = -\Delta + \kappa_0 + \delta\kappa(x), \tag{7}$$

with $\delta\kappa(x)$ being the random local transition temperature. These formulae can be re-written in terms of the density of localized states:

$$F = N/2 \int_0^\Lambda \rho(\mu) \log(\mu) d\mu. \tag{8}$$

$G_0^{-1}$ can be considered as a Hamiltonian for an impure system with the random potential $\delta\kappa(x)$. The Griffiths phase lies in the interval: $T_c < T < T_c^0$, where $T_c$ and $T_c^0$ are transition temperatures in the impure and pure samples respectively. The density of localized states can be found with the instanton approach [4]. Notice, that formally similar localized states appear if we consider fluctuations on the instanton background in the pure QP. Instantons describe the tunneling process semiclassically and can be viewed as the dynamic clusters. If the temperature is high (but low enough to stay in the quantum region, i.e. near the crossover point, density of instanton gas increases and it transformed into the instanton liquid. Isolated bound states transform into the band, similar to the Lifshiz impurity band. Most interesting things take place, if we consider an influence of the impurity effective field on the dynamic cluster. Random local fields violate the symmetry of the double-well potential. It breaks the resonance between the "left" and "right" states in wells that makes the tunneling extremely unlikely. Clusters became static and pinned.

# IV  BERRY'S PHASE IN THE DYNAMICS OF LATTICE

One of the brightly quantum phenomenon that can be important in QP is the geometric phase of Berry. There are many possible realizations of this paradigm in solids. The soft mode in cubic perovscites like $SrTiO_3, KTaO_3$ belonging to the displacive kind virtual ferroelectrics mode is doubly degenerate for a non-vanishing wave vector [8] so that for a given direction of the wave vector we have two polarization vectors, parametrizing the degeneracy. We can direct into a sample an electromagnetic wave with a circular polarization and with a frequency, corresponding to the polariton mixing with the soft mode. Thus, we can excite the phonons with rotating polarization vectors in the sample. Non-vanishing soft mode amplitude appearance means that the system is in a non-equilibrium state for a given temperature and geometrically (not thermodynamically) it can be considered as being in a state with an instantaneously broken symmetry qualitatively similar to the case of the phase transition. Suppose, that the polarization vectors draw an ellipse. The centre of this ellipse corresponds to the symmetric state of the system that inevitably leads to a degeneracy in the electronic subsystem interacting adiabatically with the phonons. Thus, we arrived at the standard situation for the Berry's phase problem: an electron wave function $\Psi_n(t)$ takes an additional geometric phase $\gamma_{gn}(T)$ apart of the dynamic phase $\gamma_{dn} = \int_0^T dt \epsilon(C(t))$. The non-equilibrium case is difficult to consider on quantitative grounds, but, on the other hand, it can be done semi-empirically using all available information concerning an influence of the spontaneous symmetry violation effect at the equilibrium phase transition on electronic states. The geometric phase can be written as [9]:

$$\gamma_{gn}(T) = -\int\int ds_i V_{in}(e_\alpha), \qquad (9)$$

where integration is carried out over a surface tightened on the circuit $C$, described by the polarization vector $e_\alpha$, and

$$V_{in}(e_\alpha) = Im \sum_{mn} <n|\nabla H|m> \wedge <m|\nabla H|n> /(E_m - E_n)^2. \qquad (10)$$

If we have initially a superposition of states $\Psi_{init} = \sum a_n |n>$ or the mixed state with a density matrix $\rho = \sum \rho_{mn} |n><m|$, the phase can be detected measuring some proper observable which is sensitive to a change of the electronic wave function. The quantum mechanical average can be written as

$$<\hat{A}> = Tr(\rho \hat{A}) = \sum \rho_{nn} <n|\hat{A}|n> +$$

$$2Re \sum_{m \neq n} \rho_{mn} <m|\hat{A}|n> \times \cos[\gamma_{dn} + \gamma_{gn} - (\gamma_{dn} + \gamma_{gn})]. \qquad (11)$$

Let us consider now a possible influence of the Berry's phase upon the lattice dynamics in QP. The obvious way to consider this problem is to use the Born-Oppenheimer adiabatic approximation. In order to take into account possible electronic degeneracies (or "almost" degeneracies), we have to classify the "trajectories" of nuclei on homotopic classes, which are distinguished by the number of degeneracy points they are encircling. Most natural way to do it is to apply the path-integral approach [10]. The path-integral in the Hamiltonian form reads:

$$K[R_0, r_0; R_1, r_1] = \int_{R_0}^{R_1} DR \int_{r_0}^{r_1} Dr \int DP \int Dp \exp[iS], \qquad (12)$$

$$S = \int_{t_0}^{t_1} dt[PdR/dt + pdr/dt - V(r, R) - P^2/2M - p^2/2m], \qquad (13)$$

where $K[R_0, r_0, R_1, r_1]$ is the Feynman probability amplitude, $r$ and $p$ are the coordinate and the momentum of the electron, $R$ and $P$ are the coordinate and the momentum of the nucleus respectively. Notice, that the number of the degrees of freedom can be arbitrary in our discussion. We can decompose the path-integral within the adiabatic approximation:

$$K = \int_{R_0}^{R_1} DRDP \exp\{i \int_{t_0}^{t_1} dt[PdR/dt - P^2/2M]\} \times$$

$$\int_{r_0}^{r_1} DrDp \exp\{i \int_{t_0}^{t_1} dt[pdr/dt - p^2/2m - V(r, R)]\}. \qquad (14)$$

The electronic amplitude can be written as: $\sum \phi_m(r_1, R_1, t_1) K_{mn}^{el}$, where

$$K_{mn}^{el}(r, R) = T \exp[-i \int_{t_0}^{t_1} dt[\epsilon_{mn}\delta_{mn} + i < \phi_m|d\phi_n/dt >] =$$

$$T \exp[-\int_{t_0}^{t_1} dt[\epsilon_m \delta_{mn} + A_{mn}(R)P/M]. \qquad (15)$$

We have used the expansion of the electron wave function on the eigenstates of the instantaneous electronic Hamiltonian:

$$\Psi(r, R) = \sum \Phi_n(R) \phi_n(r, R). \qquad (16)$$

Now we can integrate the nuclear momentum away. We obtain in result the full effective action for the nucleus within the adiabatic approximation using the path-integral formalism:

$$A_{eff} = \int_{t_0}^{t_1} dt[(-\epsilon_m \delta_{mn} + (dR/dt)^2 M/2 + i(A_{mn}/M)dR/dt + A_{mn}^2/M] \qquad (17)$$

In the work [11] the last term in the r.h.s. of (20) was lost while the effective action was derived with the Lagrangian path-integral. We see, that the Hamiltonian formulation of the path-integral is a reliable tool for a study of geometric effects in quantum theory. In the next section we shall encounter another example where the Lagrangian form of the path-integral is inapplicable.

# V QUANTUM PARAELECTRIC IN THE JOSEPHSON CIRCUIT AND IN THE MODIFIED XY-MODEL WITH CHARGING

Next to the mainstream in the modern physics of the high-temperature superconductors (HTS) is developing of different lattice-dependent models [12-14]. Unfortunately, most of them possesses by so complicatedly looking Hamiltonians, that a natural desire appears to study a simplest possible model relevant to the problem, which can be investigated in details.. By our opinion, such a "toy model" can be taken in the form of the Josephson cell with the capacitor being a quantum one:

$$H = Q^2/2C + gQ^4/4 + \Phi_0 I_c/(2\pi c)(1 - \cos(2\pi\Phi/\Phi_0)), \qquad (18)$$

where $C$ is the capacitance, $C^{-1} = \alpha C_0^{-1}$, $C_0$ is the capacitance scale, $\alpha = (T - T_c)/T_c, g > 0$. Here $\Phi_0$ is the Onsager-London magnetic flux quantum: $\Phi_0 = hc/e^*$, $c$ is the speed of light. The quartic term is introduced to provide a stability when $\alpha$ changes sign. The sample (or a cell of the large system) is assumed to be small enough so that charging effects are essential (they are taken into account with the first two terms). The last term describes the Josephson tunneling. The model (21) is a simplest possible one taking account of quantum phenomena in a system with superconducting and dielectric fluctuations. To make it quantum we have to consider $Q$ and $\Phi$ as operators commuting as

$$[Q, \Phi] = ihc. \qquad (19)$$

As one can see, the self-charging XY-model

$$H = \sum [K(Q_i)\delta_{ij} + I_c\Phi_0/(2\pi c)\cos(\theta_i - \theta_j)] \qquad (20)$$

can be reduced to our simple model with the following modification:$I_c\Phi_0/(2\pi c) \to z < \cos\theta > I_c\Phi_0/(2\pi c)$,

where $z$ is the coordination number of the lattice.The model under the consideration is really a novel dynamic system which was not yet known and studied. To see it we write the Schrodinger equation corresponding to our Hamiltonian. In $\theta$-representation we can write:

$$-\alpha d^2\Psi/d\theta^2 - (\lambda/4)d^4\Psi/d\theta^4 + a(1 - \cos\theta)\Psi = \epsilon\Psi, \qquad (21)$$

here $\epsilon = E/E_0, E_0 = (e^*)^2/2C_0, \theta = 2\pi\Phi/\Phi_0, a = (I_c\Phi_0)/E_0, \lambda = g/E_0$.
We have to take into consideration, that in the absence of a dissipation states with $\theta$ and $\theta + 2\pi n$ are indistinguishable.. It means that we have to solve this Matteue-like equation in the class of periodic functions. It can be easily done in zero order on $a$:

$$\epsilon_n = (\lambda/4)(n/2)^2(2\alpha/\lambda + (n^2/2)). \tag{22}$$

One can see that for $\alpha < 0$ the normal systematics of states is violated, that can be interpreted as a symmetry violation. In the next order on $a$ we get the degenerate levels splitted. Let us now write the same equation in the $q$-representation:

$$[\alpha q^2/2 + (\lambda/4)q^4]\Psi(q) + (a - \varepsilon)\Psi(q) - a[\Psi(q+1) + \Psi(q-1)] = 0, \tag{23}$$

where $q$ is a whole number. Being solved in the class of $\theta$-periodic functions this equation gives a discrete spectrum. In the limit of small $a$ (almost free system) the points of the spectrum are situated pairwisely that corresponds to edges of bands near the gaps which exist in the extended class of solutions.

The system under the consideration is of interest to be considered with the path-integral approach too. Our Hamiltonian is not quadratic both on the "coordinate" $\theta$ and on the "momentum" $q$. Therefore, the Lagrangian form of the path-integral can not be applied. We have shown, that the path integral for the partition function (for one cell) can be written in the following extended form, taking into account a compactness of the space:

$$Z = \sum_{m=-\infty}^{\infty} \int_0^{2\pi} d\theta \int_0^{2\pi m} D\theta Dq \exp[-S[\theta, q]/\hbar], \tag{24}$$

$$S = \int_{-\beta\hbar/2}^{\beta\hbar/2} d\tau [\alpha q^2/2 + a(1 - \cos\theta) + (\lambda/4)q^4], \tag{25}$$

where $m$ are winding numbers.

In the classical high-temperature limit the partition function is easy to find:

$$Z \propto \int_0^{2\pi} d\theta \exp[-\beta a(1 - \cos\theta)] = \exp(-\beta a)I_0(\beta a), \tag{26}$$

where $I_0(x)$ is the modified Bessel function. The following study of this system having in mind a solution of some technical problems of the theory and possible applications to the physics of HTS and to a theory of Josephson circuits including the artificial Josephson structures.

## VI  ACKNOWLEDGEMENTS

In conclusion we want to acknowledge useful discussions with M. Bednar.

# VII  REFERENCES

1. Muller,K.A., Berlinger,W., Tosatti,E., Z. Phys.B, 84, 277-281 (1991).
2. Ktitorov,S.A., .Jasdtrabik,L., Ferroelectrics,153, 137-139 (1994).
3. Chalupa,J., J.Phys.A., 12, 1257-1262 (1979).
4. Cardy,J.L., McKane,A.J.,Nucl. Phys. B,257 [FS14],383-395 (1985).
5. Vaks.V.G., Introduction to the microscopic theory of ferroelectrics, Moscow: Nauka,1973, ch.2, pp. 16-46.
6. Berry,M.V., Proc. Roy. Soc. London, A392, 45-57 (1984).
7. Popov,V.N., Functional Integral and collective excitations, Cambridge: University Press, 1987, ch. 18, pp.147-157.
8. Shapere, F., Wilczek, F., Geometric Phases in Physics, Singapore, World Scientific, 1989, p. 160.
9. Bilz, H., et al., Phys. Rev. B35, 4840-4852 (1987).
10. Bussman-Holder, A., et al., Phys. Rev. Lett. 67, 512-514 (1991).
11. Shenoy, S.R., et al, Phys. Rev. Lett. 79, 4657-4659 (1997).
12. Simanek, E., Inhomogeneous Superconductors, Oxford: University press, 1994, ch.3, pp. 40-72.

# Dynamical Model for Phase Coexistence in Proton Glass

V. Hugo Schmidt

*Physics Department, Montana State University, Bozeman, MT 59717*

**Abstract.** We describe a model for static and dynamic behavior of $KH_2PO_4$ (KDP) type crystals in the ferroelectric phase, and for proton glass crystals of this type in the temperature and composition range in which the ferroelectric and paraelectric phases coexist. Model predictions are compared with experimental results for ferroelectric $RbH_2AsO_4$ (RDA) and for the proton glass $Rb_{1-x}(NH_4)_xH_2AsO_4$ (RADA). The model is based on the statistical mechanics and dynamics of a fundamental excitation in the ferroelectric phase, namely chains consisting of nonpolar $H_2AsO_4$ groups with an $H_3AsO_4$ group at one end and an $HAsO_4$ group at the other end.

## INTRODUCTION

Proton glasses offer a rich variety of phenomena, and are of particular interest because their dynamical processes can be understood in terms of simple excitations. We reported coexistence of ferroelectric and paraelectric phases in $Rb_{1-x}(NH_4)_xH_xAsO_4$ (RADA) based on dielectric permittivity (1-3) and spontaneous polarization (3) measurements. We have employed other methods and studied other crystals as well, and have discussed the nature of such coexistence (4,5). This paper develops a model which explains some of the phenomena observed in the coexistence region of the phase diagram.

## DESCRIPTION OF MODEL

This model for static and dynamic properties of coexistence is based on the statistics of the most prevalent excitations in ferroelectric domains in KDP-type crystals. We choose $Rb_{1-x}(NH_4)_xH_2AsO_4$ (RADA) as the crystal with which we compare model and experimental results in this work. In such crystals, these excitations are chains consisting of an $H_3AsO_4$ group and an $HAsO_4$ group connected by nonpolar $H_2AsO_4$ groups. A nonpolar $H_2AsO_4$ group has energy $\varepsilon_0$, whereas a polar $H_2AsO_4$ group has zero energy, according to the Slater model (6). Each $AsO_4$ arsenate ion is connected by four O-H···O "acid" hydrogen bonds to neighboring arsenate ions. Most arsenate ions have two protons close (hence the $H_2AsO_4$ designation for such groups), in these four bonds in which protons occupy off-center positions. If both bottom (relative to the $c$ axis) protons are close, the $As^{5+}$ ion is pushed upward and the group has a positive electric dipole moment $\mu$ along $c$. If both top protons are close, the moment is negative. The remaining four $H_2AsO_4$ proton arrangements produce dipole moments in the $ab$ plane and contribute to dielectric permittivity in that plane. They are called nonpolar because they have no moment along $c$, the axis for ferroelectric polarization.

A chain of such nonpolar groups must terminate with a "Takagi" group of energy $\varepsilon_1$, typically about $5\varepsilon_0$ in KDP-type crystals. Takagi (7) incorporated them into Slater's model and obtained better agreement with experiment. The reason for existence of such chains can be understood by following the dynamics of creation and growth of such a chain. Creation occurs within, say, a positive ferroelectric domain by means of one intrabond proton transfer,

$$H_2AsO_4 + H_2AsO_4 \rightarrow H_3AsO_4 + HAsO_4. \tag{1}$$

The proton is transferred to a site close to one of the two top oxygens of an $H_2AsO_4$ group, making it an $H_3AsO_4$ group and changing one of the two groups just above it to an $HAsO_4$ group. This Takagi group pair has an *a priori* statistical weight of 2, relative to 1 for the original polar ion at the bottom of this short chain. The chain can grow upward if a proton moves close to the top of the $HAsO_4$ group, thereby converting it to a nonpolar $H_2AsO_4$ group because it has one close proton each at the top and bottom. The $HAsO_4$ group has in effect moved to one of two possible positions higher along the $c$ axis, thereby again doubling the *a priori* statistical weight of the chain, to 4. The chain energy has now increased from $2\varepsilon_1$ to $2\varepsilon_1+\varepsilon_0$. We ignore the possibility that the last proton moves away from the $HAsO_4$ group, because that would entail a large energy increase, of order $\varepsilon_1$.

In general, if no ammonium ions are encountered, the Boltzmann factor $W_{nS}$ in the Slater limit for a chain of n links, involving n proton transfers, is

$$W_{nS} = 2^n exp\{-[2\varepsilon_1 + (n-1)\varepsilon_0]/kT\}. \tag{2}$$

For each step that the chain lengthens, the Boltzmann factor increases by the factor $2exp(-\varepsilon_0/kT)$. We pointed out long ago (8) that the Slater model ferroelectric transition temperature

$$T_c(Slater) = \varepsilon_0/k\ln 2 \tag{3}$$

is exactly the temperature at which growth of the chain by one step does not change the Boltzmann factor, thereby allowing such chains to grow without limit and destroy the ferroelectric order.

This perfect agreement of the chain model with the Slater model hinges on being at the Slater limit, for which $\varepsilon_1 \rightarrow \infty$. For finite Takagi energy $\varepsilon_1$, a large number of chains of finite length will destroy the ferroelectric order at a temperature below the Slater transition temperature.

## APPLICATION OF MODEL TO RDA

As a first step in developing the chain model, without the complication of randomly placed ammonium ions, we apply it to pure $RbH_2AsO_4$ (RDA) and compare results with experiment and with Takagi model predictions. If $W_n$ is the probability that a chain of n links has its bottom end at a given arsenate anion site, and one notes that each link destroys the dipole moment of one arsenate anion (more exactly, of one formula unit), then the ferroelectric order parameter p is given by

$$p = 1 - \Sigma n W_n = 1 - \Sigma (n W_{nS}) / \Sigma W_{nS}. \tag{4}$$

These sums, and all sums over n in this work, run from zero to infinity. The $W_{nS}$ for n=0 is 1, and the others are provided by Eq. (2). Upon evaluating these sums, we find

$$p(chain) = (a-bc)(a+bc)/[(a-bc)(a+bc)+b^2/2], \tag{5}$$

whereas the Takagi result is

$$p(Takagi) = [(a-b)(a+b)]^{1/2}/a. \tag{6}$$

Here, $a=1-2exp(-\varepsilon_0/kT)$, $b=2exp(-\varepsilon_1/kT)$, and $c=exp(-\varepsilon_0/2kT)$.

The transition temperature $T_c$ is the lowest temperature at which p becomes zero. The transition temperatures for these models are given by

$$T_c(chain) \rightarrow a-bc = 1-2exp(-\varepsilon_0/kT_c) - 2exp\{-(\varepsilon_1+0.5\varepsilon_0)/kT_c\} = 0, \tag{7}$$

$$T_c(Takagi) \rightarrow a-b = 1-2exp(-\varepsilon_0/kT_c) - 2exp(-\varepsilon_1/kT_c) = 0. \tag{8}$$

The chain model result for p differs from the Takagi result in that p(chain) has finite slope at $T_c$(chain), while the Takagi model predicts infinite slope at $T_c$(Takagi) as expected for a second-order transition. This defect of the chain model occurs because intersection of chains is neglected, and this neglect becomes serious near the transition. Neither model predicts the measured weakly first-order nature of the transition. Senko (9) added a long-range interaction to the Takagi model, which was shown by Silsbee, Uehling, and Schmidt (10) to change the transition to first order if this long-range interaction is large enough.

To make numerical and graphical comparison of these model predictions with experiment, we use the experimental value $T_c$=109.75 K given by Fairall and Reese (11) for RDA, and the value $\varepsilon_1/kT_c$=4.20 provided by them, which corresponds to

$\varepsilon_1/k=460.95$ K. Because they included both the above-mentioned long-range interaction, and the tunneling effect introduced by Blinc and Svetina (12), their value for $\varepsilon_0/kT_c$ is much too low to provide the measured $T_c$ if only it and their $\varepsilon_1/kT_c$ are used in the Takagi model. Accordingly, we instead use for $\varepsilon_0/k$ the value 79.41 K which when used in the Takagi model together with $\varepsilon_1/k=460.95$ K gives the measured $T_c$. With these values for $\varepsilon_0$ and $\varepsilon_1$, which force the Takagi result for $T_c$ to coincide with the experimental value, we obtain for $T_c$ given in Eqs. (3), (7), and (8) for the various models the values

$T_c$(experimental)=$T_c$(Takagi)=109.75 K,
$T_c$(chain) =111.00 K,
$T_c$(Slater) =114.56 K.

## APPLICATION OF MODEL TO RADA

The reasonable agreement of the chain model result with the Takagi result encourages us to apply an extension of this chain model to proton glass crystals in the range of x for which coexistence of ferroelectric and paraelectric/proton glass phases occurs. This extension provides for a reduction in the chain energy wherever the chain passes by an ammonium ion. Although each cation site has probability x of containing an ammonium instead of a rubidium ion, this is not a mean field model because we specifically take into account the preference for chains to choose paths with neighboring ammonium sites. Taking this preference into account yields the non-mean-field result that even at zero temperature, the crystal contains both ferroelectric and proton glass phase regions for all x between 0 and $x_c$, the concentration at which ferroelectricity disappears at zero temperature.

To take the effect of ammonium ions into account, we use an interaction we introduced previously (13) in an extension of the Slater model. This is the "cross-cation interaction" which lowers the energy by $\varepsilon_a$ whenever the two "acid" hydrogens opposite each other across an ammonium ion occupy sites consistent with an antiferroelectric rather than a ferroelectric configuration. There are two such hydrogen pairs for each ammonium ion, so the potential for the ammonium ion to destroy ferroelectric order is not exhausted until two acid hydrogens adjacent to the ammonium ion have moved away from their ordered-phase positions. Each acid hydrogen is bonded to two oxygens, and each of these oxygens can form an N-H···O bond to a different ammonium ion if such an ion is in that cation site. If the chain has n displaced hydrogens and runs past m ammonium ions, its unnormalized Boltzmann factor $W_{nu}$, the mixed-crystal analog of $W_{nS}$ in Eq. (2), is

$$W_{nmu}=[2^n(1-y)^{2n-m}y^m(2n)!/(2n-m)!m!]f(U_{nm}). \qquad (9)$$

Here y is the effective x, taking into account the above-mentioned saturation of the tendency of ammonium ions to destroy ferroelectric order, and given by

$$y = x - 0.5 \Sigma (m W_{nmu}) / \Sigma W_{nmu}. \tag{10}$$

Here and in the following equations, $\Sigma$ signifies a sum over n from zero to infinity, and a sum over m from zero to 2n. After solving Eqs. (9) and (10) self-consistently for y, the ferroelectric order parameter is found from the analog of Eq. (4), namely

$$p = 1 - \Sigma (n W_{nm}) = 1 - \Sigma (n W_{nmu}) / \Sigma W_{nmu}. \tag{11}$$

The factor $f(U_{nm})$ in Eq. (9) depends on the chain energy $U_{nm}$, given by

$$U_{nm} = 2\varepsilon_1 + (n-1)\varepsilon_0 - m\varepsilon_a. \tag{12}$$

One might think that $f(U_{nm})$ should have the Boltzmann form seen in Eq. (2), but use of that form in this calculational method would give completely erroneous results for mixed crystals. The reason for such erroneous results is that sufficiently large m can give negative $U_{nm}$. If we would construct a finite-size crystal with a given ammonium ion distribution, the ground state corresponding to zero temperature would contain a number of such chains with negative energies of various magnitudes. Instead, we are looking at all the possible chains that can begin at one given anion site, in a crystal of infinite extent. If we would employ Boltzmann statistics, then at zero temperature only the lowest energy state (of energy negative infinity!) would be occupied. A better choice for $f(U_{nm})$ would be the Fermi-Dirac distribution function, even though we are dealing with distinguishable particles so that there is no physical necessity for using Fermi-Dirac statistics. A simpler choice which also avoids the above low-temperature catastrophe would be to choose

$$f = 1 \text{ for } U_{nm} \leq 0, \quad f = \exp(-U_{nm}/kT) \text{ for } U_{nm} \geq 0. \tag{13}$$

The f=1 choice is consistent with the unnormalized probability $W_{00u}=1$ which must be used in the above sums, and simply means that no given configuration of a chain beginning at a single site can have more than single occupancy.

The possibility of negative $U_{nm}$ in Eq. (12) justifies the above statement that a mixed crystal cannot be in a completely ordered ferroelectric state. If in Eq. (12) we would replace m by its mean-field (m-f) value 2nx, then at zero temperature the crystal would have complete ferroelectric order for $x < x_{cm-f}$, and would be completely in the proton glass state for larger x values below the antiferroelectric order range. This mean-field critical concentration is given by

$$x_{cm-f} = \varepsilon_0 / 2\varepsilon_a. \tag{14}$$

This mean-field approximation seriously overestimates $x_c$. To find the degree of overestimation, and to obtain results from the chain model for mixed crystals, we need a numerical value for the cross-cation interaction energy $\varepsilon_a$. This is obtained from our Slater-type model (13) for proton glass crystals which incorporates this interaction. For the large x values for which an antiferroelectric transition occurs, that model predicts a first-order transition with a jump from 0 to 1 for the antiferroelectric order parameter. This all-or-nothing jump is a consequence of setting the Takagi interaction energy $\varepsilon_1$ to infinity in that model. Such a jump occurs also for the ferroelectric order parameter p predicted by that model for lower x, except in the Slater model limit (x=0), where all values $0 \leq p \leq 1$ are allowed. (One can say that the Slater transition is locked to a tricritical point.) The actual transition in $NH_4H_2AsO_4$ (ADA) is strongly first-order and occurs at 216.1 K. Our Slater-type model predicted that this transition occurs for ADA at the "Néel" temperature $T_N$ for which the paraelectric and antiferroelectric free energies are equal. The condition for this equality is

$$\varepsilon_a = f_N\varepsilon_0 - kT_N[f_N\ln(2f_N) + (1-f_N)\ln(1-f_N)], \tag{15}$$

where $f_N$ is the fraction of zero-energy polar Slater groups in the paraelectric phase at temperature $T_N$, given by

$$f_N(paraelectric, T_N) = [1 + 2\exp(-\varepsilon_0/kT_N)]^{-1} = [1 + 2\exp(-T_c\ln2/T_N)]^{-1}. \tag{16}$$

Here we substitute the Slater model expression $T_c\ln2$ for $\varepsilon_0/k$, and use the measured $T_c$ of 109.75 K for RDA. These values, together with $T_N$=216.1 K, yield $\varepsilon_a/k$=116.06 K when inserted into Eq. (15) and (16). This value for $\varepsilon_a/k$, together with the value $T_c\ln2$ for $\varepsilon_0/k$, yield a mean-field critical concentration of $x_{cm-f}$=0.328, whereas the experimental value for which ferroelectric behavior disappears even at zero temperature is near $x_c$=0.16.

This large discrepancy shows that mean-field models fail badly when applied to two-phase coexistence situations in which the coexistence is partially spatial because of quenched cation disorder. These models give fairly accurate results in the paraelectric phase of RDA (Slater model) and the paraelectric phase of RADA and ADA (Slater model with cross-cation interaction added), and in the ferroelectric phase of a pure (not mixed) crystal such as RDA (Takagi model) for which there is no quenched cation disorder and consequently the coexistence of ferroelectric and paraelectric phases found at nonzero temperature is temporal rather than spatial.

Predictions of this chain model for the ferroelectric order parameter p, compared with experimental results derived from permittivity and spontaneous polarization measurements (1-3), are presented in Fig. 1 for RADA crystals with various fractional ammonium ion concentrations x. The predictions employ Eqs. (9-13) and the three energy parameters $\varepsilon_0/k$=79.41 K, $\varepsilon_1/k$=460.95 K, and $\varepsilon_a/k$=116.06 K which were defined and provided numerical values as explained above.

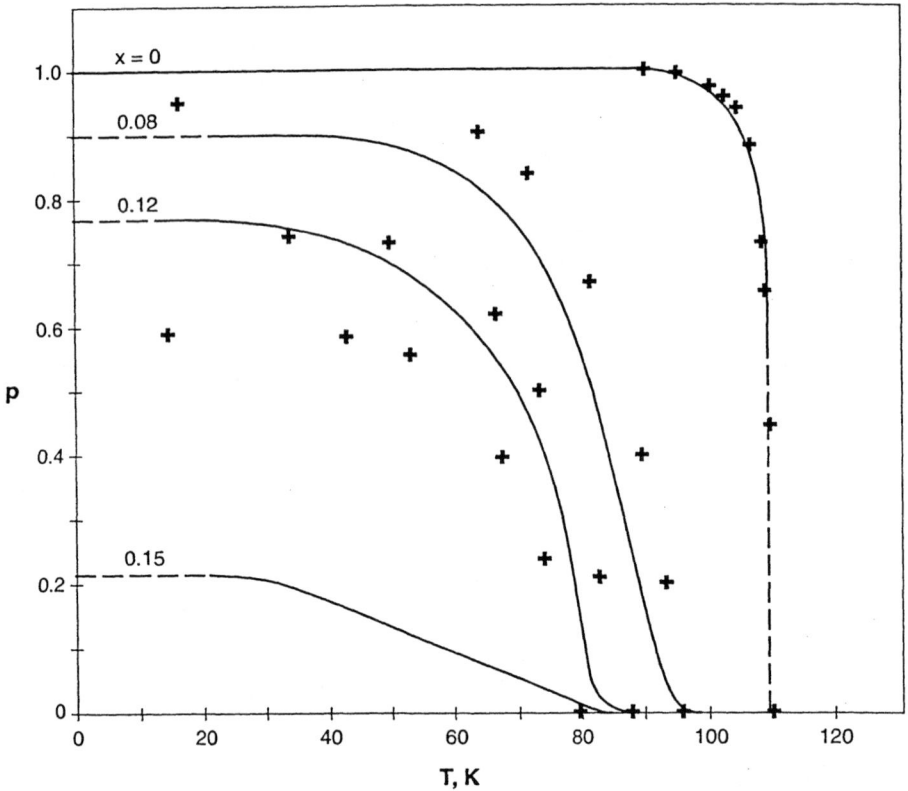

**FIGURE 1.** Comparison of temperature dependence of ferroelectric order parameter p determined experimentally (solid and dashed lines) from permittivity and spontaneous polarization results by the method outlined in Ref. 3, for $Rb_{1-x}(NH_4)_xH_2AsO_4$ (RADA) crystals with x=0, x=0.12, and x=0.15 (Ref. 1), and x=0.08 (Ref. 3), compared with predictions of the chain model (+) using Eqs. (9-13) as explained in the text.

The model fit shows promisingly good agreement with the experimental results. The x=0 fit is excellent, the x=0.08 calculated points are somewhat high, the x=0.12 fit is better, and the x=0.15 calculated points are far too high. The qualitative features are nearly all correct; the order parameter p saturates for low temperature at a level below 1 for nonzero x, the sharpness of curvature and the slopes of the straight portions are nearly right except for x=0.15, and the temperatures at which p goes to zero are nearly correct. Only the small reverse bend seen for x=0.12 and 0.08 near p=0 is not predicted. The discrepancies are due to the model and perhaps also to experiment. The model is based on paraelectric chains, and loses accuracy for low p where the paraelectric region is quite extensive. The flame atomic-absorption spectroscopy used to determine x may have been inaccurate. Model refinements and additional fits to experiment will be made.

## DIELECTRIC PERMITTIVITY AND RELAXATION

The method for finding the dielectric permittivity and the dielectric response time constant is now illustrated, using RDA as an example. The contribution of the chains to the permittivity $\varepsilon$ is

$$\varepsilon = \Delta P/\varepsilon_M E = (P_s/\varepsilon_M) dp/dE, \tag{17}$$

where $\Delta P$ is the polarization change, $\varepsilon_M$ is the MKS constant $8.85 \times 10^{-12}$ $C^2/Nm^2$, E is electric field applied along $c$, and $P_s$ is the spontaneous polarization at zero temperature, $0.036$ $C/m^2$ for RDA. We simplify the calculation without seriously affecting the result by using the Slater limit expression for p in an electric field E, which is

$$p = 1 - exp[-(2\varepsilon_1 - \varepsilon_0)/kT] \Sigma \{n2^n exp[-n(\varepsilon_0 + \mu E)/kT]\}, \tag{18}$$

where $\mu = P_s a^2 c/4$ is the electric dipole moment per polar molecular unit, and a and c are the unit cell dimensions, both near $7.5 \times 10^{-10}$ m, so $\mu$ is near $3.8 \times 10^{-30}$ C-m. We see that dp/dE is given by

$$dp/dE = -exp[-(2\varepsilon_1 - \varepsilon_0)/kT] \mu d\Sigma \{n2^n exp(-n\varepsilon_0/kT)\}/d\varepsilon_0 \tag{19}$$

$$= -exp[-(2\varepsilon_1 - \varepsilon_0)/kT] \mu d\{2exp(-\varepsilon_0/kT)/[1 - 2exp(-\varepsilon_0/kT)]^2\}/d\varepsilon_0.$$

Thus the permittivity contribution of the chains is given by

$$\varepsilon = (4P_s\mu/\varepsilon_M kT) exp(-2\varepsilon_1/kT)[1 + 2exp(-\varepsilon_0/kT)]/[1 - 2exp(-\varepsilon_0/kT)]^3. \tag{20}$$

This ignores any contribution of domain wall mobility. The permittivity appears to be heading for a singularity, but because $\varepsilon_1$ is not infinite, the transition occurs before the singularity is reached.

The dielectric relaxation time constant $\tau$ is the final change in polarization P (which is proportional to the order parameter p) induced by a small field step E, divided by the initial rate of change caused by that step,

$$\tau = \Delta p/(dp/dt) = E(dp/dE)/(dp/dt) = (\varepsilon_M \varepsilon E/P_s)/(dp/dt), \tag{21}$$

where $\varepsilon$ is the permittivity given in Eq. (20). To find dp/dt we use the common kinetic theory approximation that the jump rate $v$, for protons in this case, is

$$v = v_0 exp(-\Delta U/kT) \text{ if } \Delta U \geq 0, \ v = v_0 \text{ if } \Delta U \leq 0. \tag{22}$$

Here $\Delta U$ is the energy change, including the field-induced change, resulting from the jump. We choose $v_0 = kT/h$, where h is Planck's constant, which is applicable because we are below the Debye temperature. Because E is small, the exponential in Eq. (22) can be expanded as

$$exp(-\Delta U/kT) \cong exp(-\Delta U_0/kT)(1-\mu E/kT) \tag{23}$$

for jumps which create or lengthen chains, which require positive energy $\Delta U_0$ in the absence of a field, and which require additional energy $\mu E$ in a field E along the ferroelectric polarization direction because the jump destroys a dipole moment of magnitude $\mu$. The field-induced rate of change of p attributable to the Takagi group at the top of a chain is thus

$$(dp/dt)_n = 2W_{nS} v_0 exp(-\Delta U_0/kT) \mu E/kT. \tag{24}$$

A factor of 2 occurs because either of two protons can move in to that group and thus lengthen the chain. The $W_{nS}$ factor from Eq. (2) is the probability in the Slater limit that a chain of length n begins at a given anion site. We take into account the fact that the given anion site, which is occupied either by a polar Slater group or by the bottom of a chain, is representative of the whole crystal, so one chain lengthening step reduces p by 1. The energy change $\Delta U_0$ is $\varepsilon_0$, except for the creation process, in which case it is $2\varepsilon_1$. Summing the contributions from Eq. (3), keeping in mind that $W_{0S}=1$, yields

$$dp/dt = (2\mu E/h) exp(-2\varepsilon_1/kT)/[1-2exp(-\varepsilon_0/kT)]. \tag{25}$$

Combining Eqs. (20), (21) and (25) yields

$$\tau = (2h/kT)[1+2exp(-\varepsilon_0/kT)]/[1-2xp(-\varepsilon_0/kT)]^2. \tag{26}$$

We find that $\tau$ <u>decreases</u> with decreasing temperature. This behavior of $\tau$ contradicts the common belief that time constant increases (because motion "freezes out") as temperature decreases in the ferroelectric phase. What "freezes out" is the <u>amplitude</u> of the dielectric response $\varepsilon$, as seen in Eq. (20). As with $\varepsilon$, $\tau$ does not become infinite with increasing temperature because the noninfinite Takagi energy $\varepsilon_1$ causes $T_c$ to occur before the denominator in Eq. (26) vanishes.

If we insert our previously determined values, $\varepsilon_0/k=79.41$ K and $\varepsilon_1/k=460.95$ K, into Eq. (26), we obtain at $T_c=109.75$ K the time constant $\tau(T_c)=1.92 \times 10^{-9}$ s, which is much longer than the attempt time $\tau_0(T_c)=h/kT_c=4.37 \times 10^{-13}$ s. Again using these values and other parameters listed above, Eq. (20) for permittivity $\varepsilon$ at $T_c$ the value $\varepsilon(T_c)=673$. This large value can be swamped by the domain wall contribution.

Similar methods can be employed to find ε and τ for the *a* axis, and for both the *a* and *c* axes in the coexistence region for $0<x<x_c$. A problem for the *a* axis calculation is finding $\mu_a$, because there is no phase ordered ferroelectrically along *a*, so the method used to find μ along *c* from $P_s$ along *c* has no analog for finding $\mu_a$. The best approach may be to assume that the μ's are proportional to the Curie-Weiss constants found from measurements in the paraelectric phase.

## CONCLUSIONS

Most of the features of coexistence of the ferroelectric phase with the paraelectric/proton glass phase in $Rb_{1-x}(NH_4)_xH_2AsO_4$ (RADA) proton glass crystals are predicted by a model which employs the statistics of chainlike excitations from the ferroelectric phase. Improvements to the model are needed particularly in the region of lower ferroelectric order parameter p. Additional fits to other dielectric data, and neutron diffraction and nuclear magnetic results, are needed for RADA and other proton and deuteron glasses, also in the region of coexistence of the antiferroelectric phase with the paraelectric/proton glass phase.

## ACKNOWLEDGEMENTS

This work was supported by National Science Foundation Grant DMR-9520251. Thanks are expressed to George Tuthill for a helpful discussion.

## REFERENCES

1. Trybula, Z., Schmidt, V. H., and Drumheller, J. E., *Phys. Rev.* **B 43**, 1287-1289 (1991).
2. Howell, F. L., Pinto, N. J., and Schmidt, V. H., *Phys. Rev.* **B 46**, 13762-13766 (1992).
3. Pinto, N. J., and Schmidt, V. H., *Ferroelectrics* **141**, 207-213 (1993).
4. Schmidt, V. H., Trybula, Z., Pinto, N. J., and Shapiro, S. M., *Phase Transitions* (Proc. XVII Crystallography Congress, in press).
5. Schmidt, V. H., *J. Korean Phys. Soc.* (Proc. 9th Internat. Mtg. on Ferroelectricity, in press).
6. Slater, J. C., *J. Chem. Phys.* **9**, 16-33 (1941).
7. Takagi, Y., *J. Phys. Soc. Jpn.* **3**, 273-274 (1948).
8. Schmidt, V. H., *Deuteron Jumping in $KH_2PO_4$*, Ph.D. Thesis, University of Washington, 1961, pp. 90-91.
9. Senko, M. E., *Phys. Rev.* **121**, 1599-1604 (1961).
10. Silsbee, H. B., Uehling, E. A., and Schmidt, V. H., *Phys. Rev.* **133**, A165-A170 (1964).
11. Fairall, C. W., and Reese, W., *Phys. Rev.* **B 11**, 2066-2068 (1975).
12. Blinc, R., and Svetina, S., *Phys. Rev.* **147**, 423-429 and 430-438 (1966).
13. Schmidt, V. H., Wang, J. T., and Schnackenberg, P., *Jpn. J. Appl. Phys.* **24**, Suppl. 24-2, 944-946 (1985); see corrections in Schmidt, V. H., *J. Phys. Soc. Jpn.* **56**, 3752-3753 (1987).

# Proton Tunneling and Nonlinear Polarizability Effects in Hydrogen-Bonded Ferroelectrics

A. Bussmann-Holder* and K.-H. Michel[†]

*Max-Planck-Institut für Festkörperforschung
Heisenbergstr. 1, D-70569 Stuttgart, Germany
[†]Department Natuurkunde, Universiteit Antwerpen
B-2610 Wilrijk, Belgium

**Abstract.** Hydrogen-bonded ferroelectrics are modelled by a coupled spin/nonlinear lattice (polarizability) interaction Hamiltonian, where specifically the geometry of the hydrogen bond is included. The model leads to a structural phase transition and describes correctly the isotope effect due to the substitution H/D in hydrogen-bonded systems in terms of bond length changes.

The quantum tunneling model (QTM) [1] was long believed to be the correct approach in understanding the phase transition mechanism of hydrogen-bonded ferroelectric systems. It is based on the assumption that the structural phase transition in e.g. $KH_2PO_4$ (KDP) is driven by the direct proton-proton interaction, where the protons occupy with equal probability two equilibrium positions in a symmetric double-well potential. The large isotope effect on the structural transition temperature $T_c$ upon deuteration is attributed, in this model, to a change in the tunneling frequency due to the mass change from H to D [2].

Yet, in the last years high resolution neutron diffraction data [3] have provided increasing evidence that the QTM has to be extended to account for structural changes in the hydrogen bond geometry which are intimately related to distortions of the surrounding lattice [4,5]. This coupling between tunneling mode and the optical vibrations of the [K-PO$_4$] complex has been considered in Ref. [6], but the geometrical aspects of the hydrogen bond have not been included.

In the following a new approach to the phase transition of H-bonded ferroelectrics is given which is based on the interplay between the tunneling motion of the protons (deuterons), the concommittant polarizability changes of the neighboring $PO_4$ groups and the optical phonon mode of the [K-PO$_4$] complexes [7]. In order to have an analytically tractable problem a two-dimensional system is investigated which is thought to reproduce the essential structural features of KDP. It is important to note that the model is not only applicable to KDP but also to other hydrogen-

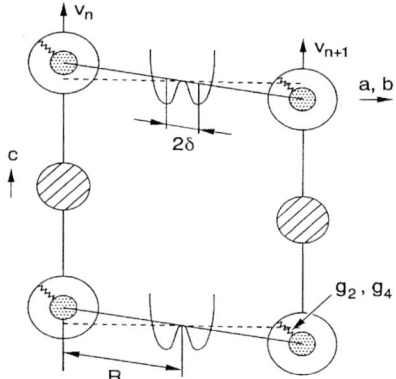

**FIGURE 1.** Model for $KH_2PO_4$.

bonded ferro- and antiferroelectric systems like e.g. squaric acid, $PbHPO_4$. The model system (Fig. 1) consists of the $K^+$ ion, the $PO_4^{2-}$-group and the proton. The $PO_4$ group is treated as a highly polarizable cluster, in analogy to the $BO_3$ group in $ABO_3$ ferroelectric perovskites [8,9]. The hydrogen bond is directed along the $y$-axes which is the line connecting two neighboring $PO_4$ groups (O-O line). The equilibrium positions of the proton with respect to the center of the bond are $1/2(\delta_y, \delta_z)$ and $1/2(-\delta_y, -\delta_z)$, where $\delta_y = \delta \cos \Psi$, $\delta_z = \delta \sin \Psi$ and $\Psi = 8.8°$, $\delta = 0.3647$ Å (KDP), $\Psi = 7.4°$, $\delta = 0.4462$ Å (DKDP), respectively [10]. The Hamiltonian reads

$$\mathbf{H} = \mathbf{H}_T + \mathbf{H}_L + \mathbf{H}_{TL} \tag{1}$$

with

$$\mathbf{H}_T = -\frac{1}{2}\Sigma J(q)S^z(q)S^z(-q) - \Omega S^x(q=0)\sqrt{N} \tag{2}$$

being the tunneling model Hamiltonian [1,11] of $N$ interacting protons which are represented by the two state pseudospin $S^z = \pm 1$, and wave vector $q$ dependent interaction $J(q)$. $S^x$ accounts for proton tunneling with tunneling frequency $\Omega$.

$\mathbf{H}_L$ refers to the host lattice Hamiltonian

$$\mathbf{H}_L = \sum_{q,i=P,K} \frac{1}{2}\Big\{-\pi_i^+(q)\pi_i(q) + m_s \dot{u}^+(q)\dot{u}(q)$$
$$+ \frac{1}{2}\omega_p^2(q)\phi_p^+(q)\phi_p(q) + \frac{2f}{m_k}\phi_k^+(q)\phi_k(q)$$
$$(2f+g)u^+(q)u(q) - 2\left[\frac{2f}{\sqrt{m_k}}\phi_k^+(q) + \frac{g}{\sqrt{M_p}}\phi_p^+(q)\right]u(q)\Big\} \tag{3}$$

$\pi_p$, $\pi_k$ and $\phi_p/\sqrt{M_p}$, $\phi_k/\sqrt{m_k}$ are momentum and conjugate displacement coordinates of the PO$_4$ center mass $M_p$ and the K$^+$ ion (mass $m_k$), and $u$, $\dot{u}$ are displacement and velocity of the PO$_4^{2-}$ shell. The lattice Hamiltonian is equivalent to the nonlinear polarizability model [8,9] as the local core-shell coupling $g$ consists of an attractive harmonic term $g_2$ and a stabilizing 4$^{th}$ order term $g_4$:

$$g = g_2 + \frac{3g_4}{N}\sum_q \left\langle \left(u^+(q) - \frac{\phi_p^+(q)}{\sqrt{M_p}}\right)\left(u(q) - \frac{\phi_p(q)}{\sqrt{M_p}}\right)\right\rangle \quad (4)$$

The coupling constant $f$ provides an indirect coupling between PO$_4$-group and K$^+$ ion through the PO$_4$ shells. $\frac{4f'}{M_p} = \omega_p^2$ is the zone boundary frequency stemming from harmonic coupling between next nearest neighbor PO$_4$ cores.

The coupling term between lattice and pseudospins is given by

$$\mathbf{H}_{TL} = C\Sigma S^z(q)u^+(q)$$

which couples the pseudospins with the PO$_4$ shells.

While in the polarizability model a displacive phase transition occurs only if $g_2 < 0$, the coupling to the tunneling proton system allows for an order-disorder driven transition also for $g_2 > 0$. Thus, if $g_2 < 0$ the considered system combines order-disorder and displacive features. As has been shown in Refs. [7,12], the coupling constant $C$ is directly proportional to $\delta_z$ and thus accounts for the specific bond geometry.

Using standard Green's functions methods [13] the coupled equations of motion are obtained:

$$\left(\omega^2 - \omega_0^2\right)\langle\langle\xi^+(q); S^z(q)\rangle\rangle_\omega = \frac{C\sqrt{\mu}}{(2f+g)}\left[\frac{g}{M_p} - \frac{2f}{m_k}\right]\langle\langle S^{z+}(q); S^z(q)\rangle\rangle_\omega$$
$$\left(\omega^2 - \tilde{\Omega}^2\right)\langle\langle S^{z+}(q); S^z(q)\rangle\rangle_\omega =$$
$$\langle S^x\rangle\Omega\left\{1 + \frac{2\sqrt{\mu}}{(2f+g)}\left[\frac{g}{M_p} - \frac{2f}{m_k}\right]\right\}\langle\langle\xi^+(q); S^z(q)\rangle\rangle_\omega \quad (6)$$

where

$$\xi(q) = \sqrt{\frac{\mu}{M_p}}\phi_p(q) - \sqrt{\frac{\mu}{m_k}}\phi_k(q);$$

$$\mu = \frac{m_k M_p}{m_k + M_p};$$

$$\omega_0^2 = \frac{2fg}{\mu}\frac{1}{(2f+g)};$$

$$\tilde{\Omega} = \Omega^2 - \Omega\langle S^x\rangle\left[J(q) + \frac{C^2}{(2f+g)}\right];$$

$$\langle S^x\rangle = \langle S^x(q=0)\rangle/\sqrt{N}.$$

In the limit $\omega = 0$ the retarded Green's functions reduce to static susceptibilities which diverge under condition

$$\tilde{\Omega}^2 \omega_0^2 = \frac{C^2 \Omega \langle S^x \rangle \mu}{(2f+g)^2} \left( \frac{g}{M_p} - \frac{2f}{m_k} \right)^2 \qquad (7)$$

which determines the occurrence of a structural phase transition. As both $g$ and $\langle S^x \rangle$ are temperature dependent, the critical temperature $T_c$ is determined by Eq. (7) where $T_c$ is defined either through $\omega_0^2 = 0$ (polarizability induced) $\tilde{\Omega}_0^2 = 0$ (proton induced) or $\omega_0^2 = \Omega_0^2 = 0$ (combined effect).

Analytically obtainable limiting cases for $T_c$ are:

i) $J \gg C^2/(2f+g)$, i.e., the transition is driven by the condensation of the proton motion:

$$T_c = \frac{1}{2} \left\{ J + \frac{C^2}{2f+g} \left[ 1 + \frac{\mu^2}{2fg^2} \left( \frac{g}{M_p} - \frac{2f}{m_k} \right)^2 \right] \right\} \qquad (8)$$

$T_c$ is enhanced as compared to QTM due to the coupling to the polarizable host lattice. Yet the effect is expected to be small as $(2f+g)$ is large.

ii) $S \approx 0$, $g < f$, i.e., the transition is polarizability induced and

$$T_c = \frac{1}{2} \left\{ \frac{|g_2|}{3g_4} - \frac{\mu}{m_k} \frac{C^2}{4f} + \sqrt{\left[ \frac{|g_2|}{3g_4} - \frac{\mu}{m_k} \frac{C^2}{4f} \right]^2 + \frac{\mu^2}{m_k^2} \frac{C^2}{3g_4}} \right\} \qquad (9)$$

which, in the absence of coupling $C$ yields the same results as the polarizability model. With $\frac{C}{f} < \frac{|g_2|}{3g_4}$, the right hand side of Eq. (9) can be expanded in terms of $C$ and

$$T_c = \frac{|g_2|}{3g_4} + \left( \frac{f}{m_k} \mu - |g_2| \right) / \left( |g_2| \frac{\mu}{m_k} f \right) \cdot \frac{\mu^2 C^2}{4m_k^2} \qquad (10)$$

Again the coupling between protons and $PO_4$ shells leads to an increase in $T_c$, whereas the direct coupling between the protons is negligible. Solving Eq. (7) numerically for $T_c$ as function of the model parameters the results shown in Fig. 2 are obtained. The variation of $C$ only, all other parameters are kept fixed, leads to a strong increase in $T_c$ which, when inserting the experimental values for $\delta$(KDP), $\delta$(DKDP) leads to a $T_{c/\text{DKDP}} = 168$ K which is in reasonable agreement with experiment. While $T_c$ is also strongly dependent on $f$, i.e., the coupling between $K^+$ ion and $PO_4$ shells, it is weakly dependent on $g_2$, $g_4$, and $J$. Thus, e.g. the substitution of $K^+$ by $Rb^+$ should induce variations in $f$ and accordingly $T_c$ is affected.

In conclusion, a new model for the phase transition mechanism of hydrogen-bonded ferroelectrics has been presented which accounts for the effect of the $H$ separation $\delta$ on the $PO_4$ shell displacements. The model combines order/disorder dynamics and displacive features and leads to the correct increase in $T_c$ upon deuteration.

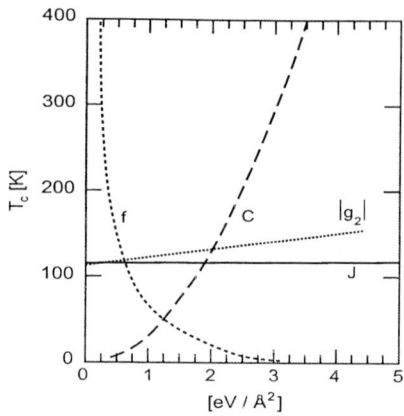

**FIGURE 2.** $T_c$ as function of model parameters.

# REFERENCES

1. R. Blinc, J. Phys. Chem. Solids **13**, 204 (1960).
2. R. Blinc, B. Zeks., Adv. Phys. **21**, 693 (1972).
3. J.E. Tibbals, R.J. Nelmes, G.J. McIntyre, J. Phys. C: Solid State Phys. **15**, 37 (1982); Z. Tun, R.J. Nelmes, W.F. Kuhs, R.D.F. Stansfield, J. Phys. C: Solid State Phys. **21**, 245 (1988).
4. M.J. McMahon, R.J. Nelmes, W.F. Kuhs, R. Dorwarth, R.O. Piltz, Z. Tun, Nature **348**, 317 (1990).
5. J.A. Krumhansl, Nature **348**, 285 (1990).
6. K.K. Kobayashi, J. Phys. Soc. Japan **24**, 497 (1968).
7. A. Bussmann-Holder, K.H. Michel, Phys. Rev. Lett., in press.
8. R. Migoni, H. Bilz, D. Bäuerle, Phys. Rev. Lett. **37**, 1155 (1976).
9. H. Bilz, G. Benedek, A. Bussmann-Holder, Phys. Rev. B **35**, 4840 (1987); A. Bussmann-Holder, H. Bilz, G. Benedek, Phys. Rev. B **39**, 9214 (1989).
10. A. Katrusiak, Phys. Rev. B **51**, 589 (1995).
11. P.G. de Gennes, Solid State Commun. **1**, 132 (1963).
12. K.H. Michel, Phys. Rev. B **24**, 3998 (1981); R.M. Lyndon-Bell, K.H. Michel, Rev. Mod. Phys. **66**, 721 (1994).
13. D.N. Zubarev, Sov. Phys. Uspekhi **3**, 320 (1960).

# First-principles and semiempirical Hartree-Fock calculations for $F$ centers in KNbO$_3$ and Li impurities in KTaO$_3$

R. I. Eglitis[a,b], E. A. Kotomin[b,c], A. V. Postnikov[a],
N. E. Christensen[c], and G. Borstel[a]

[a] *Universität Osnabrück – Fachbereich Physik, D-49069 Osnabrück, Germany*
[b] *Institute of Solid State Physics, University of Latvia, 8 Kengaraga, Riga LV-1063, Latvia*
[c] *Institute of Physics and Astronomy, University of Aarhus Aarhus C, DK-8000, Denmark*

(Received 4 February 1998)

**Abstract.** The linear muffin-tin-orbital method based on the density-functional theory and the semi-empirical method of the Intermediate Neglect of the Differential Overlap based on the Hartree–Fock formalism are used for a supercell study of $F$ centers in cubic and orthorhombic ferroelectric KNbO$_3$ crystals. Two electrons are found to be considerably delocalized even in the ground state of the defect. The absorption energies are calculated by means of the INDO method using the $\Delta$SCF scheme after a relaxation of atoms surrounding the $F$ center.

As an example of another type of point defect in perovskite, an isolated Li impurity in KTaO$_3$ as well as interacting Li pairs are considered in the supercell approach, using supercells of up to 270 atoms. The off-center Li displacement, reorientational energy barriers and the lattice relaxation around impurities are calculated. The results are compared with those obtained earlier within the shell model, revealing the relaxation pattern somehow different from the shell model estimations.

## INTRODUCTION

It is well understood now that point defects play an important role in the electro-optic and non-linear optical applications of KNbO$_3$ and related materials [1]. In particular, the light frequency doubling in KNbO$_3$ is seriously affected by unidentified defects responsible for the induced IR absorption [2]. The photorefractive effect, important in particular for holographic storage, is also known to depend on the presence of impurities and defects.

One of the most common defects in oxide crystals is the so-called $F$ center, an O vacancy $V_O$ which traps two electrons [3]. In electron-irradiated KNbO$_3$, a broad absorption band is observed around 2.7 eV at room temperature and tentatively ascribed to $F$ centers [4,5]. The motivation behind the present study of this defect by means of theory is twofold. First, due to the lowering of the local point symmetry at the O site in ferroelectric phases of $KNbO_3$ (for instance, in the orthorhombic phase which is the room-temperature one), the degeneracy of the $2p$-type excited state may be lifted, resulting in several split absorption bands. Second, there were claims in the literature in favor of the symmetry breaking between two Nb atoms neighboring the O vacancy, resulting in an asymmetric electron density distribution [6]. In order to clarify these questions and to check the assignment of the 2.7 eV absorption band, we study in the present paper the $F$ center in $KNbO_3$ using the supercell model and two different theoretical techniques: the full-potential linear muffin-tin orbital (FP-LMTO) and the semiempirical Intermediate Neglect of the Differential Overlap (INDO) methods. In the present contribution we essentially summarize the results presented recently in more detail in Ref. [7].

The same techniques, also in the supercell approach, have been applied to the study of another point defect, inducing lattice relaxation in a incipient ferroelectric crystal: Li substituting K in $KTaO_3$. As is experimentally known [8], such substitutional impurity gets spontaneously displaced along one of six possible [100]-type directions. The magnitude of this displacement and (in some cases) the lattice relaxation related to it have been estimated by empirical models [9,10], the shell model [11,12], first-principles FP-LMTO calculations [13] and recently by the INDO method [14]. As an extension of the latter study, we discuss in the present contribution also the effects of interaction between Li impurities by the same method. The INDO method has the advantage of being relatively compact in what regards the necessary computer resources. This allows in the study of impurity systems to treat relatively large supercells which remain well beyond the range of *ab initio* total-energy methods. At the same time, once the essential parameters of the INDO method are known for the system in question, the description of excitation energies and/or total-energy trends with the INDO method is much better than within any model schemes.

The INDO method is described in detail in Ref. [15]. We used its practical implementation in the computer code CLUSTERD. The technical details related to the application of the method (e.g. parametrisation) for $KNbO_3$ can be found in Ref. [16], for Li-doped $KTaO_3$ – in Ref. [14]. As a benchmark for the INDO parametrisation in the previous study [14,16], as well as for an independent investigation of the $F$-center in Ref. [7] (the essential results of which are discussed below), we used the FP-LMTO method in the implementation by M. Methfessel [17].

# TECHNICAL DETAILS OF CALCULATION AND RESULTS FOR F-CENTERS

In both LMTO and INDO calculations of the oxygen vacancy in $KNbO_3$, we used $2 \times 2 \times 2$ supercells, including 39 atoms, for the geometry of an ideal cubic perovskite lattice. A more detailed study of $F$-center system within the local density approximation, using e.g. the atomic sphere approximation along with the FP-LMTO, the electronic structure of the defect system and the aspects of correcting the band gap are discussed in Ref. [7] and skipped here. Summarizing, the band gap estimated from the ground-state band structure in the local density approximation (LDA), as is not justified but commonly used, is known to be underestimated in dielectrics. This does not present a problem for the total-energy studies (even structure optimizations) in pure oxides, but in case of the $F$-center the impurity band formed in the band gap exhibits too strong dispersion (due to limited supercell size) and overlaps with the states in the conduction band, resulting in a metallic behavior of the impurity system. In the INDO calculation which is essentially a Hartree-Fock scheme and hence tends to overestimate the band gap, no such problem arises. On the other hand, we would like to keep the LDA results as a useful reference point for the electronic structure and as a reliable benchmark of the total energy-based structure optimization. It can be achieved either by an artificial operator shifting the Nb$4d$ states in the conduction band upwards (that is useful for the analysis of the electronic structure but makes the total energy results unreliable), or by performing the total energy calculation with only one k-point in the Brillouin zone, in order to suppress the dispersion of the defect states. In the latter case, the calculation setup becomes somehow resembling that of INDO, where also only the $\Gamma$ point is traditionally used for the k-space sampling, in the spirit of the "large unit cell" LUC scheme [18]. Since the structure optimization normally needs good convergency in the number of k-points, the total-energy result of such LDA calculation should be considered as an approximate one, giving rather an error bar when taken together with that of INDO. According to the LDA calculation, the relaxation of two Nb atoms neighboring to the $F$ center is outwards by 4.8% of the lattice constant, resulting in the energy lowering by 1.2 eV. From the INDO calculation, both values are somehow larger but in qualitative agreement with the LDA results: outward relaxation of two Nb neighbors by 6.5% and the energy gain of $\sim$3.7 eV.

In addition to analyzing this most important aspect of relaxation, we optimized in the INDO calculation the positions of more distant neighbors (14 atoms in total) to the O vacancy as well. The 0.9% outward displacement of K atoms and the 1.9% inward displacement of O atoms gives the total relaxation energy of $\sim$4.7 eV, mostly due to the contributions from the Nb and O displacements.

The analysis of the effective charges of atoms surrounding the $F$ center shows that of the two electrons associated with the removed O atom, only $\approx -0.6|e|$ is localized at the vacancy, and about a similar extra charge is localized on the two nearest Nb atoms. The $F$ center produces a local energy level $\approx 0.6$ eV above the top of the valence band. Its molecular orbital contains primarily contribution from the atomic orbitals of the two nearest Nb atoms.

The structure optimization based on the INDO calculation has also been done for a low-symmetry geometry corresponding to a room-temperature (ferroelectric) orthorhombic phase of KNbO$_3$. This phase is stable in a broad temperature range (263 to 498 K) and hence subject to most studies and practical applications. The

**TABLE 1.** Calculated absorption energy for the $F$ center ($E_{abs}$) and the energy of the nearest-neighbors Nb relaxation ($E_{rel}$) in cubic and orthorhombic phases of KNbO$_3$.

| Symmetry, phase | $E_{abs}$ (eV) | | | $E_{rel}$ (eV) |
|---|---|---|---|---|
| $C_{4v}$, cubic | 2.73 | 2.97 | – | 3.7 |
| $C_s$, orthorhombic | 2.56 | 3.03 | 3.10 | 3.6 |
| $C_{2v}$, orthorhombic | 2.72 | 3.04 | 3.11 | 3.6 |

structure parameters in pure KNbO$_3$ have been optimized with the INDO method in Ref. [16] and are in quite good agreement with the experimental measurements. In the orthorhombic phase, there are two inequivalent positions of oxygen and hence the possibility to form two different kinds of $F$-center, possibly with different optical properties. The displacements of Nb atoms nearest to $V_O$ were calculated for these both types of defects and found to be very close to those found for the cubic phase. The relevant relaxation energies (3.6 eV) are also nearly the same as for the Nb relaxation found in the cubic phase.

Because of different local symmetry at the O site (either $C_{2v}$ or $C_s$, in contrast to $C_{4h}$ in the cubic phase), the energetics of the impurity levels changes. In the cubic phase, the $V_O$ excited state splits into two levels, one of which remains two-fold degenerate. Our $\Delta$SCF calculations predict the two relevant vacant bands, the absorption energies of which are given in Table 1. In the orthorhombic phase, the degeneracy of the impurity level is completely lifted (Table 1). The relaxation energies (associated with the displacement of Nb neighbors only) are also listed in Table 1 for comparison.

## OFF-CENTER DISPLACEMENT OF Li IN KTaO$_3$ AND RELATED LATTICE RELAXATION

The off-center displacement of substitutional Li in KTaO$_3$ is known to induce a considerable long-range polarization of crystal. Therefore the supercell size needs to be larger than for the study of $F$ centers. The size of the polarized region associated with the [100]-displaced Li ion was estimated in the shell model calculation by Stachiotti and Migoni [11] to be about 5 lattice constants along the direction of displacement, with ~99% of the 'effective dipole' polarization being confined to nearest Ta–O chains, that go parallel to the displacement. As is discussed below, the magnitudes of the atomic displacements and polarization in our present calculation is considerably smaller than those found in Ref. [11], and the relaxed neighbors to the Li impurity are well within the 3×3×3 supercell. In order to check our approach, we performed as well the calculations for a supercell doubled in the direction of Li displacement, i.e., 6×3×3, with a single [100]-displaced Li atom. The equilibrium displacement in this case is 0.62 Å, exactly as for the 3×3×3 supercell, with the energy lowering 57 meV. The difference from the result for a 3×3×3 supercell

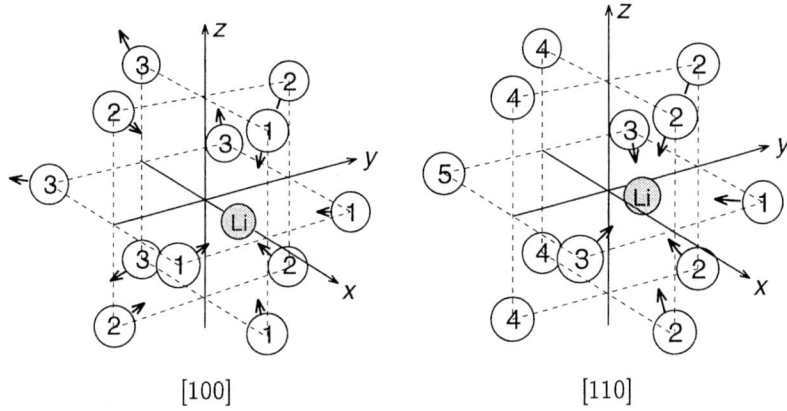

**FIGURE 1.** [100] and [110] Li off-center displacements and relaxation pattern of neighboring oxygen atoms.

(62.0 meV) roughly represents the uncertainty related to the supercell size in our calculations.

The parametrisation of the INDO method for the KTaO$_3$:Li system has been done in Ref. [14] based on the comparison with the results of earlier FP-LMTO calculations [13] in what regards the magnitude of the Li off-center displacement (0.61 Å [13]) as well as the energy gain due to the Li displacement. The total energy as function of [100] and [110] Li off-center displacements as calculated by INDO is shown in Fig. 2. Our equilibrium [100] off-center displacement of 0.62 Å is smaller as calculated by Stachiotti and Migoni within the shell model [11]. On the other hand, our value is in good agreement with a more recent, and apparently more elaborately parameterized, shell model calculation by Exner et al. (0.64 Å, Ref. [12]).

The displaced Li ion with its neighboring oxygen atoms is shown schematically in Fig. 1 for the cases of [100] and [110] off-center displacements. The [100] displacement is known to occur in reality [8] whereas the [110] displacement represents a saddle point between two such adjacent and equivalent displaced configurations. Oxygen atoms are numbered according to their separation into several inequivalent groups.

The energy gain due to the Li off-center displacement is not directly measurable in an experiment, but there are estimations for the 90$^0$-energy barrier via the saddle point between [100] and [010]-displaced positions to be 86 meV [10]. Our estimate of the energy difference between [100] and [110] minima is ~30.2 meV, roughly two times larger than in the FP-LMTO calculation [13], but much less than the experimental estimate. The origin of this discrepancy, as has been mentioned in Ref. [13], is most probably related to the lattice relaxation around the displaced Li ion, that makes the net energy gain from the displacement larger, and the 90$^0$-activation energy (involving now the displacement of many atoms) correspondingly higher.

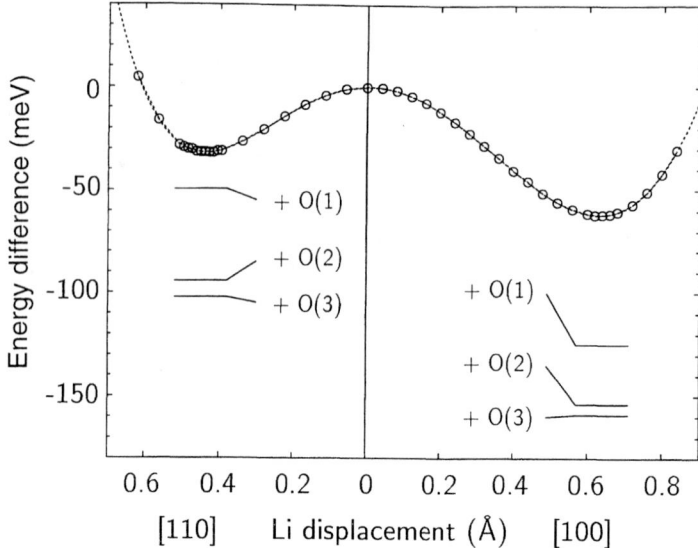

**FIGURE 2.** Total energy gain as function of [100] and [110] Li displacements without lattice relaxation (dashed line with open circles), and the total energy values after including the relaxation of three groups of nearest oxygen atoms.

Indeed, the second harmonic generation-based estimates of the activation barrier [19] reveal two types of processes, apparently one involving the lattice relaxation (with the barrier height 86.2 meV) and another one that is too fast for the lattice to follow, with the barrier 14.7 meV.

In order to clarify this point, we performed a lattice relaxation of several shells

**TABLE 2.** Relaxed atomic positions for the Li displacement as calculated by INDO.

| Atom | Lattice coordinates | | | Displacement |
|---|---|---|---|---|
| | | Li [100] displacement | | |
| Li | $\Delta_x$ | 0 | 0 | $\Delta_x = 0.1550$ |
| 4×O(1) | $\frac{1}{2}+\Delta_x$ | $\frac{1}{2}+\Delta_y$ | 0 | $\Delta_x = -0.0045; \Delta_y = -0.0105$ |
| 4×O(2) | $\Delta_x$ | $\frac{1}{2}+\Delta_y$ | $\frac{1}{2}+\Delta_y$ | $\Delta_x = 0.0070; \Delta_y = -0.0026$ |
| 4×O(3) | $-\frac{1}{2}+\Delta_x$ | $\frac{1}{2}+\Delta_y$ | 0 | $\Delta_x = -0.0020; \Delta_y = 0.0020$ |
| | | Li [110] displacement | | |
| Li | $\Delta_x$ | $\Delta_x$ | 0 | $\Delta_x = 0.0760$ |
| 1×O(1) | $\frac{1}{2}+\Delta_x$ | $\frac{1}{2}+\Delta_x$ | 0 | $\Delta_x = -0.0090$ |
| 4×O(2) | $\frac{1}{2}+\Delta_x$ | $\Delta_y$ | $\frac{1}{2}+\Delta_z$ | $\Delta_x = -0.0060; \Delta_y = 0.0030; \Delta_z = -0.0080$ |
| 2×O(3) | $\frac{1}{2}+\Delta_x$ | $-\frac{1}{2}+\Delta_y$ | 0 | $\Delta_x = -0.0020; \Delta_y = 0.0060$ |
| 4×O(4) | $-\frac{1}{2}+\Delta_x$ | $\Delta_y$ | $-\frac{1}{2}+\Delta_z$ | $\Delta_x = 0.0003, \Delta_y = 0.0001, \Delta_z = 0.0003$ |
| 1×O(5) | $-\frac{1}{2}+\Delta_x$ | $-\frac{1}{2}+\Delta_x$ | 0 | $\Delta_x \sim 0$ |

of neighbors to the displaced Li ion, for the cases of [100] and [110] displacements. The relaxed coordinates of atoms are given in Table 2, where the oxygen atoms are numbered consistently with Fig. 1. The total energy values resulting from the gradual inclusion of neighbor relaxation are shown in Fig. 2. We found the relaxation of twelve nearest oxygen atoms essential, and the effect of relaxing nearest Ta and more distant atoms to be negligible, in what regards the effect on the total energy. The energy gain in the fully relaxed [100]-displaced configuration, with respect to a non-relaxed central Li position, is 158.9 meV; the energy gain in the relaxed [110]-configuration is 102.3 meV. Therefore, the enhancement of the excitation barrier due to relaxation effects is by a factor of two, but still not sufficient to reach experimentally expected ~86 meV. This discrepancy may be due to the fact that in reality the $90^0$-reorientation process of the impurity does not necessarily occur via the fully relaxed saddle-point configuration. Depending on the actual degree of relaxation around the saddle-point Li position, the barrier height is expected from Fig. 2 to be between ~57 meV (full relaxation at the saddle point) to ~127 meV (no relaxation).

## INTERACTING Li IMPURITIES IN KTaO$_3$

The experimental investigations of the diluted $K_{1-x}Li_xTaO_3$ system are numerous and include e.g. the nuclear magnetic resonance studies of relaxational dynamics associated with dipole reorientations [20], ultrasound attenuation measurements [21] and the measurements of the low-frequency shear modulus [22]. Due to different technical limitations, none of these methods allows to attain the ground state of the system in the concentration range $x \leq 7\%$.

Up to now, there are reported only few theoretical studies of interacting Li impurities in KTaO$_3$, using mainly analytical approaches [23] or oversimplified shell model calculations [24], but there are no *ab initio* studies reported to our knowledge at this field. In order to get some theoretical predictions, to check results of shell model calculations [24], and to answer the question about the nature of the low-temperature phase of $K_{1-x}Li_xTaO_3$ at small concentrations of Li spin glasses or ferroelectrics, INDO calculations of Li-Li interaction in KTaO$_3$ may be of certain interest.

In the preliminary calculations done by now, we concentrated on two following subjects. First, we wanted to know how the interaction between Li impurities which substitute two neighboring K sites affects the energy characteristics and the lattice relaxation associated with each impurity. For this purpose, we allowed the simultaneous adjustment of the structure coordinates as listed in Table 3 (affecting both impurities and 20 oxygen neighbors). The labelling of atoms in Tab. 3 and the qualitative scheme of the relaxation pattern is shown in Fig. 3. The relaxation of Ta and K atoms was found to be much smaller than that of O neighbors. The energy gain that was 57 meV due to the [100] displacement of a single Li impurity and 151 meV for the oxygen relaxation taken into account, makes correspondingly

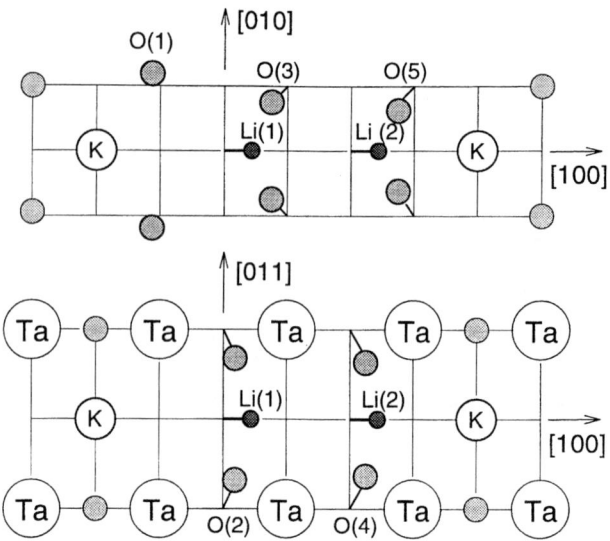

**FIGURE 3.** Off-center displacements of two nearest Li atoms and the relaxation pattern of 20 neighboring oxygen atoms.

176 meV and 407 meV per two Li impurities. It indicates that a substantial Li-Li interaction of the magnitude $E(2\text{Li}) - 2 \times E(\text{Li}) = 62$ meV for bare Li and 105 meV for Li with oxygen "cloud" indeed occurs and is enhanced by lattice polarization.

The second important characteristic is the strength and spatial distribution of the dipole field created by each Li impurity. In order to study it, we used the "probe" Li impurity at different lattice positions and displaced in different directions with respect to the "central" one. The "central" impurity was displaced along [100] by 0.62 Å, i.e. the equilibrium displacement for the single off-center Li ion. The "probe" impurity was allowed to relax along the given direction, and the interaction energy was extracted as a measure of the dipole field in crystal. The tested positions of the "probe" impurities in crystal are indicated in Fig. 4, and the

**TABLE 3.** Lattice relaxation around two nearest Li impurities

| Atom | Lattice coordinates | | | Displacement |
|---|---|---|---|---|
| Li(1) | $\Delta_x$ | 0 | 0 | $\Delta_x = 0.1550$ |
| Li(2) | $1 + \Delta_x$ | 0 | 0 | $\Delta_x = 0.178$ |
| O(1) | $-\frac{1}{2} + \Delta_x$ | $\pm(\frac{1}{2} + \Delta_{yz})$ | 0 | $\Delta_x = -0.0022;\ \Delta_{yz} = 0.0022$ |
| O(2) | $\Delta_x$ | $\pm(\frac{1}{2} + \Delta_{yz})$ | $\pm(\frac{1}{2} + \Delta_{yz})$ | $\Delta_x = 0.0075;\ \Delta_{yz} = -0.0028$ |
| O(3) | $\frac{1}{2} + \Delta_x$ | $\pm(\frac{1}{2} + \Delta_{yz})$ | 0 | $\Delta_x = -0.0074;\ \Delta_{yz} = -0.0082$ |
| O(4) | $1 + \Delta_x$ | $\pm(\frac{1}{2} + \Delta_{yz})$ | $\pm(\frac{1}{2} + \Delta_{yz})$ | $\Delta_x = 0.0080;\ \Delta_{yz} = -0.0032$ |
| O(5) | $\frac{3}{2} + \Delta_x$ | $\pm(\frac{1}{2} + \Delta_{yz})$ | 0 | $\Delta_x = -0.0057;\ \Delta_{yz} = -0.0125$ |

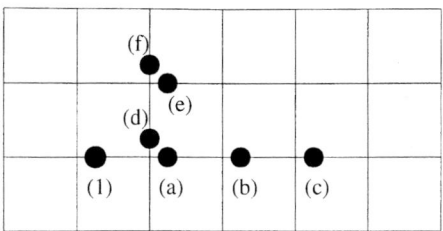

**FIGURE 4.** Distribution of interacting Li impurities in 3×3×6 extended KTaO$_3$ supercell. The interaction between Li(1) and another Li placed in a,b,c,d,e,f position is considered.

**TABLE 4.** Effective interaction energy $E_{int}$ and equilibrium displacement $\Delta$ of the second Li atom (a–f) for different mutual configurations displacements. The displacement of Li(1) is fixed at 0.62 Å along [100].

| Li pair | $E_{int}$ (meV) | $\Delta$ (Å) | Li pair | $E_{int}$ (meV) | $\Delta$ (Å) |
|---|---|---|---|---|---|
| 1–a | 62.0 | 0.72 | 1–d | 28.52 | 0.677 |
| 1–b | 10.29 | 0.633 | 1–e | 5.93 | 0.633 |
| 1–c | 7.56 | 0.629 | 1–f | 17.91 | 0.657 |

resulting interaction energies – in Table 4. One can clearly see the anisotropy of the interaction field. The numerical values may be somehow affected by the choice of even larger supercell in future calculations. It is noteworthy, however, that the interaction strength decreases with the distance faster than it was found in the shell model calculations [24].

## SUMMARY

In the present study of two different point defects in perovskites by means of a semiempirical INDO method, based on the comparison with *ab initio* calculation results, we established the following: The ground state of the $F$ center in KNbO$_3$ is associated with a strong symmetrical relaxation of two nearest Nb atoms outwards relative to the O vacancy. We presented a strong argument that the 2.7 eV absorption band observed in electron-irradiated crystals could be due to the $F$-type centers, and predicted the existence of two additional absorption bands (at 3.04 eV and 3.11 eV) for the same defect in the orthorhombic phase of KNbO$_3$. In the analysis of single and interacting off-center Li impurities in KTaO$_3$ we estimated the lattice relaxation around impurities and the characteristic interaction energies, depending on the distance between the defects and their mutual orientation. The interaction energies are lower and less long-ranged than it was estimated from earlier shell-model calculation.

## Acknowledgments

Financial support of the Deutsche Forschungsgemeinschaft (SFB 225, Graduate College) and by the Volkswagen-Stiftung is greatly acknowledged.

## REFERENCES

1. P. Günter and J.-P. Huignard (eds.) *Photorefractive Materials and Their Application* (Topics in Applied Physics, **61, 62**), Berlin, Heidelberg: Springer-Verlag, 1988.
2. L. Shiv, J. L. Sørensen, E. S. Polzik, and G. Mizell, *Optics Letters* **20**, 2271 (1995).
3. J. H. Crawford, Jr., *Nucl. Inst. Meth. B* **1** 159 (1984); J.-M. Spaeth, J. R. Niklas, and R. H. Bartram, *Structural Analysis of Point Defects in Solids* (Springer series in Solid State Sciences, vol. 43), Berlin, Heidelberg: Springer-Verlag, 1993.
4. E. R. Hodgson, C. Zaldo, and F. Agullo-López, *Solid State Commun.* **75**, 351 (1990).
5. E. A. Kotomin, R. I. Eglitis, and A. I. Popov, *J. Phys.: Condens. Matter* **9**, L315 (1997).
6. S. A. Prosandeyev, A. V. Fisenko, A. I. Riabchinski, A. I. Osipenko, I. P. Raevski, and N. Safontseva, *J. Phys.: Condens. Matter*, **8**, 6705 (1996).
7. R. I. Eglitis, N. E. Christensen, E. A. Kotomin, A. V. Postnikov, and G. Borstel, *Phys. Rev. B* **56**, 8599 (1997).
8. Y. Yacoby and S. Just, *Solid State Commun.* **15**, 715 (1974).
9. F. Borsa, U. Höchli, J. J. van der Klink, and D. Rytz, *Phys. Rev. Lett.* **45**, 1884 (1980); U. T. Höchli, K. Knorr, and A. Loidl, *Adv. Phys.* **39**, 405 (1990).
10. J. J. van der Klink and S. N. Khanna, *Phys. Rev. B* **29**, 2415 (1984).
11. M. G. Stachiotti and R. L. Migoni, *J. Phys.: Condens. Matter.* **2**, 4341 (1990).
12. M. Exner, C. R. A. Catlow, H. Donnerberg, and O. F. Schirmer, *J. Phys.: Condens. Matter* **6**, 3379 (1994).
13. A. V. Postnikov, T. Neumann, and G. Borstel, *Ferroelectrics*, **164**, 101 (1995).
14. R. E. Eglitis, A. V. Postnikov, and G. Borstel, *Phys. Rev. B* **55**, 12976 (1997).
15. A. Shluger, *Theoret. Chim. Acta* (Berl.) **66**, 355 (1985); E. Stefanovich, E. Shidlovskaya, A. Shluger, and M. Zakharov, *Phys. Status Solidi B* **160**, 529 (1990); A. Shluger and E. Stefanovich, *Phys. Rev. B* **42**, 9664 (1990).
16. R. I. Eglitis, A. V. Postnikov, and G. Borstel, *Phys. Rev. B* **54**, 2421 (1996).
17. M. Methfessel, *Phys. Rev. B* **38**, 1537 (1988); M. Methfessel, C. O. Rodriguez, and O. K. Andersen, *ibid.* **40**, 2009 (1989).
18. R. A. Evarestov and L. A. Lovchikov, *Phys. Status Solidi B* **93**, 469 (1977).
19. P. Voigt and S. Kapphan, *J. Phys. Chem. Solids* **55**, 853 (1994).
20. J. J. van der Klink and F. Borsa, *Phys. Rev. B* **30**, 52 (1992); S. Rod, F. Borsa, and J. J. van der Klink, *ibid.* **38**, 2267 (1988).
21. P. Doussineau, C. Frenois, A. Lavelut, and S. Ziolkievicz, *J. Phys.: Condens. Matter* **3**, 8369 (1991).
22. U. Höchli, J. Hessinger, and K. Knorr, *J. Phys.: Condens. Matter* **3**, 8377 (1991).
23. B. E. Vugmeister and M. D. Glinchuk, *Rev. Mod. Phys.* **82**, 993 (1990).
24. M. G. Stachiotti, R. L. Migoni, H. M. Christen, J. Kohanoff, and U. Höchli, *J. Phys.: Condens. Matter* **6**, 4297 (1994).

# LAPW vs. LMTO full-potential simulations and anharmonic dynamics of KNbO$_3$

## A. V. Postnikov and G. Borstel

*University of Osnabrück – Department of Physics, D-49069 Osnabrück, Germany*

(Received February 4, 1998)

**Abstract.** With the aim to get an insight in the origin of differences in the earlier reported calculation results for KNbO$_3$ and to test the recently proposed "NFP" implementation of the full-potential linear muffin-tin orbital (FP-LMTO) method by M. Methfessel and M. van Schilfgaarde, we perform a comparative study of the ferroelectric instability in KNbO$_3$ by FP-LMTO and full-potential linear augmented plane-wave (LAPW) method. It is shown that a high precision in the description of the charge density variations over the interstitial region in perovskite materials is essential; the technical limitations of the accuracy of charge-density description apparently accounted for previously reported slight disagreement with the LAPW results. With more accurate description of the charge density by sufficiently fine real-space grid, the results obtained by both methods became almost identical.

In order to extract additional information (beyond the harmonic approximation) from the total energy fit obtainable in total-energy calculations, a scheme is proposed to solve the multidimensional vibrational Schrödinger equation in the model of noninteracting anharmonic oscillators via the expansion in hyperspherical harmonics. Preliminary results are given for the $t_{1u}$ vibrational modes in cubic KNbO$_3$.

## INTRODUCTION

KNbO$_3$ is one of benchmark systems for *ab initio* analysis of ferroelectric perovskites. It has been extensively studied by the whole spectrum of numerical methods from an apparently ultimately accurate full-potential linear augmented plane waves (LAPW) [1–5] through pseudopotentials [6] and tight-binding *ab initio* schemes [7] to semiempirical and model-based techniques [8].

Modern state-of-art simulations of ferroelectric correlations, lattice dynamics and phase transitions are dependent on reliable and accurate description of the total energy as function of displacements and strain variables. The full-potential linear muffin-tin-orbitals (FP-LMTO) method [9] was proven able to provide a reasonable balance of accuracy and low computational effort even when applied to supercells of up to 40 atoms [10–13]. However, in sensitive benchmark calculations of Γ phonons, due to certain technical limitations, FP-LMTO provided apparently lower accuracy for phonon eigenvectors, as compared to FP-LAPW [11,3,5]. A new version of the FP-LMTO code by Methfessel and van Schilfgaarde [14], in contrast to the previous one used in our earlier calculations [9], is unsensitive to sphere packing and uses much more efficient, albeit more mathematically involved, basis of augmented "smooth Hankel functions" that enables one to drastically reduce the size of diagonalization problem without loss of accuracy. We compare the results obtained with the new FP-LMTO and with the WIEN97 implementation of the FP-LAPW method [15], concentrating on the accuracy and performance.

In order to get an additional insight into lattice dynamics of $KNbO_3$, we present vibrational frequencies as calculated quantum-mechanically in the assumption of uncoupled multidimensional anharmonic oscillators, based on the total energy data obtained from first-principles calculations, and give a preliminary estimate of the lowest vibration frequency within this approach.

## COMPARISON OF CALCULATION METHODS

The linear augmented plane-wave (LAPW) and the linear muffin-tin orbital (LMTO) methods are closely related and originate in the same work by Andersen [16]. In the modern stage at their development, they share the advantage of being all-electron methods (in contrast to pseudopotential ones and to those depending on the frozen-core approximation). As compared to tight-binding schemes with fixed (e.g. Gaussian-type) bases, the basis in LAPW and LMTO is optimized in the course of iterations, as the heads of wavefunctions are recalculated inside the (arbitrarily) chosen muffin-tin spheres from one iteration to another. The tails of basis functions that span the interstitial region between the spheres are constructed by an augmentation procedure which matches the numerical solutions inside the MT-spheres either to the plane waves of different $\mathbf{k}$ (in LAPW) or to the spherical Hankel functions of different energy (in LMTO). This difference accounts for relative advantages and disadvantages of these two calculation schemes. In LAPW, the number of augmented plane waves with different $\mathbf{k}$ can be safely saturated, until the desirable convergence of results with respect to the completeness of basis is achieved. This however has its price in terms of computational effort.

The advantage of LMTO lies in the fact that its basis functions may be tuned to resemble true atom-centered wavefunctions in a crystal. This means, in the ideal case, quite efficient and compact basis set and hence computational speed and the ability to treat larger systems than is possible with LAPW, given certain amount of computer resources. The weak point of LMTO is that the saturation of the basis set is not as straightforward as in the LAPW, because adding more atom-centered tail functions of certain angular symmetry needs care to prevent linear dependences within the basis set. In order to overcome this problem and at the same time to maintain the LMTO as, ideally, the 'minimal basis' method of competitive accuracy, one can try to experiment with more sophisticated but hopefully more efficient envelope functions. As some examples of proposed alternatives to conventional spherical Hankel functions one can single out the tight-binding LMTO method [17] or the "exact muffin-tin orbital theory" development [18], both of them being not yet implemented, to our knowledge, in workable full-potential total-energy codes.

Recently, yet another extension within the general LMTO formalism has been proposed and implemented by M. Methfessel and M. van Schilfgaarde and referred to as the "NFP" LMTO code [14]. While in many aspects a development of the earlier described [9] and widely used FP-LMTO formalism, the NFP algorithm incorporates an essential new element, that is, using the "smooth" Hankel functions rather than the "standard" ones for the augmentation of numerical radial solutions in the interstitial region. Whereas the standard spherical Hankel function satisfies the differential equation

$$(\Delta + \epsilon)h_0(r) = -4\pi\delta(\mathbf{r})$$

for $l = 0$, the equation for the smooth function contains the smeared $\delta$ function $g_0(r) = C \exp(-r^2/R_{\text{sm}}^2)$ as a source term

$$(\Delta + \epsilon)\tilde{h}_0(r) = -4\pi g_0(r) \, ,$$

and the smooth functions of higher orders are generated by applying a differential operator $\mathcal{Y}(-\nabla)$, defined by $\mathcal{Y}(\mathbf{r}) = r^l Y_L$, to the function of the 0th order. These new envelope functions can be tuned by a proper choice of the smoothing parameter $R_{\text{sm}}$ to imitate the actual shape of the wavefunction in crystal.

Cubic perovskite ferroelectrics provide an excellent benchmark system for the high-precision total-energy calculation scheme. Whereas one typically cannot pinpoint any essential disagreement in the band dispersions calculated by different full-potential schemes, the energy differences on the ~1 mRy scale related to ferroelectric instability, soft mode phonon frequencies and eigenvectors in these materials may be quite differently estimated by different computation schemes, result in qualitatively different predictions and thus dramatize the competition between different numerical approaches. For $KNbO_3$, a number of calculations has been already done using different methods. Full-potential LAPW calculations were performed by Singh [1–3] and Krakauer et al [4,5], LMTO calculations by us [10–13]. In spite of the overall agreement (the ability of both methods to account for the ferroelectric instability, consistent results for the Γ-frozen phonon frequencies), some disagreements prevailed in the description of ferroelectric instability, as well as in the estimation of the soft mode phonon eigenvector [11,3,5].

What makes perovskites generally (and $KNbO_3$ as an example) a hard test for any computational scheme relying on a site-centered basis, like LMTO, is their relatively loose structural packing (if one thinks in terms of nominal ionic radii). The electron density is unevenly distributed between compact $NbO_6$ octahedra and intermediate large cavities which host small K ions. In our earlier LMTO calculations using the "old" LMTO code by Methfessel [9], good sphere packing was essential for accurate integration over the interstitial region, but could not be guaranteed in a completely satisfactory way (see, e.g. the discussion in Ref. [13]).

**TABLE 1.** NFP-LMTO calculation setup ($\kappa^2$ in Ry, $R_{MT,SM}$ in a.u.)

| \multicolumn{3}{c}{K ($R_{MT}$=1.95)} | | | Nb ($R_{MT}$=1.85) | | | O ($R_{MT}$=1.55) | |
|---|---|---|---|---|---|---|---|---|
| $n,l$ | $\kappa^2$ | $R_{SM}$ | $n,l$ | $\kappa^2$ | $R_{SM}$ | $n,l$ | $\kappa^2$ | $R_{SM}$ |
| 4s | −0.5 | 3.0 | 5s | −0.1 | 2.0 | 2s | −0.5 | 0.79 |
| 3p | −0.5 | 1.3 | 4p | −1.9 | 0.9 | 2p | −0.15 | 0.71 |
| 3d | −0.2 | 3.4 | 4d | −0.5 | 1.2 | s,p,d | −0.2 | 2.0 |
| s,p | −0.2 | 2.0 | s,p,d | −1.0 | 1.5 | | | |

In the present implementation of LMTO, the aspect of good sphere packing is no more sensitive, therefore we tried to figure out finally whether the mentioned disagreement with the LAPW results was due to problems of technical inefficiency (unadequate integration scheme etc.) or has to do with some basic limitations of the LMTO formalism.

In our present benchmark calculations, we used the implementation of the full-potential LAPW method knows as the WIEN97 code [15]. Sphere sizes were chosen as shown in the Table 1 (the same for LAPW and LMTO). The **k**-space integration was performed in an identical way in both schemes, using the sampling on a mesh of 18 inequivalent points, corresponding to 6×6×6 divisions of the full Brillouin zone. This mesh was found to be sufficiently dense for the estimations of ferroelectric instability, based on previous experience [2]. The exchange-correlation was treated either in the local density approximation (LDA) or in the generalized gradient approximation (GGA).

In the construction of the LMTO setup, the usual procedure is to optimize the possibly minimal basis, using the freedom in the choice of basis parameters, and then to expand the basis in order to ensure its sufficient completeness. In the NFT formalism, the quality of the basis depends on both energies of the smooth Hankel functions and their smoothing radii, the latter apparently having more pronounced effect. In contrast to earlier FP-LMTO scheme [9] which favored the basis functions with at least three tail energies per orbital for sufficient accuracy, the NFP provides a reasonable description of the valence band states with augmenting just one smooth Hankel function to each; the setup is then refined by adding some other tail functions. The calculation setup we used in the present LMTO calculation is shown in Table 1. It includes 70 basis functions, that can be either somehow more extended, or reduced, to get the desirable compromise between the accuracy and performance. For comparison, the basis size for a LAPW calculation of comparable accuracy should include at least ∼800 augmented plane waves.

The present version of NFP incorporates only the LDA treatment of the exchange/correlation. Another technical drawback is the impossibility to treat the states with different principal quantum number and the same orbital quantum number within the valence band. For KNbO$_3$, we had to include Nb4p states and neglect Nb5p in the valence panel. This seems to be an acceptable compromise, however. More discussion related to the LMTO setup may be found in Ref. [10].

# CALCULATION RESULTS AND DISCUSSION

Fig. 1 shows the energy/volume curves as calculated by LAPW in the LDA and in the GGA, and by LMTO in the LDA. First two curves essentially reproduce previous results by Singh [2] aimed at the comparison of LDA and GGA. The energy/volume curve generated now with LMTO (crosses in Fig. 1) practically coincides with that obtained with the LAPW. Absolute energy values lie by ~0.9 Ry lower with LAPW, understandably due to more complete basis.

For the study of ferroelectric instability, we concentrated on the displacement pattern compatible with the $t_{1u}$ TO phonon modes, i.e. the $z$-displacements of K(0 0 0), Nb($\frac{1}{2}\frac{1}{2}\frac{1}{2}$) and O$_{II}$ ($\frac{1}{2}\frac{1}{2}0$) with respect to two equivalent O$_{I}$ ($0\frac{1}{2}\frac{1}{2}$) and ($0\frac{1}{2}\frac{1}{2}$) atoms. In Fig. 2, the energy differences are shown as function of the displacement pattern which roughly corresponds to that in the soft mode, ultimately resulting in the equilibrium structure of tetragonal ferroelectric phase: Nb is displaced twice farther as K relatively to the oxygen cage. We found the calculated energy differences to be extremely sensitive to the quality of the charge density representation in the unit cell. In both methods we used, this expansion is done by the fast Fourier transformation; in WIEN97, the magnitude of the largest reciprocal-space vector $G$ is specified whereas in the NFP-LMTO the number of divisions $N$ along each unit cell edge for a real-space uniform grid has to be explicitly provided. For a perovskite, both cutoffs need to be relatively large in order to achieve a convergency in

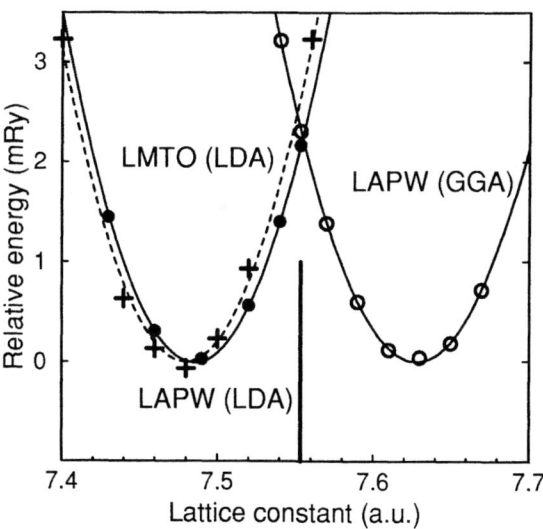

**FIGURE 1.** Total energy difference (with respect to the equilibrium value) depending on the lattice constant in KNbO$_3$ as calculated by LMTO-LDA (crosses), LAPW-LDA (filled circles) and LAPW-GGA (open circles). The parabolic fit is shown for each case. Experimental lattice constant extrapolated to zero temperature is indicated by a vertical line.

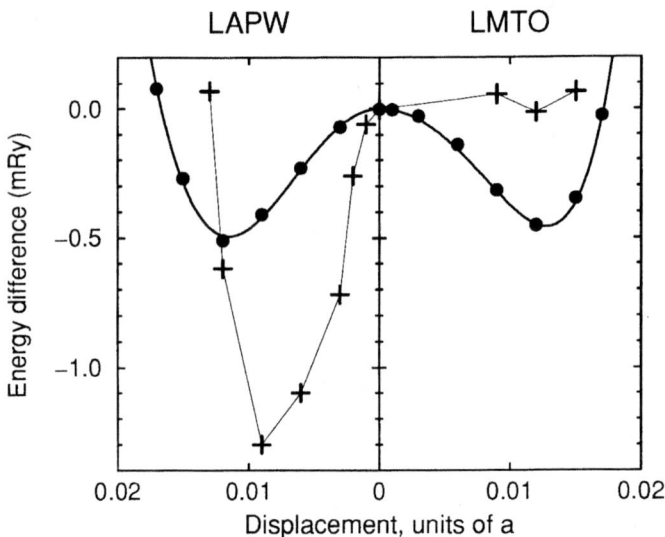

**FIGURE 2.** Total energy difference depending on atomic displacements $\Delta_z(K) + 2\Delta_z(Nb)$ as calculated by LAPW and LMTO for different values of $\Delta_z$ and different cutoffs in the charge density expansion. Crosses: $G=12.0$ in LAPW and $N=18$ in LMTO; dots: $G=14.0$ in LAPW and $N=24$ in LMTO (essentially converged results). Polynomial fit is a guide to eye.

this parameter. As is seen in Fig. 2, the value $G = 10$ in LAPW overestimates the ferroelectric instability whereas the LMTO with $N = 18$ finds yet no trace of the instability. At $G = 12$ and $N = 24$, the trends are about comparable, becoming even closer for $G = 14$ and $N = 32$.

As an additional test for the proper balance of the energetics of different displacement patterns, we calculated the $\Gamma$-TO phonons in the cubic phase of KNbO$_3$. The insufficient accuracy of previous LMTO calculations [11,12] manifested itself as a noticeable deviation of the vibration eigenvectors from those obtained by LAPW

**TABLE 2.** Calculated frequencies and eigenvectors of $\Gamma$-TO phonons of the $t_{1u}$ symmetry from LAPW and LMTO

| $\omega$ (cm$^{-1}$) | Eigenvectors | | | | $\omega$ (cm$^{-1}$) | Eigenvectors | | | |
|---|---|---|---|---|---|---|---|---|---|
| | K | Nb | O$_I$ | O$_{II}$ | | K | Nb | O$_I$ | O$_{II}$ |
| | LAPW, Ref. [1] | | | | | LMTO, Ref. [11] | | | |
| 115$i$ | 0.04 | −0.60 | 0.40 | 0.56 | 203$i$ | 0.32 | −0.67 | 0.29 | 0.53 |
| 168 | −0.88 | 0.35 | 0.19 | 0.16 | 193 | −0.81 | 0.12 | 0.36 | 0.27 |
| 483 | 0.03 | −0.09 | 0.46 | −0.75 | 459 | 0.13 | −0.14 | 0.45 | −0.75 |
| | LAPW, Ref. [5] | | | | | NFP-LMTO, present results | | | |
| 197$i$ | 0.01 | −0.59 | 0.42 | 0.55 | 106$i$ | 0.07 | −0.61 | 0.45 | 0.48 |
| 170 | −0.88 | 0.37 | 0.18 | 0.15 | 179 | −0.88 | 0.33 | 0.21 | 0.17 |
| 473 | 0.02 | −0.08 | 0.46 | −0.76 | 518 | 0.03 | −0.03 | 0.41 | −0.81 |

[1,5]. Table 2 shows the results of some preliminary calculations with the NFP code (obtained with the 2d-order total energy fit over the results for several combined displacements) in comparison with the LAPW data. One can see that the correct displacement pattern within the soft mode is now restored, and the overall agreement with the LAPW eigenvectors is quite satisfactory.

## TREATMENT OF ANHARMONIC VIBRATIONS

With the total-energy fit generally available from first-principles calculations, one may tend to extract some additional information than is possible within the harmonic approximation. The treatment of anharmonic effects in crystal is rather complicated (see, e.g., Ref. [19] for a review). In principle, the modes of different symmetry and related to different **q**-values couple beyond the harmonic approximation. Nevertheless, in the study of ferroelectrics there have been several attempts to single out any particular mode, which is believed to be principally associated with anharmonic effects, and to solve the vibrational Schrödinger equation related to it. This has been done e.g. for $LiTaO_3$ and $LiNbO_3$ by Inbar and Cohen [20] and by Bakker et al. [21] (for an empirical potential well, in the latter case) as well as for a one-dimensional $A_2$ mode in orthorhombic $KNbO_3$ by Postnikov and Borstel [12]. This approach was referred to as non-interacting anharmonic oscillators [20], meaning the oscillators related to a particular Γ TO-mode in crystal. It is assumed that the displacement potential is separable into components with different **q**-dependence. Such separation is less valid for several symmetry coordinates which mix already in the harmonic approximation, therefore the solution of a multidimensional oscillator problem is necessary in this case. A straightforward treatment by, e.g., a finite-difference method on a multidimensional grid rapidly becomes prohibitive with a number of dimensions (see [12]). Therefore, we propose a scheme which uses the expansion in hyperspherical harmonics. This approach is known in the calculation of vibration spectra of three-atomic molecules [22], however, for an *ad hoc* constructed system of variables. In the following, we describe the formalism for an arbitrary number of symmetry coordinates.

We start with an arbitrary convenient set of symmetry-adapted displacement coordinates (see, e.g., Ref. [23]): $S_t = \sum_i B_{ti} x_i$ ($x_i$ are conventional cartesian displacements), which form a complete basis within a particular irreducible representation, but do not need to be orthonormal. The Schrödinger equation then acquires a form:

$$\left[ -\frac{\hbar^2}{2} \sum_{tt'} \frac{\partial}{\partial S_t} G_{tt'} \frac{\partial}{\partial S_{t'}} + V(\{S_t\}) \right] \Psi = E\Psi, \tag{1}$$

with the kinetic-energy matrix $G_{tt'} = \sum_i B_{ti} m_i^{-1} B_{t'i}$. Mixed derivatives can further be excluded by the following orthogonalizing transformation:

$$Q_t = \sum_{t'} \frac{X_{t't}}{\sqrt{\lambda^t}} S_{t'},$$

where $X_{t't}$ is the $t$-th eigenvector, corresponding to the eigenvalue $\lambda^t$, of the kinetic-energy matrix. In $n$-dim. space of generalized coordinates $Q_t$, we use spherical coordinates $(r, \vartheta_1, \cdots, \vartheta_p, \varphi)$ for $p = n-2$,

$$Q_1 = r \cos \vartheta_1$$
$$Q_2 = r \sin \vartheta_1 \cos \vartheta_2$$
$$\cdots$$
$$Q_{p+1} = r \sin \vartheta_1 \sin \vartheta_2 \cdots \sin \vartheta_{p-1} \sin \vartheta_p \cos \varphi$$
$$Q_{p+2} = r \sin \vartheta_1 \sin \vartheta_2 \cdots \sin \vartheta_{p-1} \sin \vartheta_p \sin \varphi$$
$$0 \leq \vartheta_j \leq \pi \quad (j = 1, 2, \cdots, p), \quad 0 \leq \varphi \leq 2\pi.$$

There are $(2N+p)(N+p-1)!/(p!\,N!)$ harmonic polynomials of degree $N$ numbered by $n-1$ integers $m_0, \cdots, m_p$ such that $N = m_0 \geq m_1 \geq \cdots \geq |m_p| \geq 0$, $m_p = \pm|m_p|$. The explicit form of the polynomials of is the following:

$$H(N, m_1, ..., m_p; Q_1, ..., Q_{p+2}) = r^N Y_{N, m_1, ..., m_{p-1}}^{\cdot m_p}(\vartheta_1, ..., \vartheta_p, \varphi).$$

The hyperspherical harmonics are chosen either as complex functions

$$Y_{m_0,..,m_{p-1}}^{m_p}(\vartheta_1,..,\vartheta_p,\varphi) = e^{im_p\varphi} \prod_{k=0}^{p-1} (\sin\vartheta_{k+1})^{m_{k+1}} C_{m_k - m_{k+1}}^{m_{k+1} + \frac{p-k}{2}}(\cos\vartheta_{k+1}),$$

or in the real form, $\sim \cos m_p$ for $m_p \geq 0$ and $\sim \sin m_p$ for $m_p < 0$. $C_n^p(z)$ are Gegenbauer polynomials [24,25], generated by the following recursion:

$$C_0^p(z) = 1; \quad C_1^p(z) = 2pz; \quad (n+1)C_{n+1}^p(z) = 2(n+p)\,z\,C_n^p(z) - (n+2p-1)\,C_{n-1}^p(z).$$

$Y_{m_0,..,m_{p-1}}^{m_p}(\vartheta_1,..,\vartheta_p,\varphi)$, orthogonal on a unit sphere, are eigenfunctions of the multi-dimensional Laplace operator:

$$\Delta Y = -\frac{m_0(m_0 + p)}{r^2} Y.$$

With the potential and the wave function expanded in hyperspherical harmonics

$$V(\vec{Q}) = \sum_{m_0,...m_p} V_{m_0,..,m_p}(r) Y_{m_0,..,m_p}(\vartheta_1,...,\vartheta_p,\varphi),$$

$$\Psi(\vec{Q}) = \sum_{m_0',...m_p'} R_{m_0',...,m_p'}(r) Y_{m_0',...,m_p'}(\vartheta_1,...,\vartheta_p,\varphi),$$

the multidimensional Schrödinger equation (1) transforms into a system of $\sum_{N=0}^{N_{\max}} (2N + p)(N+p-1)!/(p!\,N!)$ coupled 1-dimensional equations:

$$-\frac{\hbar^2}{2} \frac{1}{r^{p+1}} \frac{d}{dr}\left[r^{p+1} \frac{dR_{[m]}(r)}{dr}\right] + \frac{m_0(m_0+p)}{r^2} R_{[m]}(r) +$$
$$+ \sum_{[m''][m']} V_{[m'']}(r) R_{[m']}(r) \int Y_{[m'']} Y_{[m']} Y_{[m]} d\Omega = E\, R_{[m]}(r). \qquad (2)$$

The expansion of the potential (provided in a polynomial form by a fit to total-energy values) is finite whereas for the wave function a cutoff value $N_{max}$ has to be introduced.

As a practical example of this approach, we considered the solution of a 3-dimensional oscillator problem corresponding to the vibration pattern within the $t_{1u}$ mode in cubic $KNbO_3$. For the symmetry coordinates as discussed above, we included the 4th power of the Nb displacement into the total energy fit. (For real applications, one should of course consider some other degrees of freedom beyond the harmonic approximation). The system of coupled equations (2) was solved by a finite difference method, with 50 points in the equidistant radial mesh from up to $r = 5.0$ where a boundary condition $R(r) = 0$ was imposed on radial wavefunctions (this scheme may be somehow refined in more precise calculations, incorporating a nonuniform mesh). For the maximal degree of polynome $N = 8$ in the wave function expansion, the energy difference between two lowest oscillator levels practically converged to 70 cm$^{-1}$. The convergence is more slow for higher levels.

## CONCLUSIONS

We compare in the present paper the results obtained for the ferroelectric instability in $KNbO_3$ with two methods, LAPW and LMTO, both of which have been applied to this system before but apparently never underwent a thorough comparison with the, as far as possible, identical calculation setup. The result of this comparison is that not only the energy/volume curves are identical in the LDA, but the description of the ferroelectric instability, involving equilibrium displacements of ~0.1 a.u. and energy differences of ~0.5 mRy, is practically identical by both schemes, provided the sufficient accuracy in the description of the charge density variations over the unit cell is guaranteed. The LAPW method provides understandably lower absolute values of the total energy, but the new formulation of LMTO has the advantage of much more compact basis set (about 10 times smaller than that of LAPW) and some resources to expand the basis somehow for even better controllable accuracy without running into numerical problems of overcompleteness. As a useful tool for the analysis of total energy data obtained from any first-principle calculation, we describe the scheme to solve the multi-dimensional vibrational Schrödinger equation in the approximation of non-interacting anharmonic oscillators. The preliminary results for the lowest energy difference are presented for $KNbO_3$.

### Acknowledgments

We thank M. Methfessel for providing us with his new version of the LMTO code and for numerous helpful discussions. We are grateful to K. Schwarz and P. Blaha for making the WIEN97 code available to us. C. O. Rodriguez valuably contributed in working discussions, e.g., on applying the LAPW method to $KNbO_3$. A.P. is grateful to D. Vanderbilt for stimulating discussions on lattice dynamics. Financial support of the Deutsche Forschungsgemeinschaft (SFB 225) is gratefully acknowledged.

# REFERENCES

1. D. J. Singh and L. Boyer, *Ferroelectrics* **136**, 95 (1992).
2. D. J. Singh, *Ferroelectrics* **164**, 143 (1995).
3. D. J. Singh, *Ferroelectrics* **194**, 299 (1997).
4. R. Yu and H. Krakauer, *Phys. Rev. Lett.* **74**, 4067 (1995).
5. Ch.-Zh. Wang, R. Yu, and H. Krakauer, *Phys. Rev. B* **54**, 11161 (1996).
6. W. Zhong, R. D. King-Smith, and D. Vanderbilt, *Phys. Rev. Lett.* **72**, 3618 (1994); R. D. King-Smith and D. Vanderbilt, *Phys. Rev. B* **49**, 5828 (1994).
7. S. Dall'Olio, R. Dovesi, and R. Resta, *Phys. Rev. B* **56**, 10105 (1997).
8. M. Sepliarsky, M. G. Stachiotti, and R. L. Migoni, *Phys. Rev. B* **52**, 4044 (1995); R. I. Eglitis, A. V. Postnikov, and G. Borstel, *Phys. Rev. B* **54**, 2421 (1996).
9. M. Methfessel, *Phys. Rev. B* **38**, 1537 (1988); M. Methfessel, C. O. Rodriguez, and O. K. Andersen, *ibid.* **40**, 2009 (1989).
10. A. V. Postnikov, T. Neumann, G. Borstel and M. Methfessel, *Phys. Rev. B* **48**, 5910 (1993).
11. A. V. Postnikov, T. Neumann, and G. Borstel, *Phys. Rev. B* **50**, 758 (1994).
12. A. V. Postnikov and G. Borstel, *Phys. Rev. B* **50**, 16403 (1994).
13. A. V. Postnikov and G. Borstel, *Ferroelectrics* **194**, 69 (1997).
14. M. Methfessel, *NFP Manual 1.00*, Institute for Semiconductor Physics, Frankfurt (Oder), 1997.
15. P. Blaha, K. Schwarz, and J. Luitz, *WIEN97*, Vienna University of Technology, 1997. [Improved and updated UNIX version of the original copyrighted WIEN-code by P.Blaha *et al.*, *Comput. Phys. Commun.* **59**, 399 (1990)].
16. O. K. Andersen, *Phys. Rev. B* **12**, 3060 (1975).
17. O. K. Andersen and O. Jepsen, *Phys. Rev. Lett.* **53**, 2571 (1984).
18. O. K. Andersen, O. Jepsen, and G. Krier, *Exact Muffin-Tin Orbital Theory*, unpublished.
19. G. Leibfried and W. Ludwig, *Solid State Phys.* **12**, 276 (1961).
20. I. Inbar and R. E. Cohen, *Ferroelectrics* **164**, 45 (1995).
21. H. J. Bakker, S. Hunsche, and H. Kurz, *Phys. Rev. B* **48**, 9331 (1993); *ibid.* **50**, 914 (1994).
22. J. Makarewicz, *J. Phys. B: At. Mol. Opt. Phys.* **21**, 1803 (1988); R. W. Whitnell and J. C. Light, *J. Chem. Phys.* **90**, 1774 (1989); T. R. Horn, R. B. Gerber, and M. A. Ratner, *ibid.* **91**, 1813 (1989); S. Carter and W. Meyer, *ibid.* **100**, 2104 (1994).
23. R. E. Cohen and H. Krakauer, *Phys. Rev. B* **42**, 6416 (1990).
24. A. Erdélyi (Ed.), *Higher Transcendental Functions*, Vol .I, p. 175; Vol. II, p. 280, New York: McGraw-Hill, 1953, 1954.
25. M. Takeuchi, *Modern Spherical Functions*, Providence, Rhode Island: American Mathematical Society, 1994.

# Self-Consistent Atomic Deformation Calculations for Strontium Titanate

## L. L. Boyer

*Complex Systems Theory Branch, Naval Research Laboratory, Washington, D.C. 20375-5345*

## H. T. Stokes

*Department of Physics and Astronomy, Brigham Young University, Provo, Utah 84602*

## M. J. Mehl

*Complex Systems Theory Branch, Naval Research Laboratory, Washington, D.C. 20375-5345*

January 29, 1998

**Abstract.** The self-consistent atomic deformation method employs density functional theory by expressing the total electronic density as a sum over "atomic" densities determined self-consistently from a variational treatment of the total energy. We use this method to compute electronic and vibrational properties of $SrTiO_3$. Certain modes are found to be quite sensitive to the kinetic energy resulting from overlapping densities. We find that enhancing the Thomas-Fermi energy by $\sim 4\%$ gives good agreement with experimental results for phonon frequencies and lattice parameter.

## I INTRODUCTION

We have developed a self-consistent atomic deformation (SCAD) method for applying density functional theory. In this approach the electronic density is expressed as a sum over "atomic" densities,

$$n(\mathbf{r}) = \sum_i n_i(\mathbf{r} - \mathbf{R}_i), \qquad (1)$$

and the total energy is given by

$$E[n(\mathbf{r})] = \sum_i T_0[n_i(\mathbf{r})] + T_k[n(\mathbf{r})] - \sum_i T_k[n_i(\mathbf{r})] + F[n(\mathbf{r})]. \qquad (2)$$

Each atomic-like density $n_i$ is determined from the solutions of a one electron Schrodinger's equation with a potential

$$v_i(\mathbf{r}) = v_F[n(\mathbf{r})] + v_k[n(\mathbf{r})] - v_k[n_i(\mathbf{r})],  \qquad (3)$$

derived variationally from the total energy. [1] In the above expressions $T_0[n_i(\mathbf{r})]$ is the kinetic energy of non-interacting electrons centered about the site at $\mathbf{R}_i$, $T_k$ is a functional to account for the kinetic energy due to overlapping densities, $F$ denotes all non-kinetic (i.e. exchange-correlation and electrostatic) contributions to the total energy and $v_F$ ($v_k$) are the functional derivatives of $F$ ($T_k$). The self-consistent solution for $v_i$ (and hence, $n_i$ and $n$), obtained by occupying the lowest one electron energy levels for the entire system, allowing for charge transfer when indicated, minimizes the total energy in accord with Janak's theorem [2]. In this sense the SCAD method follows the more familiar Kohn-Sham [3] formulation of density functional theory. It can also be considered an extension of the Gordon-Kim [4] method. In fact, some of the earliest density functional formulations of total energy were done in terms of overlapping atomic densities. [5]

The SCAD method has been discussed previously [6] and a detailed account of the method for spherical atoms has been presented [7].

## II  APPLICATION

In principle, the procedure for obtaining a self-consistent solution for the total energy is straight forward. Given $n_i$, the potentials $v_i$ are determined from Eqs. (1) and (3). New densities are obtained by solving the one electron Schroedinger equations for the $v_i$, and the process is repeated until convergence is achieved. In practice, many complicated details must be worked out in order to obtain total energies sufficiently free of numerical noise to use in frozen phonon calculations. A complete account is currently being written [8]. Here we give a brief discussion of the method as we have applied it to $SrTiO_3$.

Since $SrTiO_3$ is ionic, the potentials have important contributions from all ions in the crystal. Contributions from ions with no significant charge overlap are included as point monopoles, dipoles and quadrupoles using an Ewald-type procedure. In our case, it is sufficient to treat neighbors with separation greater than the lattice parameter $a$ as point poles.

The calculations reported here assume a kinetic energy functional of the Thomas-Fermi form,

$$T_k = A \frac{\pi^{4/3} 3^{5/3}}{10} \int n^{5/3}(\mathbf{r}) d\mathbf{r}, \qquad (4)$$

where $A$ is a scaling parameter for the electron gas [9] and $T_k = T_{TF}$ for $A = 1$. We also have begun to consider functionals of the form studied by Lacks and Gordon [10]. The local density exchange-correlation functional of Hedin and Lundqvist [11] is used.

We solve Schrodinger's equation for a given potential $v_i$ using a basis with radial functions given by Clementi and Roetti [12]. Angular dependence of the wave function is obtained by multiplying the radial functions by spherical harmonics ($Y_{lm}$) with $l \geq l_r$ where $l_r$ identifies the radial function. Extra basis functions, up to $l = 2 > l_r = 1$ for the Sr p basis, and, up to $l = 3$ for the oxygen p basis, are included. No additional $Y_{lm}$'s are needed for the Ti basis. Inclusion of higher $l$ spherical harmonics in the basis was tested and found to unimportant.

## III  RESULTS

The results presented here illustrate two main points. 1) The energetics of certain structural distortions of cubic $SrTiO_3$ are quite sensitive to the particular kinetic energy functional. We illustrate this by presenting results obtained using two different values for $A$. 2) The SCAD method can provide a good overall agreement with experimentally known properties of $SrTiO_3$. Particularly, when $A$ is chosen to be about 4% larger than the Thomas-Fermi value we obtain good overall agreement with experimental phonon frequencies and their dependence on the lattice parameter $a$ is consistent with the observed low temperature structural transformation.

### A  Electronic Structure

The SCAD results give a completely ionic picture for the electronic structure, with net monopole charges of +2, +4 and -2, respectively for the strontium, titanium and oxygen ions. The uppermost energy levels, obtained for $A = 1$ and 1.042, with $a = 7.3$ Bohr (which is near the experimental value), are listed in Table 1.

TABLE 1. Uppermost energy levels (in Hartree) for $Sr^{+2}$, $Ti^{+4}$ and $O^{-2}$ ions in the cubic perovskite structure with $a = 7.3$ Bohr obtained using kinetic energy parameters $A = 1$ and 1.042.

| Level ID | $A = 1$ | | $A = 1.042$ | |
|---|---|---|---|---|
| Ti 3p | -1.343 | | -1.369 | |
| Sr 4s | -1.274 | | -1.262 | |
| O 2s | -0.759 | | -0.741 | |
| Sr 4p | -0.657 | | -0.645 | |
| O 2p | -0.319 | -0.220 | -0.300 | -0.201 |
| Ti 3d | -0.021 | 0.014 | -0.042 | -0.006 |

The Ti d levels are well above the highest occupied (oxygen p) levels. This result, which leads to the completely ionic monopole charges, holds for any reasonable values of $A$ and $a$. On the other hand, band structure methods produce some charge with Ti d character that has been interpreted as covalent bonding between the Ti and O ions. [13] In principle, the SCAD model could have some charge occupying the Ti d states, but the energetics simply do not allow it. If we employ the spherical SCAD method (SSCAD) the oxygen p and Ti d levels are brought into coincidence with a small amount of charge transfer from oxygen to the Ti d states. This happens as well for $BaTiO_3$. [6,7] Fortunately, this does not happen for the general SCAD results. Partially occupied levels would imply metallic behavior – in effect, providing a nonzero density of states at the Fermi level. As such, electric fields would be completely screened, and therefore, no splitting would occur between longitudinal and transverse optic modes. In the SCAD model the $Sr^{+2}$ and $Ti^{+4}$ ions are nearly spherical while the oxygen ions are highly deformed by large quadrupole moments that put additional charge density between the titanium and oxygen ions.

## B  $E$ vs. $a$

We have investigated the dependence of the total energy $E(a)$ on the kinetic energy parameter $A$. As expected, increasing $A$ increases the total energy and moves the minimum to larger values of $a$. An approximately 4% increase from the Thomas-Fermi value gives an equilibrium lattice parameter close to the $a = 7.36$ Bohr experimental value [14]. Results for two selected values of $A$ are shown in fig.1

Except for the vertical and horizontal shifts the two curves are nearly the same, with bulk moduli of 2.36 and 2.17 Mbar respectively for $A = 1$ and $A = 1.042$. For comparison, the room temperature experimental value is 1.77 Mbar. [15]

## C  Frozen Phonons

Strontium titanate provides a good test of the SCAD method because it has competing structural instabilities and the lattice dynamics of this compound has been widely studied: experimentally, [16,17] by models fit to experimental results [16–18] and by first principles band structure methods [19,20] A structural transition occurs at 90 K that corresponds to a soft mode at the R-point. It also exhibits mode softening at $\Gamma$ from a would-be ferroelectric instability.

Frozen phonon calculations have been carried out for selected values of $a$ and $A$. The procedure involves computation of the energy change for small displacements consistent with the symmetry of the mode studied. Tabulated symmetry data [21] were used to facilitate and automate this aspect of the calculation. For each distortion the energy was determined for two amplitudes, corresponding to displacements of $\sim 0.04$ and $\sim 0.08$ Bohr to be sure we were investigating the harmonic part of the total energy surface. These are small displacements that require highly precise energy differences. Calculations were carried out for modes at $\Gamma$, R, M and X of the Brillouin zone. To identify modes in the subsequent discussion we use labeling convention of Miller and Love [22], subsequently employed by Stokes and Hatch [21]. These labels are matched in table 2. with those of Cowley [16].

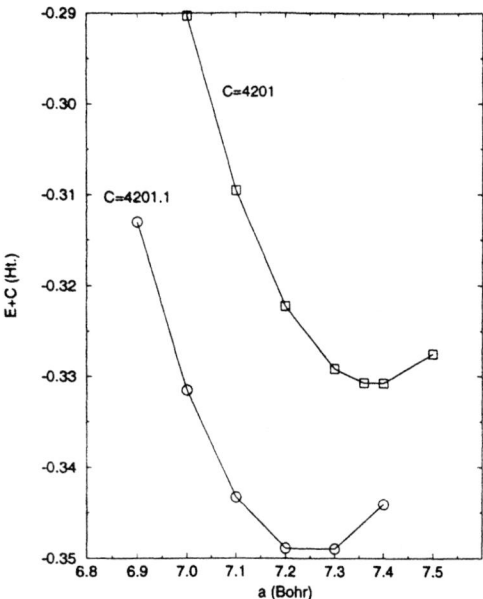

**FIGURE 1.** Total energy $E$ as a function of lattice parameter $a$ for $SrTiO_3$ with kinetic energy parameters $A = 1$ (circles) and $A = 1.042$ (squares). $C$ indicates the amount the numbers are shifted on the vertical axis.

At $\Gamma$ there is one triply degenerate Raman active mode ($\Gamma_5^-$) and three (or four, if you count the zero frequency translation mode) $\Gamma_4^-$ modes which, in the absence of macroscopic electric fields, are also triply degenerate. In fact, they are split by the macroscopic electric field that accompanies the longitudinal polarization wave associated with these modes. The splitting between longitudinal (L) and transverse (T) is produced by a term in the dynamical matrix that depends on the Born effective charges ($Z^*$) for each atom and the high frequency dielectric constant ($\epsilon_\infty$). [23] The $Z^*$'s give the polarization produced by a small displacement of the ions while $\epsilon_\infty$ derives from the polarization induced by an external macroscopic field with ions in their undisplaced positions. Either quantity is straight forward to calculate within the SCAD framework because charge is localized on ions and the moments of charge are determined uniquely in the calculation. SCAD calculations of $Z^*$ have already been discussed. [6] Calculation of $\epsilon_\infty$ is accomplished simply by adding a small term to each potential $v_i$, proportional to $Y_{1,0}$, for example, for an electric field in the $z$ direction. The resultant induced dipole moments give the susceptibility, which in turn, yields the dielectric constant. Results obtained for

**TABLE 2.** Comparison of the symmetry notation of Stokes and Hatch(SH), used here, with that of Cowley(C).

| Γ | | R | | M | | X | |
|---|---|---|---|---|---|---|---|
| SH | C | SH | C | SH | C | SH | C |
| 4− | 15 | 2− | 2' | 1+ | 1 | 1+ | 1 |
| 5− | 25 | 3− | 12' | 2+ | 3 | 2+ | 3 |
| | | 4− | 15 | 2− | 3' | 3− | 2' |
| | | 5+ | 25' | 3+ | 2 | 5+ | 5 |
| | | 5− | 25 | 3− | 2' | 5− | 5' |
| | | | | 4+ | 4 | | |
| | | | | 5+ | 5 | | |
| | | | | 5− | 5' | | |

Born effective charges and high frequency dielectric constant are shown in Fig. 2. O(I) (II) identifies $Z^*$ for displacement of oxygen toward (perpendicular to) a Ti neighbor. Our values of $Z^*$ are in reasonably good agreement with those derived from band structure calculations. For comparison, Zhong et. al. [19] (Lasota et. al. [20]) find values of 2.54(2.55), 7.12(7.56), -5.66(-5.92) and -2.00(2.12) respectively for Sr, Ti, O(I) and O(II). Our values for Ti and O(I) are somewhat smaller in magnitude than the band structure results. Our calculated $\epsilon_\infty$ is 30 to 35% smaller than the experimental value and nearly 50% smaller than the band structure value [20].

Results for frozen mode frequencies at the Γ-point are shown in figure 3. The frequency of the lowest $\Gamma_4^-(T)$ (ferroelectric) mode, is strongly dependent on $a$, with the ferroelectric instability developing with increasing $a$. This behavior is consistent with pressure dependent studies [24]. Even though increasing the kinetic energy parameter $A$ increases the equilibrium volume (fig.1), tending to promote ferroelectric instability, it has an even greater direct effect in stabilizing the ferroelectric mode. For example, as indicated in figure 3, the ferroelectric mode frequency is shifted from $245i$ to 160 at $a = 7.2$ by changing $A$ from 1 to 1.042. The stabilizing effect of the larger $A$ value is just enough to stabilize the ferroelectric mode at the computed equilibrium lattice parameter, which happens to correspond to the measured value, $a = 7.36$ Bohr. This agrees qualitatively with experiments which find a soft, but stable, ferroelectric mode in strontium titanate [17].

Our results for frozen mode frequencies at the $R$-point are shown in Fig. 4. The ground state structure of $SrTiO_3$ is characterized by a distortion with $R_5^-$ symmetry. Thus we expect to find an unstable $R_5^-$ mode at the equilibrium value of $a$. In fact, we do, and, as can be seen from figure 4, this result is relatively independent of small variations in $A$. In addition, this instability is much less sensitive to the precise value of $a$ and has the opposite sign for pressure dependence compared to the ferroelectric mode. On the other hand, the $R_3^-$ mode, like the ferroelectric mode, is very sensitive to the kinetic energy parameter $A$.

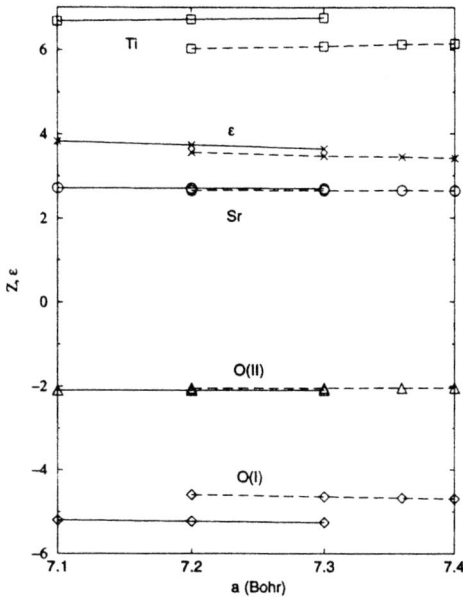

**FIGURE 2.** Results for $Z^*$ and the high frequency dielectric constant as a function of lattice parameter $a$ with kinetic energy parameter $A = 1$ (solid lines) and $A = 1.042$ (dashed lines).

The frequencies of all modes at $\Gamma$, $R$, $M$ and $X$, obtained for $A = 1.042$ and $a = 7.36$ are listed in table 3 along with Sterling's [17] model 5 (SM5), a shell-type model with parameters fit to experimental results, and the band structure (BS) results of Lasota et. al., [20] for comparison.

With the exception of two modes, $R_3^-$ and $M_3^+$, the overall agreement between our frequencies and either Sterling's Model 5 results or the band structure derived results is rather good. Another shell model of Sterling (Model 4) gives a similarly good fit to measured frequencies but with substantially higher values for the highest frequency modes, for which the only experimental value is the highest $\Gamma_4^-(L)$. Specifically, the Model 4 frequencies for $\Gamma_4^-(L)$, $R_2^-$, $M_4^+$ and $X_3^-$ are 812, 871, 841 and 752 cm$^{-1}$ respectively, in better agreement with SCAD and BS values (table 3) and the experimental $\Gamma_4^-(L)$ value (815).

The largest discrepancies between the SCAD model and the other results are for the $R_3^-$ and $M_3^+$ modes. This is understandable because these are the ones most affected by the kinetic energy parameter. (It is interesting to note that Cowley found these particular modes to be problematic as well. [25]) These discrepancies could be reduced by selecting a larger value of $A$. For example, the frequencies of the

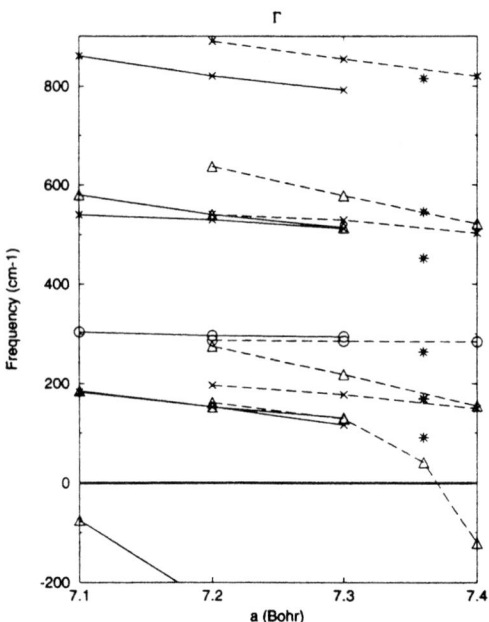

**FIGURE 3.** Results for phonon frequencies at the $\Gamma$-point as a function of lattice parameter $a$ obtained using kinetic energy parameters $A = 1$ (solid lines) and $A = 1.042$ (dashed lines). The symbols denote: $\Gamma_5^-$, o; $\Gamma_4^-(T)$, $\triangle$; $\Gamma_4^-(L)$, $\times$; and experimental values [17], $*$. Imaginary frequencies are plotted as negative real numbers.

$R_3^-$ and $M_3^+$ modes are increased to 280 and 248 cm$^{-1}$, respectively, by increasing $A$ to 1.052. These modes, along with the ferroelectric mode, have displacements that are dominated by movement of oxygen ions toward a neighboring Ti. [26] Given this similarity, it is not surprising that these modes have a similar dependence on $A$ and $a$.

## IV  SUMMARY

We have shown that the SCAD method can provide a reasonably good description of the lattice dynamics of SrTiO$_3$ if we are willing to scale the electron gas expression to increase the overlap kinetic energy by about 4% from the Thomas-Fermi value. This is crucial for stabilizing the ferroelectric mode and other modes that are dominated by motion of the oxygen ions toward their neighboring titanium ions. More complicated kinetic energy functionals [10] tend to reduce this energy,

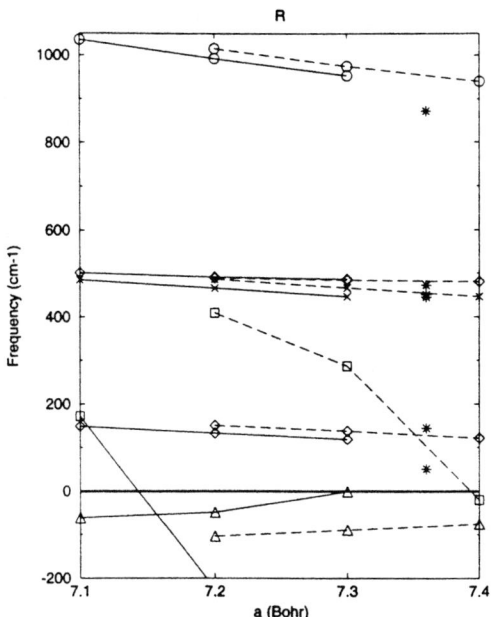

**FIGURE 4.** Results for phonon frequencies at the $R$-point as a function of lattice parameter $a$ obtained using kinetic energy parameters $A = 1$ (solid lines) and $A = 1.042$ (dashed lines). The symbols denote: $R_2^-$, ○; $R_3^-$, □; $R_4^-$, ◇; $R_5^+$, ×; $R_5^-$, △; experimental values and the $R_2^-$ Model 4 value [17], ∗. Imaginary frequencies are plotted as negative real numbers.

producing a "correction" in the wrong direction. The $R_5^-$ instability, corresponding to the observed ground state, is correctly given by the SCAD method.

We find it interesting that the ferroelectric mode and the zone boundary modes, $R_3^-$ and $M_3^+$, have similar dependences on the overlap kinetic energy and lattice parameter. These modes are driven toward instability primarily by contributions from electrostatic forces that are common to both the band structure and SCAD approaches. The overlap kinetic energy provides the stability for these modes in the SCAD method. The $R_3^-$ and $M_3^+$ modes seem to be more effectively stabilized by the band structure method. Since the major difference between the SCAD and band structure approach is in the accounting of kinetic energy, we believe a more detail analysis of these particular modes may provide greater insight into the nature of fundamental forces governing the properties of perovskite ferroelectrics.

By increasing the kinetic energy parameter a small amount from its Thomas-Fermi value $A = 1$, we obtained much better results for both the equlibrium lattice parameter and phonon frequencies. A ∼ 4% increase moved the equilibrium $a$

**TABLE 3.** Frequencies (in cm$^{-1}$) for SrTiO$_3$ at the Γ-, R-, M- and X-points of the Brillouin zone: from the SCAD model with $A = 1.042$ and $a = 7.36$ Bohr: from results of Sterling's Model 5 (SM5); and from band structure (BS) calculations of Lasota et. al.

|   | Γ SCAD | SM5 | BS |     | R SCAD | SM5 | BS |
|---|---|---|---|---|---|---|---|
| T | 41 | 87 | 90$i$ | 2− | 953 | 730 | 830 |
| T | 174 | 170 | 155 | 3− | 185 | 410 | 480 |
| T | 556 | 553 | 525 | 4− | 129 | 123 | 130 |
| L | 161 | 168 | 150 | 4− | 481 | 414 | 420 |
| L | 530 | 487 | 445 | 5+ | 454 | 478 | 430 |
| L | 833 | 723 | 750 | 5− | 83$i$ | 58 | 80$i$ |
| 5− | 285 | 264 | 220 |    |     |     |     |

|    | M SCAD | SM5 | BS |     | X SCAD | SM5 | BS |
|---|---|---|---|---|---|---|---|
| 1+ | 516 | 439 | 435 | 1+ | 289 | 297 | 270 |
| 2+ | 52$i$ | 66 | 30$i$ | 1+ | 635 | 595 | 530 |
| 2− | 170 | 107 | 100 | 2+ | 307 | 304 | 265 |
| 2− | 438 | 531 | 445 | 3− | 166 | 134 | 160 |
| 3+ | 130 | 413 | 510 | 3− | 774 | 659 | 745 |
| 3− | 101 | 120 | 95 | 5+ | 154 | 118 | 90 |
| 4+ | 883 | 712 | 780 | 5+ | 191 | 199 | 155 |
| 5+ | 231 | 302 | 310 | 5+ | 512 | 549 | 495 |
| 5− | 108 | 108 | 105 | 5− | 103 | 120 | 100 |
| 5− | 283 | 295 | 250 | 5− | 320 | 307 | 305 |
| 5− | 524 | 480 | 460 |    |     |     |     |

to near the experimental value and just stabilized the ferroelectric mode, consistent with observed softness of the ferroelectric mode at low temperature. A more fundamental justification for this kind of scaling was presented by Waldman and Gordon [9]. They determined scaling factors for rare gas systems ranging from 1.037 to 1.112, by requiring the kinetic energy functional give the correct result for the atom. Applying this approach to the valence density of $O^{-2}$ in a stabilizing charged shell of radius 2 Bohr, we obtain a value of $A = 1.056$, which is consistent with the value we selected empirically.

# REFERENCES

1. M. J. Mehl, L. L. Boyer and H. T. Stokes, J. Phys. Chem. Solids, **57** 1405 (1996).
2. J. F. Janak, Phys. Rev. B, **18** 7165 (1978).
3. W. Kohn and L. J. Sham, Phys. Rev., **140** A1133 (1965).
4. R. G. Gordon and Y. S. Kim, J. Chem. Phys., **56** 3122 (1972).
5. M. Born and K. Huang, "Dynamical Theory of Crystal Lattices" (Oxford Press, London, 1954); chapter 1 and references therein.

6. L. L. Boyer, H. T. Stokes and M. J. Mehl, Ferroelectrics, **194** 173 (1997).
7. H. T. Stokes, L. L. Boyer and M. J. Mehl, Phys. Rev. B, **54** 7729 (1996).
8. L. L. Boyer, H. T. Stokes and M. J. Mehl, in preparation.
9. M. Waldman and R. G. Gordon, J. Chem. Phys., **71** 1325 (1979).
10. D. L. Lacks and R. G. Gordon, J. Chem. Phys., **100** 4446 (1994).
11. L. Hedin and B. I. Lundqvist, J. Phys. C, **4** 2064 (1971).
12. E. Clementi and C. Roetti, Atomic Data and Nuclear Data Tables **14** 177 (1974).
13. R. E. Cohen, Nature, **358**, 136 (1992).
14. Value estimated from a room temperature value and thermal expansion data, both expressed in "Thermal Expansion of NonMettalic Solids", Vol. 13 of Thermophysical Properites of Matter, edited by Y. S. Touloukian and C. Y. Ho (Plenum, New York, 1977).
15. R. O. Bell and G. Rupprecht, Phys. Rev. **129**, 90 (1963).
16. R. A. Cowley, Phys. Rev. **134**, A981 (1964).
17. W. G. Sterling, J. Phys. C: Solid State Phys. **5**, 2711 (1972).
18. A. Bussmann-Holder, H. Bilz and G. Benedek, Phys. Rev. B. **39**, 9214 (1989).
19. W. Zhong, R. D. King-Smith and D. Vanderbilt, Phys. Rev. Lett. **72**, 3618(1994).
20. C. Lasota, C. Z. Wang, R. Yu and H Krakauer, Ferroelectrics, **194** 109 (1997).
21. H. T. Stokes and D. M. Hatch, "Isotropy Subgroups of the 230 Crystallographic Space Groups" (World Scientific, Singapore, 1988).
22. S. C. Miller and W. F. Love, "Tables of Irreducible Representations Space Groups and Co-Representations of Magnetic Space Groups" (Pruett, Boulder, 1967)
23. R. M. Pick, M. H. Cohen and R. M. Martin, Phys. Rev. B **1**, 910(1970).
24. G. A. Samara, Phys. Rev. **151**, 378(1966).
25. Cowley [16] derived shell-type models based on experimental results that included knowledge of the phonon dispersion only along the (q,0,0) direction. Without experimental values at $R$ and $M$ to temper the fit, Cowley found frequencies for the $R_3^-$ mode that changed from $\sim 350i$ to $\sim 250$ cm$^{-1}$ depending on whether the Sr and Ti charges were held at +2 and +4, or allowed to adjust to achieve a better fit. In another model, which included anharmonic effects, he found the frequencies of these modes changed by $\sim 150$ cm$^{-1}$ upon changing $T$ from 90K to room temperature. The key experimental result from Sterling's work [17] that forces a high frequency value on $R_3^-$ (and hence on $M_3^+$) appears to be the presence of two nearly degenerate modes with frequencies of $446 \pm 13$ and $449 \pm 7$ cm$^{-1}$.
26. Both the $R_3^-$ and $M_3^+$ modes are described by a 10-atom supercell in which the only displaced atoms are the oxygens at $(a/2, -\delta, a/2)$, $(-a/2, \delta, a/2)$, $(\delta, a/2, a/2)$ and $(-\delta, \pm a/2, \mp a/2)$, where the upper (lower) signs correspond to the R (M) mode, $\delta$ is the mode amplitude and the primitive lattice vectors are $[(-a, 0, a), (a, 0, a), (0, a, -a)]$ and $[(a, a, 0), (-a, a, 0), (0, 0, a)]$ respectively for R and M.

# Pressure - Induced Changes in the Local Structure of KNbO$_3$

Anatoly I. Frenkel,* Edward A. Stern† and Yizhak Yacoby‡

*Materials Research Laboratory,[1]
University of Illinois at Urbana - Champaign, Urbana, Illinois 61801
†Department of Physics Box 351560, University of Washington, Seattle, WA 98195,[2]
‡Racah Institute of Physics, Hebrew University, Jerusalem, Israel, 91904.[3]

**Abstract.**
The local structure of the perovskite KNbO$_3$ at 77 K and 300 K under high pressure, up to 15.8 GPa, has been investigated using the X-ray absorption fine structure (XAFS) technique. We found that local distortions exist throughout the measured range which are peaked off-center at the rhombohedral symmetry sites with a width narrower than the separation between the rhombohedral and orthorhombic sites. On the other hand, diffraction indicates an orthorhombic average structure at 300K, and optical measurements suggest a transition to a cubic phase at high pressures. To explain the difference between pressure - induced changes in KNbO$_3$ structure as measured by XAFS and optical techniques, a pressure - dependent hopping rate in the framework of the eight - site model is proposed.

## INTRODUCTION

Phase transitions in potassium niobate (KNbO$_3$) have been extensively studied, both experimentally and theoretically, since the discovery of its ferroelectric activity in 1949 [1]. Along with other perovskites, KNbO$_3$ has long been considered a classical example of a system undergoing purely displacive phase transitions between three low temperature ferroelectric (FE) phases with different symmetries and a high temperature paraelectric (PE) phase [2].

The question which transition mechanism, displacive or order-disorder, dominates in a particular ferroelectric, is of central importance. Within a purely displacive model, the atoms in each unit cell vibrate quasi-harmonically about their average positions in the cell. Thus the peak of the position probability distribution

---
[1] *Mailing address*: Building 510 E, Brookhaven National Laboratory, Upton, NY 11973. Electronic address: frenkel@bnl.gov. Supported by the DOE grant DEFG02-96ER45439.
[2] Supported by the DOE grant DEFG06-90ER45425.
[3] Supported by the German-Israeli binational science foundation.

function (PDF) of each atom is at its average position. This means that in the paraelectric phase the peak of the PDF is at a centrosymmetric point in the unit cell and at off center position in the ferroelectric (FE) phase. The peak position is expected to be displaced to [100], [110] and [111] directions in the tetragonal orthorhombic and rhombohedral phases, respectively. In contrast, within a pure order-disorder model, the atoms are locally displaced to off-center position in a disordered fashion even in the paraelectric phase. Namely, the atoms vibrate quasi-harmonically about off center positions and may hop among symmetry equivalent off-center positions. The hopping time is long relative to the ordinary quasi-harmonic vibration period even relative to the soft mode period. Below the ferroelectric transition temperature, $T_c$, the symmetry breaks and the probability of occupying different off-center positions is no longer equal. Thus, within an order-disorder model the PDF has peaks at off-center positions both in the PE and FE phases. In the PE phase the peaks in symmetry equivalent positions have equal intensities. As $T$ decreases below the FE transition, some peaks intensify at the expense of others.

As seen above, the comparison between the local and average structures plays a very important role in our understanding of structural phase transitions. Various experiments provide different kinds of structural information. X-ray and neutron diffraction provide detailed information on the periodic part of the structure. However if one wants to obtain information on local deviations from the average structure one needs to include this possibility explicitly in the data analysis. However, in most cases, the spatial resolution in the pair correlation function provided by this technique is insufficient in order to distinguish between a broad PDF peak at a high symmetry point and a PDF with multiple peaks at off-center positions. Resonant techniques such as NMR and ESR provide local structural information provided the distortions are slow enough to avoid motional narrowing. Thus whenever local structural distortions are observed by these techniques one also obtains an upper limit of the hopping rate. In contrast, if the existence of the local distortions is established by another, faster technique and is not observed by the resonance technique, one can set the lower limit of the hopping rate. In this sense Raman scattering is also a resonance method. If the local distortions are slow compared to a few cm$^{-1}$ the effect of the local distortions will show up either as a violation of the selection rules producing distortion induced first order Raman scattering and/or a central peak. Diffuse X-ray scattering can provide very valuable information on the distortion coherence lengths. Unfortunately these measurements can in certain cases be interpreted in different ways. Finally X-ray Absorption Fine Structure (XAFS) measurements provide local structural information and will detect local deformations even if the hopping rate is almost in the phonon range. Furthermore, the crystal momentum range involved in XAFS experiments ($q = 30\,\text{Å}^{-1}$) is so large that it allows one to determine the pair distance PDF with enough resolution to distinguish between a singly peaked broad PDF and a PDF with multiple peaks at off-center positions. Thus it can be used to determine unequivocally the existence of local structural distortions which cannot be described by quasi-harmonic vibrations.

Temperature has been the most commonly used parameter in a variety of experiments. Hydrostatic pressure has not been often used but it can provide very valuable information. Hydrostatic pressure is known to suppress the ferroelectric phase transition temperatures in perovskite crystals [2]. Although no experimental data is available for $KNbO_3$, it is expected to behave in a way similar to $BaTiO_3$ and obey the modified Curie-Weiss law: $1/\epsilon \propto (P - P_c)$ at constant temperature [2]. If the displacive model holds, the effect of pressure will solely be to change the temperature dependence of the soft mode. On the other hand if the transition is of the order-disorder type or combined displacive and order-disorder it may affect both the off-center displacements in the paraelectric phase and the soft mode. The most suitable method to study the effect of pressure on the off-center displacements is XAFS.

The next Section contains a brief review of previous experimental results in $KNbO_3$. The various experimental results are compared to each other taking into account their different time and length scales. In the following Sections, we report results of XAFS measurements in $KNbO_3$ at 77 K and 300 K and pressures up to 10.2 GPa and 15.8 GPa, respectively. At ambient pressure, the average structure is rhombohedral and orthorhombic, respectively. On the other hand, the experimental XAFS results show that at both temperatures and all the pressures studied, the local structure is rhombohedrally distorted thus supporting the existence of an important element of order-disorder in the pressure-induced phase transitions similar to the one induced by temperature.

# REVIEW OF PREVIOUS EXPERIMENTS WITH $KNBO_3$

The properties of $KNbO_3$ at different temperatures and ambient pressure have been thoroughly investigated. The sequence of rhombohedral-orthorhombic-tetragonal-cubic phase transitions in $KNbO_3$ at 263 K, 498 K and 708 K, respectively, was obtained from neutron diffraction experiments [3,4]. Hewat found that the oxygen octahedra vibrate almost as rigid bodies around Nb atoms. The anomalous anisotropy of the oxygen atom mean square displacements has been attributed to the oxygen octahedra librations [4].

Infrared [5], Raman [6], and inelastic neutron scattering [7] experiments, showed that $KNbO_3$ has a transverse optic mode that softens with decreasing temperature. However, the frequency of this mode does not extrapolate to zero at the PE to tetragonal FE phase transition as would be expected in a displacive like model but at a temperature which is hundreds of degrees lower [8]. This result indicates that a simple displacive like model cannot account for the properties of $KNbO_3$.

The Curie-Weiss constant of $KNbO_3$ is about $2.8 \times 10^5$ K. This value is consistent with values calculated theoretically for displacive like ferroelectric perovskites and is approximately two orders of magnitude larger than the values calculated in order-disorder like models, thus supporting a displacive model.

Comes, *et al.* [9], observed the existence of X-ray diffuse scattering planes in the reciprocal space of KNbO$_3$, in the orthorhombic, tetragonal and cubic phases. The planes are perpendicular to the [100] type axes. In the cubic phase the only planes missing are the ones that go through the origin. In the lower temperature phases the planes which are not parallel to the vector order parameter disappear. These results have been later confirmed by Holma, *et al.* [10], using synchrotron radiation and electronic detection. Comes, *et al.* [11], proposed that the observed diffuse scattering is due to disordered spontaneous displacements of the Nb atoms in [111] type directions (commonly referred to as the eight-site model). In the paraelectric phase the displacement component in the [100] direction has a relatively long correlation length along the [100] direction and much shorter correlation lengths in the other directions. In the lower temperature phases the correlation lengths along the directions of the order parameter become infinite and the corresponding planes disappear. Later on Comes and Shirane [12] proposed an alternative explanation based on the very flat dispersion of the transverse acoustic phonons along [100] type directions. Holma, *et al.* [10], measured the temperature dependence of the diffuse X-ray scattering and found that the width of the diffuse scattering planes and their intensity vary with temperature. In particular the width seems to extrapolate to zero as $T$ approaches the transition temperature from above. From these results they conclude that a static eight site model cannot account for the observations. They show that their results are consistent with the theoretical quasi harmonic model of Hüller [13] which is qualitatively similar to the alternative suggestion of Comes and Shirane [12]. These models do not require the existence of disorder in the paraelectric phase.

The most direct evidence that Nb atoms are displaced in [111] type directions comes from XAFS experiments. XAFS provides the pair distance PDF between a probe atom and its neighbors. It can be shown that in the quasi-harmonic approximation the interatomic distance PDF would be expected to peak at the average crystal site. The XAFS results show that in both pure KNbO$_3$ [14,15] and in mixed KTa$_x$Nb$_{1-x}$O$_3$ [16], the Nb-oxygen distance PDF has peaks at distances corresponding to Nb atoms displaced in [111] type directions in all phases including the paraelectric phase.

The existence of local off-center displacements in the paraelectric phase has been recently reproduced theoretically in KNbO$_3$ by molecular-dynamics simulations [17]. The simulation was done in the frame-work of the nonlinear oxygen polarizability model. The order-disorder behavior was observed in the paraelectric phase over several hundred degrees above the FE-PE transition temperature and the inclusion of acoustic phonon coupling led to the correct sequence of phase transitions in this material [17]. Numerous theoretical calculations were carried out in KNbO$_3$ to explain its properties [18–21]. LAPW linear response calculations of the lattice dynamics of cubic KNbO$_3$ by Krakauer, *et al.* [22], succeeded in reproducing the picture of dynamic linear chains directed along the principal cubic axes, thus confirming the eight-site model.

Recently, Girshberg and Yacoby [23] presented a model of ferroelectrics based on the existence of both a soft mode and spontaneous local off-center ion displacements, and the interaction between the two. Their model accounts quantitatively for both displacive and order-disorder like properties. In $KNbO_3$, for example, their model explains why its Curie-Weiss constant ($2.8 \times 10^5$ K) is as large as the constant predicted for purely displacive type crystals and at the same time its soft mode does not vanish at $T_c$ but extrapolates to zero at a temperature hundreds of degrees below $T_c$. They also fully reproduce the temperature and frequency dependence of the imaginary part of the dielectric constant as measured by Vogt [24]. This model also explains the breakdown of the selection rules in the PE phase Raman spectra and the temperature dependence of the diffuse X-ray scattering results. According to this model the eight site off-center displacements are coupled to the soft mode producing a renormalized relaxation mode. As the temperature decreases the relaxation mode slows down and the coherence length increases.

Contrary to temperature dependent measurements, there are only few experiments that studied the effect of pressure on the properties of $KNbO_3$ at room temperature, and almost none that studied them at low temperatures. Recent birefringence and Raman scattering experiments by Gourdain, et al. [25], in the pressure range from 0 to 33 GPa, suggest that $KNbO_3$ undergoes a FE-PE transition at 9 - 10 GPa. The authors describe the structure of the paraelectric phase as cubic and the phase transition as weakly first order. The frequencies of most TO modes soften and their Raman intensity decreases as the transition pressure is approached from below. However, first-order Raman scattering persists in the paraelectric phase all the way up to 33 GPa, the upper limit of the pressure range studied. Other Raman spectroscopy measurements by Shen, et al. [26], reported "three new crystalline phases and an amorphous phase" in the pressure range from 0 to 20 GPa. According to the authors, the displacive phase transitions occur at 2, 6, 9, and 15 GPa, respectively. A phase transition near 2 GPa was obtained by measuring the low frequency dielectric constant under pressure up to 3.2 GPa [27]. X-ray diffraction experiments were performed with $KNbO_3$ under pressure up to 12 GPa by Moya, et al. [28]. No structural transformation was observed in the lattice cell parameters and the structure was characterized as orthorhombic at least up to 12 GPa. The unit cell volume vs pressure behavior was found to be in a good agreement with the Murnaghan equation of state [29]:

$$V = V_0 \left[1 + \left(\frac{B'_0 P}{B_0}\right)\right]^{-\frac{1}{B'_0}}, \tag{1}$$

where $V_0$ is the volume per unit formula at 300 K, $B_0$ and $B'_0$ are the isothermal bulk modulus and its pressure derivative, respectively, evaluated at $P = 0$. The best fit of Eq. (1) to the $P - V$ data provided $V_0 = 132.8$ Å$^3$ and $B_0 = 143$ GPa, in good agreement with the bulk modulus obtained from the dielectric measurements [30]. XAFS experiments with $KNb_{0.87}Ta_{0.13}O_3$ at pressures up to 7.5 GPa and room temperature were recently performed at the Nb $K$ edge [31]. The results showed no changes in the local structure within this pressure range.

The discussion in the preceding paragraph shows that the pressure phase diagram at room temperature is highly controversial. However, almost all authors agree that at least one phase transition does take place at pressures below 15 GPa. The question is whether this transition is reflected in the local structure. In the next section we present the results of our XAFS analysis of the $KNbO_3$ powder which was measured at high pressure and constant temperatures: 77 K and 300 K, where the ambient pressure phases are rhombohedral and orthorhombic, respectively. Besides providing new experimental information on the effect of pressure on the local structure, the low temperature measurement will allow us to compare the effects of pressure in two different phases with (orthorhombic phase) and without (rhombohedral phase) structural disorder.

## XAFS ANALYSIS AND RESULTS

Pressure XAFS experiments were performed with $KNbO_3$ powder on the Nb $K$ absorption edge at liquid nitrogen and room temperatures. Details of the sample preparation, experimental setup, synchrotron XAFS measurements and pressure calibration can be found in a separate article [32]. In the present paper, we summarize our modelling procedure which allowed us to solve the local structure around Nb atom within 5 nearest neighbor shells and minimize uncertainties in the obtained parameters.

Data analysis was performed by fitting a theoretical XAFS signal calculated with computer code FEFF6 to the experimental data in $r$ space by Fourier transforming both data and theory. Based on the results of the previous works [14–16,31] we chose the rhombohedrally distorted prototype cubic structure as a model of the local lattice distortion in $KNbO_3$ under pressure. At 77 K, the ambient pressure phase is rhombohedral, therefore this model is a natural choice. At 300 K, the ambient pressure phase is orthorhombic but it is composed of the disordered, locally rhombohedral domains, therefore the same model is a reasonable choice for our data analysis at the both temperatures. Other models, (*e.g.*, orthorhombic distortions) were also tried but the fit quality was much worse than for the rhombohedral model described below.

The independent structural parameters allowed to be varied in the fits to the data at each pressure were the displacements of the oxygen octahedron ($\Delta(O)$) and potassium cube ($\Delta(K)$) in the [111] direction from their positions in the ideal $Pm3m$ perovskite cubic structure, and the isotropic lattice contraction factor $\epsilon = a(P)/a(0)$, where $a(P)$ is the lattice parameter. Hence our model is general enough to accommodate a pure displacive behavior (where Nb atom off-center displacement $d$ vanishes with pressure), a pure order-disorder behavior (where $d$ remains constant under pressure, similar to what was obtained at different temperatures (Ref. [14])), and any combination of these two scenarios.

Five nearest coordination shells around absorbing Nb atom have been analyzed (Fig. 1). Rhombohedral distorted, the oxygen octahedron splits into 2 subshells

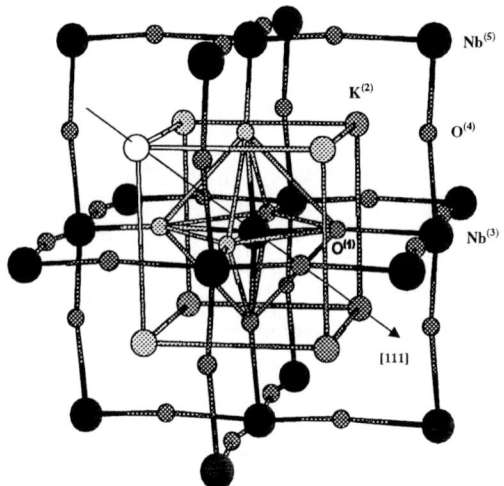

**FIGURE 1.** Rhombohedral structure of KNbO$_3$. Shown are 5 coordination shells around central Nb atom. Different shades of grey indicate inequivalent O$^{(1)}$ and K$^{(2)}$ sites relative to central Nb.

with 3 atoms each. The potassium cube splits into 4 subshells with 1, 3, 3 and 1 atoms. The 3rd nearest coordination shell consists of 6 niobium atoms. The 4th shell consists of 24 oxygen atoms from neighboring octahedra. They form 4 subshells with 6 atoms each. Finally, the 5th shell consists of 12 niobium atoms.

In our computer program FEFFIT [34], which fits an XAFS theory to the data, physically reasonable constraints between the structural parameters were applied to break their correlations in the fit and reduce the total number of variables.

All the distances $R_i$ between the Nb$^{(0)}$ atom and its $i$ neighbor in each of these 5 shells including inequivalent O$^{(1)}$ and K$^{(2)}$ subshells were expressed as a function of the three independent structural distortion parameters: $\Delta$(O), $\Delta$(K) and a lattice contraction factor $\epsilon$ using a linear approximation [32].

Effective scattering amplitude of the multiple - scattering photoelectron paths between the central Nb, its neighboring O$^{(1)}$ atoms, and the Nb$^{(3)}$ in the same [100] direction, which are very sensitive to the small deviations from collinearity, were related to $\Delta^2$(O) via the quadratic term in Taylor expansion.

Besides $\epsilon$, $\Delta$(O), and $\Delta$(K), the mean square disorder in the half path length, $\sigma^2$, and the muffin-tin energy reference corrections, $\Delta E_0$, were also varied in the non-linear least square fit of theory to data. Additional constraints were applied to relate the $\sigma^2$ and $\Delta E_0$ of the multiple scattering paths to the single scattering paths connecting the same atoms [32].

Numerical results for $\Delta$(O), the lattice parameter $a$ and the Nb off-center displacement $d$ relative to the oxygen octahedra at all pressures measured at 77 K and 300 K, respectively, have been reported in Ref. [32]. The values of $\Delta$(K) and the Nb$^{(0)}$-K$^{(2)}$ $\sigma^2$ are comparable to the uncertainties at all pressures and temperatures,

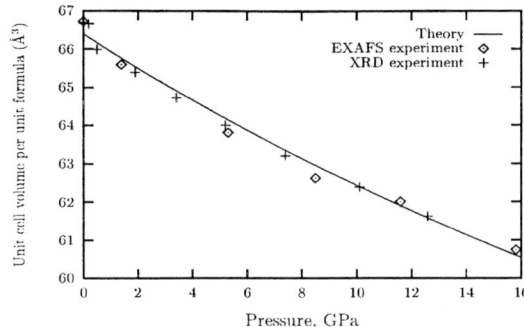

**FIGURE 2.** Unit cell volume per unit formula vs pressure at 300 K (symbols) calculated from the lattice parameter determined by XAFS (this work) and XRD (Ref. [28]) Solid line shows the behavior predicted by theory (Ref. [29]). Uncertainties in the experimental data are smaller than the symbols and are not shown.

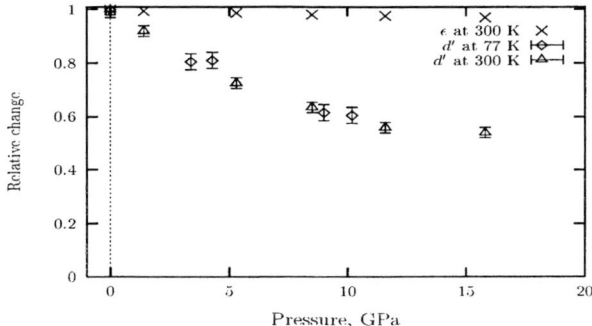

**FIGURE 3.** Changes in lattice parameter (at 300 K), Nb off-center displacements at 77 K and 300 K, normalized by their values at ambient pressure and respective temperatures.

and are, therefore, omitted.

The pressure dependence of the unit cell volume per unit formula calculated using the obtained $Nb^{(0)}$-$Nb^{(3)}$ distances (see Discussion Section) are shown in Fig. 2. The pressure dependent, Nb off-center displacements $d'(P) = d(P)/d(0)$ normalized to ambient pressure are shown in Fig. 3 for both 77 K and 300 K, along with the normalized unit cell dimension $\epsilon(P) = a(P)/a(0)$ measured at room temperature.

Mean square relative deviations of the distances between the $Nb^{(0)}$ atom and its 1st, 3rd, 4th and 5th NN have been reported in Ref. [32] for all temperatures and pressures. We obtained that $\sigma^2$ of the nearer neighbor pairs $Nb^{(0)}$ - $O^{(1)}$ and $Nb^{(0)}$ - $Nb^{(3)}$ decrease with pressure as expected. $\sigma^2$ of the longer $Nb^{(0)}$-$O^{(4)}$ and $Nb^{(0)}$ - $Nb^{(5)}$ pairs, however, have a completely different trend. These $\sigma^2$ increase with pressure at both temperatures. This indicates that pressure increases the oxygen octahedra librations and distortions, and decreases the Nb displacement -

displacement correlation length [32].

## DISCUSSION

Before discussing the new experimental results obtained for the Nb off-center displacements under pressure, we describe several important cross-checks made between our results and those obtained by other techniques.

The ambient pressure Nb off-center displacements along [111] obtained at 77 K with XAFS [32], are in good agreement with the 0.218 Å displacement measured at 230 K by neutron diffraction [4]. Although the XAFS and diffraction techniques generally measure different structures (XAFS measures local relative atomic positions while diffraction measures the average crystal structure), at this temperature, the rhombohedral structure of $KNbO_3$ is ordered, and the results of these two techniques should agree with each other. The fact that we obtained the same displacement as neutron diffraction, is, therefore, expected, and adds confidence to our other results.

Another important cross-check is the behavior of the unit cell volume with pressure. It is known from the X-ray diffuse scattering results [35,9] that the off-center displacements of the $Nb^{(0)}$ and its nearest neighbor $Nb^{(3)}$ atoms strongly correlate in [100] cubic directions. Eight nearest neighbor Nb atoms (in [100] directions) define the prototype cubic cell (the rhombohedral angle is 89.93° (Ref. [4]) and its deviation from 90° can be neglected). The local disorder of Nb displacements should not distort the rigid cubic frame of the neighboring Nb atoms due to the [100] correlations between their displacements. The small value of $\sigma^2$ between the $Nb^{(0)}$ and $Nb^{(3)}$ atoms as measured by XAFS confirms this high correlation. The volume of the unit cell per formula, therefore, can be approximated as cube of the distance between $Nb^{(0)}$ and $Nb^{(3)}$ atoms, obtained in our XAFS analysis for all pressures: $V_{XAFS} = R^3$. The average unit cell volume per formula, $V_{XRD}$, measured by X-ray diffraction, is expected to be equal to $V_{XAFS}$ in this case. The plot of the $P-V$ data (Fig. 2) for $V_{XAFS}$ (this work) and $V_{XRD}$ (Ref. [28]) demonstrates that these pressure behaviors of the unit cell volume, independently measured by XAFS and X-ray diffraction, are indeed in a very good agreement with each other and with Eq. (1).

We obtained that, at ambient pressure, the direction of the Nb displacements of almost the same magnitude in both orthorhombic and rhombohedral phases remains along the [111] cubic axis thus confirming previous XAFS results [14–16]. This indicates that the local structure unit cells in the orthorhombic phase of $KNbO_3$ retain their rhombohedral distortions. The fact that the X-ray diffraction (XRD) measures the structure of $KNbO_3$ at room temperature as orthorhombic does not contradict but complements our XAFS results. Since XRD is a fast measurement, with the same time scale as XAFS ($\approx 10^{-15}$ s), the difference between the two results should be attributed to a small correlation length of the Nb atom [111] displacements. As follows from the analysis of the behavior of $\sigma^2$ of the

$Nb^{(0)}$-$Nb^{(5)}$ bonds, high pressure decreases the already small correlation length even further [32]. Diffraction averages over the disorder of the local unit cells to obtain the orthorhombic phase.

It is worth noting that in this study we obtained that the distribution function of the $Nb^{(0)}$-$O^{(1)}$ distances remains convincingly peaked off the center of the first neighbor oxygen octahedron along the [111] directions for all temperatures and pressures since the displacement of the peak persists to be about three times the corresponding vibrational amplitude $\sigma$ of the $Nb^{(0)}$-$O^{(1)}$ bonds [32]. Because of the good spatial resolution of the XAFS data (0.04 Å in our case), and small uncertainties in both the Nb-$O^{(1)}$ distances (0.01 Å, or better) and $\sigma^2$ (0.0010 - 0.0016 Å$^2$), it is possible to distinguish this off center distribution, characteristic of a multi-well potential, from a single-well anisotropic vibration peaked about the center which has its largest displacement along the [111] directions. It is also possible to distinguish between the [111] (rhombohedral) and [110] (orthorhombic) directions of the local Nb displacements, since our spatial resolution is much better than the 0.11 Å required. To verify this we checked a fit to the model with orthorhombic distortions. As expected the fit quality was clearly significantly worse than for the rhombohedral model. Unless the measurement has sufficient spatial resolution it is not possible to make this distinction. For example, this distinction could not be made in measurements of $KNbO_3$ [4] and $PbTiO_3$ [36] using the neutron diffraction technique.

In disagreement with our XAFS results, Shuvaeva, et al. [37,38], using polarized XAFS and X-ray diffraction measurements of the single crystal $KNbO_3$ at room temperature, obtained that the direction of the local Nb atom off-center displacement agrees closely with the direction of the [110] polar axis in the orthorhombic phase. The error analysis, however, was not performed. The quality of the fits reported in Ref. [37] indicates that the uncertainties in the obtained parameters (due to noise in the data or lack of spatial resolution) may have been too large to convincingly discriminate between the [110] and [111] directions of Nb atom displacements.

As to the pressure effects, where a displacive transition to the cubic phase at 9 - 10 GPa has been proposed from the birefringence and Raman spectroscopy results [25,26] from the fact that those effects vanish or become small above 9 - 10 GPa [25], XAFS measurements show conclusively that the local distortions do not vanish. As Fig. 3 demonstrates, the Nb displacement does not vanish but remains fixed in direction and changes continuously through the 9 - 10 GPa region with no visible slope change.

This apparent contradiction between the two local probes of XAFS and Raman can be reconciled by taking into account the different time scales of the optical and x-ray measurements as discussed in Sects. I and II. In the eight-site model [11] the local displacements are in the [111] directions as XAFS observes. However, these local displacements become progressively more disordered as $KNbO_3$ transforms from the rhombehedral to the orthorhombic, tetragonal, and then completely disordered in the cubic phase. These disordered regions are small enough to average out in diffraction and birefrigence measurements which probe over relatively larger dimensions. To account for the difference between the two local probes (Raman and XAFS) the dynamics of the disordering must be considered. At room temperature and below 9 - 10 GPa the hopping time of the Nb atoms between equivalent [111] sites is longer than the lifetime of the Raman measurement and Raman senses the displacement. However, above 9 -10 GPa the hopping lifetime becomes shorter than the lifetime of the Raman measurement which then senses the average over the disordered [111] sites, giving higher symmetry and a weaker first order line. The XAFS measure is always much faster than any hopping time and always senses the local displacement. The theory of Girshberg and Yacoby [23] demonstrated that a temperature - dependent hopping rate among equivalent off-center positions is a crucial element in giving a quantitative explanation of the properties of the temperature - induced FE-PE transitions in $KNbO_3$.

What remains to explain is the cause of the increase of the hopping rate with increasing pressure. The hopping rate $\nu$ is given by $\nu = \nu_0 e^{-u/k_B T}$ where $\nu_0$ is some attempt frequency, $u$ is the barrier height between equivalent sites, and $k_B$ is the Boltzmann constant. Since $T$ is constant, the increase in $\nu$ must occur because of a decrease of $u$ with pressure. As Fig. 3 shows, pressure has two effects: to decrease slightly the lattice constant and to decrease much more the displacement $d$. Decreasing the lattice constant would be expected to increase $u$. However, the much larger concomitant decrease of $d$ should decrease $u$ because it decreases the distance required to hop to neighboring equivalent sites. For example the barrier vanishes when $d = 0$. It is reasonable to expect that the much greater change in $d$ would dominate, explaining the increased rate of hopping with increasing pressure.

The measured decrease of the local distortion $d$ with increasing pressure can be understood theoretically using the pseudo Jahn-Teller distortion ideas of Bersuker [39]. As Bersuker has shown, $d$ depends inversely on the energy gap between the ground and excited states coupled by the distortion. With the decreasing lattice constant with increasing pressure the splitting of the excited states will increase exponentially because of the corresponding increase in overlap of neighboring atom wave-functions, explaining the greater rate of decrease of $d$ compared to that of the lattice constant as shown in Fig. 3.

## CONCLUSIONS

Nb K-edge XAFS data of $KNbO_3$ were measured at 77 K and 300 K under high pressure, up to 15.8 GPa. We observed a pressure-induced gradual displacement of Nb atom towards the center of oxygen octahedra at both temperatures, 77 K and 300 K. The Nb off-center displacements are in the [111] type directions at all temperatures and pressures, in spite of the fact that the crystal has different crystallographic structures.

Our results show, that, while the average structure reportedly exhibits one or more pressure-induced phase transitions, the local structure remains rhombohedrally distorted at all pressures up to 15.8 GPa, indicating a significant order - disorder element in the phase transitions.

The difference between XAFS and other relevant experimental techniques as a function of pressure is explained by their different spatial resolution, time and length scales. To reconcile these differences it is concluded that the hopping rate of Nb atoms among their equivalent sites increases with increasing pressure due to the concomitant relatively large decrease in the Nb displacements.

## ACKNOWLEDGEMENTS

We would like to thank Prof. E. Moya for providing his X-ray diffraction results (Fig. 2) and helpful discussions. EAS is grateful to Victor Polinger for pointing out how the Bersuker theory can explain the decrease of $d$ with increasing pressure.

## REFERENCES

1. Matthias B. T., *Phys. Rev. B* **75**, 1771 (1949).
2. Lines M. E. and Glass A. M., *Principles and Applications of Ferroelectrics and Related Materials*, Oxford: Clarendon Press, 1977.
3. Perry C. H., Hayes R. R., Tornberg N. E., *Proceedings of the International Conference on Light Scattering of Solids*, Ed. M. Balkanski.
4. Hewat A. W., *J. Phys. C: Solid State Phys.*, **6**, 2559 (1973).
5. Fontana M. D., Metrat G., Servoin J. L., and Gervais F., *Ferroelectrics* **38**, 797 (1981).
6. Fontana M. D., Kugel G. E., Vamvakas J., and Carabatos C., *Solid State Commun.*, **45**, 873 (1983).
7. Fontana M. D., Dolling G., Kugel G. E., and Carabatos C., *Phys. Rev. B* **20**, 3850 (1979).
8. Fontana M. D., Metrat G., Servoin J. L., and Gervais F., *J. Phys. C* **16**, 483 (1984).
9. Comes R., Lambert M., and Guiner A., *Acta. Cryst.*, **A26**, 244 (1970).
10. Holma M., Takesue N., and Chen H., *Ferroelectrics* **164**, 237 (1995).
11. Comes R., Currat R., Denoyer F., Lambert M. and Quittet A. M., *Ferroelectrics* **12**, 3 (1976).
12. Comes R. and Shirane G., *Phys. Rev. B* **5**, 1886 (1972).
13. Hüller A., *Solid State Commun.*, **7**, 589 (1969).
14. Kim K. H., Elam W. T., and Skelton E. F., *Mat. Res. Soc. Symp. Proc.*, **172**, 291 (1990).
15. Mathan N., Prouzet E., Husson H., and Dexpert H., *J. Phys.: Condens. Matter* **5**, 1261 (1993).
16. Hanske-Petitpierre O., Yacoby Y., Mustre de Leon J., Stern E. A., and Rehr J. J., *Phys. Rev. B* **44**, 6700 (1991).

17. Sepliarsky M., Stachiotti M. G., and Migoni R. L., *Phys. Rev. B* **56**, 566 (1997); Stachiotti M. G., Sepliarsky M., Tinte S., Migoni R. L., and Rodrigues C. O., Proceedings of the Fifth Williamsburg Worskhop on First-Principles Calculations for Ferroelectrics, Williamsburg, VA, 1998.
18. Yu R. and Krakauer H., *Phys. Rev. Lett.*, **74**, 4067, (1995); Krakauer H., Yu R., Wang C.-Z., Rabe K., and Waghmare U. V., Proceedings of the Fifth Williamsburg Worskhop on First-Principles Calculations for Ferroelectrics, Williamsburg, VA, 1998.
19. Kvyatkovskii O. E., *Phys. Solid State* **39**, 602 (1997).
20. Dorfman S., Fuks D., Gordon A., Postnikov A. V., and Borstel G., *Phys. Rev. B* **52**, 7135 (1995).
21. Bussmann-Holder A., *J. Phys. Chem. Solids* **57**, 1445 (1996).
22. Krakauer H., Yu R., and Wang C. Z., *J. Phys. Chem. Solids* **57**, 1409 (1996).
23. Girshberg Ya. and Yacoby Y., *Solid State Commun.*, **103**, 425 (1997).
24. Vogt H., *Phys. Rev. B* **51**, 8046 (1995).
25. Gourdain D., Moya E., Chevrin J. C., Canny B., and Pruzan Ph., *Phys. Rev. B* **52**, 3108 (1995).
26. Shen Z. X., Hu Z. P., Chong T. C., Beh C. Y., Tang S. H., and Kuok M. H., *Phys. Rev. B* **52**, 3976 (1995).
27. Errandonea D. and Moya E., *Phys. Stat. Sol. (B)* **203**, R1 (1997).
28. Moya E., Tinoco T., Polian A., and Itie J. P., *Ferroectric Letters* **22**, 59 (1997).
29. Murnaghan F. D., *Am. J. Math.*, **49**, 235 (1937).
30. Wiesendanger E., *Ferroelectrics* **6**, 263 (1974).
31. Wang F., Ravel B., Yacoby Y., Stern E. A., and Ingalls R., *J. Phys. IV* **7**, C2-1225 (1997).
32. Frenkel A. I., Wang F. M., Kelly S., Ingalls R., Haskel D., and Stern E. A., *Phys. Rev. B* **56**, 10869 (1997).
33. Zabinsky S. I., Rehr J. J., Ankoudinov A., Albers R. C., and Eller M. J., *Phys. Rev. B* **52**, 2995 (1995).
34. Stern E. A., Newville M., Ravel B., Yacoby Y., and Haskel D., *Physica B* **208 & 209**, 117 (1995).
35. Comes R., Lambert M., and Guiner A., *Solid State Commun.*, **15**, 715 (1974).
36. Nelmes R. J. and Kuhs W. F., *Solid State Commun.*, **54**, 721 (1980); Nelmes R. J., Piltz R. O., Kuhs W. F., Tun Z., and Restori R., *Ferroelectrics* **108**, 165 (1990).
37. Shuvaeva V. A., Yanagi K., Sakaue K., and Terauchi H., *J. Phys. Soc. Japan* **66**, 1351 (1997).
38. Shuvaeva V. A. and Antipin M. Y., *Crystallography Reports* **40**, 466 (1995).
39. Bersuker I. B. and Polinger V. Z., *Vibronic Interactions in Molecules and Crystals*, Springer Series in Chemical Physics, Vol. 49, New York: Springer, 1989.

# Weighted Density Functionals for Ferroelectric Materials

I.I. Mazin[1,2] and D. J. Singh[1]

[1] *Code 6691, Naval Research Laboratory, Washington, DC 20375*
[2] *CSI, George Mason University, Fairfax, VA 22030*

**Abstract.** The weighted density approximation, its implementation and its application to ferroelectric materials is discussed. Calculations are presented for several perovskite oxides and related materials. In general the weighted density approximation is found to be superior to either the local density or generalized gradient approximation for the ground state. Electronic structures are little changed. The linear response of the weighted density approximation is calculated for the homogeneous electron gas, and found to be improved relative to the local density result, but not in full agreement with existing Monte Carlo data. It is shown that the agreement can be further improved by a simple modification. Calculations of the ferroelectric soft mode in $KNbO_3$ suggest that the low temperature distortion is approximately 20% smaller than indicated by existing experiments.

## I INTRODUCTION

Piezoelectric, ferroelectric and related perovskites are both technologically important and physically interesting systems and, as such, have been the subject of considerable recent interest. Several groups have performed first principles calculations based on density functional theory. These calculations, primarily based on the local density approximation (LDA) have elucidated important aspects of the underlying physics of these materials as well as providing quantitative information about phonons, polarizations, crystal structure, elastic and other properties. [1]

Two of the many themes that have emerged are (1) Ferroelectric and piezoelectric properties are due to a delicate balance between large competing interactions, [2,3] so very high accuracy is required for predictive calculations and (2) The lattice instabilities behind these phenomena are extraordinarily sensitive to volume. In fact, in important materials, like $KNbO_3$ [4–10] and $BaTiO_3$ [2,3,11] ferroelectricity, while predicted at the experimental volume, is absent, or at least strongly suppressed, at the predicted LDA volume.

Generally excellent agreement with experimental knowledge has been obtained when calculations are performed at the experimental volume. This includes even transition temperatures, polarizations and transition temperatures [1]. In antiferroelectric $PbZrO_3$, LDA calculations [12] at the experimental volume yielded a crystal structure that differed from existing experimental measurements. However, the LDA structure has now been confirmed by two independent neutron investigations, [13,14] providing yet another demonstration of the utility of the first principles approach. On the other hand, even at the experimental volume, the LDA consistently overestimates the dielectric constant, $\varepsilon_\infty$. [15–17] Exploration of novel systems that are of necessity less well studied experimentally than current materials will require an ability to calculate properties independent of experimental measurements of the volume. This may also be important for determination of piezoelectric properties in materials like $PbTiO_3$ where the LDA gives an incorrect prediction of the strain associated with the deviation of the $c/a$ ratio from unity even with the volume constrained. [18] This raises the question of whether there is a practical way of correcting the LDA errors, in order to construct a more predictive first principles approach for these materials.

The LDA errors can sometimes be trivially fixed. For example, if the experimental volume is known, the calculations may be done constrained to this volume. In some cases, e.g. $KNbO_3$, simple modification of the LDA functional through use of the Wigner form rather than an electron gas-like form, [7,8] or the use of generalized gradient approximations (GGA) to density functional theory are sufficient to correct the volume (although the dielectric constant remains too high). However, in general these approaches do not solve the LDA problem. For example, in $BaTiO_3$ the GGA overcorrects the LDA volume, so that the error is as large but in the opposite direction. [9] The implication is that a more sophisticated, presumably non-local, density functional is needed to fully circumvent the problems noted above with the LDA.

The first efforts at developing practical non-local functionals date from the 1970's when the average density approximation [19] (ADA) and weighted density approximation [20–22] (WDA) were proposed. However, over most of the intervening period the field has been relatively dormant, in part because of the success of the simpler LDA and GGA schemes and in part because it was widely thought that such schemes could not be implemented in a computationally tractable fashion. However, at least for the WDA, computationally efficient algorithms are now known [10,23–25] and benchmark calculations have been reported. The ADA has attracted less interest.

In the cases that have been studied ground state properties of solids are generally improved over the LDA. [26] These include tests for several simple elements and compounds and $KNbO_3$. Significantly, this latter test showed that the WDA predicts an equilibrium volume for $KNbO_3$ in almost perfect agreement with experiment. [10] Unlike the GGA, which also greatly improves the equilibrium volume of $KNbO_3$, the WDA is a truly non-local density functional, in that the exchange correlation potential at a point $\mathbf{r}$ incorporates information about the charge density $n(\mathbf{r'})$ over a finite region of space.

Here, following a brief overview of the method, we report investigations of a range of perovskite and related oxides within the WDA. These show much improved volumes relative to either the LDA or GGA. Phonon frequencies and the ferroelectric soft mode in $KNbO_3$ are calculated within the WDA and compared with LDA results and experiment. An analytical expression for the linear response of the uniform electron gas in the WDA is derived.

## II  THE WEIGHTED DENSITY APPROXIMATION

The WDA and the ADA are both based on the general expression for the exchange correlation energy of a general electron gas $E_{xc}$, in density functional theory (DFT).

$$E_{xc} = \frac{e^2}{2} \int \frac{n(\mathbf{r})n(\mathbf{r}')}{|\mathbf{r}-\mathbf{r}'|} G(\mathbf{r},\mathbf{r}')\{n(\mathbf{r})\} d\mathbf{r} d\mathbf{r}', \qquad (1)$$

where the function $G(\mathbf{r},\mathbf{r}')$ is also a functional of the total electronic density $n(\mathbf{r})$. A rigorous expression for $G$ can be derived [27] in terms of coupling constant averaged pair correlation function:

$$G(\mathbf{r},\mathbf{r}') = \int_0^1 [\bar{g}(\mathbf{r},\mathbf{r}';\lambda)\{n(\mathbf{r})\} - 1] d\lambda. \qquad (2)$$

Were the actual pair correlation function used instead of the coupling constant averaged one, this expression would give the interaction energy of each electron with its exchange correlation hole. The difference due to the averaging reflects the fact that kinetic energy of interacting electrons differs from that of a non-interacting system with the same density. It is this difference, also known as exchange-correlation kinetic energy, which is responsible for the high $q$, short wavelength behavior of the response, and is also implicated in usual underestimation of semiconducting gaps in Kohn-Sham DFT.

For the uniform gas this function, $G_0(|\mathbf{r}-\mathbf{r}'|,n)$, is known with high accuracy [28], but for an arbitrary inhomogeneous system, like occurs in a real density functional calculation for a solid, $G$ is not known and there is therefore no practical way to use this formula without making some approximation.

The LDA instead of Eq. (1) uses $(e^2/2) \int d\mathbf{r} d\mathbf{r}' n^2(\mathbf{r}) G_0[|\mathbf{r}-\mathbf{r}'|, n(\mathbf{r})]/|\mathbf{r}-\mathbf{r}'|$, so that $E_{xc}$ becomes $E_{xc}^{LDA} = \int n(\mathbf{r}) \epsilon_{xc}[n(\mathbf{r})] d\mathbf{r}$, $\epsilon_{xc}$ being the density of exchange-correlation energy of the uniform gas. The LDA is incorrect in both most important limits: the fully localized, *i.e.*, a one electron system, and the fully delocalized limit, *i.e.*, homogeneous electron gas. In the former case the LDA gives a spurious self-interaction with energy $(e^2/2) \int d\mathbf{r} d\mathbf{r}' n(\mathbf{r}) n(\mathbf{r}')/|\mathbf{r}-\mathbf{r}'| + \int n(\mathbf{r}) \epsilon_{xc}[n(\mathbf{r})] d\mathbf{r}$, which is widely thought [29] to be a key problem with the LDA. In the homogeneous limit, the LDA gives the correct exchange-correlation energy, but the *changes* of this energy upon small perturbations are not properly described; the second variation of $E_{xc}$ with density, *i.e.*, the exchange-correlation part of the dielectric response, $K_{xc}(\mathbf{r}-\mathbf{r}') = \delta^2 E_{xc}/\delta n(\mathbf{r}) \delta n(\mathbf{r}')$, is a delta function, which is incorrect. The Fourier transform of $K_{xc}(r)$ in LDA is independent of the wave vector. Since LDA is exact for the uniform gas, $K_{xc}^{LDA}$ corresponds to the correct $K_{xc}$ at $q=0$. GGAs also give correct behavior at $q=0$, but become even worse than the LDA at high $q$'s.

The two nonlocal expressions for $E_{xc}$, WDA and ADA, are aimed at correcting one or the other of these two limits. The former uses the general expression (1), but instead of the actual function $G$ uses a model function, defined so that the one electron limit is honored. This begins by choosing a generic expression for $G$, which depends on one parameter $\bar{n}$, to be defined later. In the original papers it was suggested that $G(\mathbf{r}, \mathbf{r}', \bar{n}) = G_0(\mathbf{r}, \mathbf{r}', \bar{n}) = \int_0^1 [\bar{g}(\mathbf{r}, \mathbf{r}'; \lambda, \bar{n}) - 1] d\lambda$, where $\bar{g}$ is the averaged pair correlation function of the homogeneous electron gas. Later it was realized [30] that other choices of $G$ may be better than $G_0$, and that there is no physical reason to prefer $G_0$ over many other choices. In the WDA $\bar{n}$ is a function of $\mathbf{r}$, but differs from $n(\mathbf{r})$, and is chosen so that $\int G[\mathbf{r}, \mathbf{r}', \bar{n}(\mathbf{r})] n(\mathbf{r}') d\mathbf{r}' = -1$. This assures that for a one electron system $E_{xc}$ cancels the self-interaction exactly.

Within the WDA, non-local information about the charge density is incorporated into $E_{xc}$ both through the construction of $\bar{n}$ via the sum rule, and through the non-local Coulomb integral.

We note that $G$ need not neccesarily be the actual pair correlation function of the system: although Eq.(1) has the same functional form as the WDA energy, the fact that $G^{WDA}$ is a function of an averaged density $\bar{n}$, not a functional of the true density $n(\mathbf{r})$, means that it is possible that the best approximations for $G^{WDA}$ could be different from the physical function $G$ defined in Eq. (2) even for the uniform electron gas. However, the few reported calculations suggest that this function, while perhaps not optimal, does yield results much superior to the LDA. Besides, use of the homogeneous electron gas averaged pair correlation function has an appealing simplicity and calculations of ground state properties, below, use this ansatz. The results confirm that at least for the structural properties considered, the WDA with this choice of $G$ yields results that are uniformly superior to the LDA.

## III GROUND STATE PROPERTIES IN THE WDA

As mentioned, one of the basic problems with application of the LDA to ferroelectric materials is the high sensitivity of ferroelectric properties to the volume, combined with the several percent errors in predicted LDA equilibrium volumes. One of our main goals is to determine whether the WDA can repair this problem while retaining the good features of the LDA. We begin by calculating the equilibrium volumes of several cubic or near cubic oxides, namely CaO, SrO, BaO, BaTiO$_3$, KNbO$_3$ and KTaO$_3$. Some results for KNbO$_3$ were reported previously. [10] The LDA underestimates the cubic lattice parameters of these materials by 1-2%, while at least in BaO and BaTiO$_3$ GGA calculations seriously overestimate the volume. [9]

The calculations were done using a planewave basis set pseudopotential method, [31,32] with hard Troullier-Martins pseudopotentials [33] and the implementation of the WDA discussed in Ref. [23]. The Perdew-Wang form of $G$, i.e. that from the uniform electron gas coupling constant averaged pair correlation function was

**TABLE 1.** LDA, WDA and experimental lattice parameters in Å for some cubic materials. Values for the elements C, Si, Mo and V and are from Ref. [23]. BaTiO$_3$ and KNbO$_3$ are really rhombohedral at low temperature, but the rhombohedral strain is small. Calculations were done for the cubic structure. Numbers in parentheses are the percentage deviation from experiment.

| Material | LDA | WDA | Expt. | Material | LDA | WDA | Expt. |
|---|---|---|---|---|---|---|---|
| C | 3.53 (-1.1) | 3.56 (-0.3) | 3.57 | Si | 5.36 (-1.3) | 5.40 (-0.6) | 5.43 |
| Mo | 3.11 (-1.3) | 3.14 (-0.3) | 3.15 | V | 2.93 (-3.0) | 2.99 (-1.0) | 3.02 |
| CaO | 4.71 (-2.0) | 4.81 (-0.0) | 4.81 | SrO | 5.06 (-1.8) | 5.16 (-0.0) | 5.16 |
| BaO | 5.46 (-1.4) | 5.56 (+0.3) | 5.54 | BaTiO$_3$ | 3.95 (-1.2) | 4.00 (-0.1) | 4.00 |
| KNbO$_3$ | 3.96 (-1.4) | 4.02 (+0.1) | 4.02 | KTaO$_3$ | 3.92 (-1.6) | 3.98 (-0.2) | 3.98 |

used for the WDA. The pseudopotentials included the semi-core states as valence states. In particular, 3s and 3p states of K, Ti and Ca, 4s and 4p states of Nb and Sr and 5s and 5p states of Ba and Ta were treated in the electronic structure calculations, while lower lying states were pseudized. Well converged basis sets, including planewaves up to 121 Ry, were used, except for CaO where a higher cut-off of 132 Ry was used.

One complication in the WDA is the need to use shell partitioning to avoid unphysical exchange-correlation interactions between valence and core electrons. It is used to prevent the core electrons from contributing to the exchange correlation hole seen by valence electrons, since it is unreasonable for tightly bound core electrons to dynamically screen valence electrons. In shell partitioning, the interaction between the valence electrons is treated with the WDA, but core-core and core-valence interactions are treated within the LDA as discussed in Refs. [23,10]. In the present calculations, the semi-core states on the metal ions and the O 2s states are treated with the LDA, lower states are pseudized with an LDA pseudopotential, and higher states are treated with the WDA.

The LDA and WDA lattice parameters of these oxides are summarized in Table 1 along with some previous results for elements. As may be seen, the WDA lattice parameters are quite dramatically improved relative to the LDA, including BaTiO$_3$ and BaO, for which GGA calculations yield overcorrected volumes.

Next we turn to phonons and the ferroelectric soft mode. The above results imply that the WDA can generally correct LDA errors in the equilibrium volume. The question then arises as to the extent that the WDA is able to reproduce the desirable features of the LDA with the volume corrected. LDA calculations at the experimental lattice parameter have demonstrated very good agreement with experimental data for phonons, crystal structures and even derived quantities like transition temperatures. Moreover, in ferroelectric/piezoelectric materials these quantities are governed by delicate balances between competing terms. In KNbO$_3$, very different results are obtained for the ferroelectric soft mode if the lattice parameter is varied by 1-2 percent, and in fact the WDA does change the LDA prediction of the lattice parameter by an amount of this order. Thus one is led to ask whether the WDA will drastically change LDA results for phonons or the ferroelectric mode

**TABLE 2.** LDA and WDA Γ-point TO phonon frequencies for cubic perovskite structure KNbO$_3$. The lattice parameter is fixed at its experimental value, which is the same as the WDA value.

| Γ-point mode | LDA frequencies (cm$^{-1}$) | WDA frequencies (cm$^{-1}$) |
|---|---|---|
| Γ$_{15}$(*unstable*) | 195 $i$ | 176 $i$ |
| Γ$_{15}$ | 170 | 143 |
| Γ$_{15}$ | 473 | 497 |
| Γ$_{25}$ | 247 | 257 |

or otherwise degrade the description of the material.

To begin addressing this issue we performed frozen phonon calculations of the zone center transverse modes in KNbO$_3$. Parallel LDA and WDA planewave pseudopotential were performed as described above using a 6x6x6 special k-point sampling of the Brillouin zone.

Calculations of atomic forces were performed for small (less than 0.05Å) displacements of the atoms from the ideal cubic sites in a rhombohedral symmetry consistent with the Γ$_{15}$ modes. A total of seven such force calculations were performed. The Γ$_{15}$ dynamical matrix was then least squares fit to these force calculations, and diagonalized to obtain the frequencies and eigenvectors. The LDA and WDA eigenvectors were very similar to each other and to previously reported [4,10] linearized augmented planewave (LAPW) [34] results, and are not reproduced here. There is only a single Γ$_{25}$ frequency, which was calculated separately, using force calculations with O displacements according to this mode. Calculated LDA and WDA frequencies of the Γ$_{15}$ and Γ$_{25}$ modes in cubic perovskite structure KNbO$_3$ are given in Table 2. The LDA results are similar to those obtained previously by all electron LAPW, [4] pseudopotential linear response LAPW methods [7,8] and planewave pseudopotential calculations for the Γ$_{15}$ modes. [35] The agreement of the LDA results with the linear response calculations of Yu and co-workers *et al.* [7,8] is almost exact, with a maximum deviation of 4 cm$^{-1}$, while somewhat larger differences from the earlier all-electron results of Singh and Boyer [4] are present. Tests [3,9] have shown that most of this latter difference is due to the use of a 4x4x4 k-point mesh in the calculations of Ref. [4] leading to an underestimation of the ferroelectric instability. Fontana *et al.* [36] have performed measurements of phonon frequencies in KNbO$_3$. These should not be uncritically compared to the present ground state calculations because the KNbO$_3$ undergoes three structural transitions as the temperature is raised from 0K, and only becomes cubic at approximately 700K. Nonetheless, as discussed in Ref. [4], they are consistent with the LDA results for the stable modes.

The WDA phonon frequencies are very similar to the LDA results. The main difference is that the splitting between the ferroelectric soft mode and the middle (lowest stable) Γ$_{15}$ mode is smaller in the WDA. However, differences between the LDA and WDA frequencies are comparable to the numerical accuracy of the calculations as estimated from the spread between the recent results of various groups. In any case, the present results do not provide any indication that the WDA degrades the favorable agreement of LDA phonon frequencies with experiment in these materials although further tests are clearly in order.

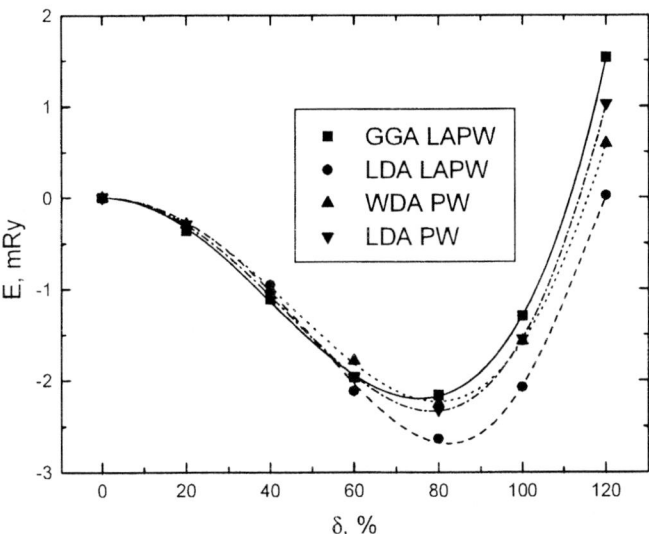

**FIGURE 1.** Total energy as a function of ferroelectric displacement in $KNbO_3$ within the WDA, LDA and GGA using both planewave pseudopotential (LDA and WDA) and LAPW approaches (LDA and GGA). The displacement $\delta$ is in percent of the experimental displacement of Ref. [37]. Calculations are at the experimental volume, which effectively coincides with the WDA and GGA volumes.

Next we turn to the ferroelectric instability. The present LDA and WDA calculations done with a planewave pseudopotential method, and earlier LAPW calculations both within the LDA and with GGA density functionals are in agreement with the experimental distortion [37] regarding the soft mode eigenvector. However, both LDA and GGA calculations give a soft mode amplitude that is 20% smaller than the existing experimental value, which was determined by powder neutron diffraction. Interestingly, both LDA and GGA calculations [9,10] for $BaTiO_3$ give a distortion that agrees very closely, (within 5%) with experiment. Fig. 1 shows the result of WDA calculations of this instability for $KNbO_3$ at the experimental volume, in terms of the reported experimental distortion. As may be noted, there is some scatter between the curves, which represent three different exchange correlation potentials and two different band structure methods. However, all the calculations are in agreement that there is an instability of order 2 mRy/f.u. and that the distortion is only 80% of the reported experimental distortion. In particular, the WDA does not significantly change the distortion relative the the LDA value. Taken together the results strongly suggest an experimental re-examination of the low temperature structure of $KNbO_3$.

## IV  LINEAR RESPONSE IN THE WDA

As mentioned, LDA systematically overestimates the static dielectric constant, $\varepsilon_\infty$. Not only is this itself an important characteristic of ferroelectric materials, it also underscores the fact that the dielectric response in general is not accurately reproduced, which could lead to difficulties calculating small energy differences associated with phonons and lattice distortions. The problem has been discussed in terms of the differences between the real one-electron excitation spectrum and the LDA band structure, primarily to underestimation of the band gap. An apparent paradox here is that the exact DFT theory should reproduce the static response functions, but not the excitation spectrum. The solution of this paradox is that the RPA (random phase) dielectric function, which is directly defined by the one-electron spectrum, is enhanced by the so-called exchange-correlation local field corrections. Correspondingly, a small gap and small local field corrections result in the same $\varepsilon_\infty$ as a large gap and large corrections. These corrections, in turn, are defined by the second variation of the exchange-correlation energy, $K_{xc}(\mathbf{r},\mathbf{r}') = \delta^2 E_{xc}/\delta n(\mathbf{r})\delta n(\mathbf{r}')$. The LDA provides correct $K_{xc}(\mathbf{r},\mathbf{r}')$ only in metals and only in the long range limit.

One of the first attempts to correct this was the formulation of ADA, where $n(\mathbf{r}')$ in Eq. (1) is substituted by $n(\mathbf{r})$, so that $E_{xc}^{ADA} = \int n(\mathbf{r})\epsilon_{xc}[\tilde{n}(\mathbf{r})]d\mathbf{r}$. Then $\tilde{n}(\mathbf{r})$ is defined as $\tilde{n}(\mathbf{r}) = \int w[|\mathbf{r}-\mathbf{r}'|, \tilde{n}(\mathbf{r})]n(\mathbf{r}')d\mathbf{r}'$, and the universal function $w$ is chosen so that $\delta^2 E_{xc}^{ADA}/\delta n(\mathbf{r})\delta n(\mathbf{r}')$ gives the correct $K_{xc}(\mathbf{r},\mathbf{r}')$ for the uniform gas. Contrary to the WDA, the ADA is not self-interaction free in one electron systems, and thus was never as popular as WDA.

From the beginning there was substantial interest in the behavior of WDA in the delocalized limit [27]. Williams and von Barth [38] suggested that the WDA should give substantial improvement over the LDA in this limit, but till now no systematic study has been reported. If this conjecture is true, the WDA has a great advantage over any other known approximation to the DFT in the sense that it would accurately reproduce two key physical limits. At least *some* improvement over LDA is to be expected: in the short range $K_{xc}^{WDA}(\mathbf{r},\mathbf{r}')$ remains finite, and correspondingly decay with $q \to \infty$ in reciprocal space. Smaller $K_{xc}$ will result in weaker local field corrections and in an improvement in $\varepsilon_\infty$. It is not clear *a priori*, though, how much $K_{xc}^{WDA}(q)$ is improved over the LDA over the whole $q$ range and, if the improvement is only modest, whether or not an approximation based on the WDA exists that does provide proper behavior. Here we derive an expression for $K_{xc}$ in the WDA, calculate $K_{xc}$ for popular flavors of WDA, and discuss construction of a WDA method with improved $K_{xc}$.

We start by deriving a closed expression for $K_{xc}$ in the WDA for arbitrary $G$. First some notation: denote the product $(e^2/r)G(r)$ as $W(r)$, use atomic units ($e = 1$, $\hbar = 1$), and use primes for the derivative with respect to the density argument, e.g. $G' = dG/dn$. We also introduce two functions, reflecting implicit dependence of the weighted density $\bar{n}$ on variations of the real density:

$$d(\mathbf{r'}-\mathbf{r}) = \delta\bar{n}(\mathbf{r'})/\delta n(\mathbf{r}) \qquad (3)$$

$$f(\mathbf{r'}-\mathbf{r},\mathbf{r'}-\mathbf{r''}) = \frac{\delta^2 \bar{n}(\mathbf{r'})}{\delta n(\mathbf{r})\delta n(\mathbf{r''})} = \frac{\delta d(\mathbf{r'}-\mathbf{r})}{\delta n(\mathbf{r''})}. \qquad (4)$$

Using the WDA expression for the exchange-correlation energy,

$$E_{xc} = (1/2) \int n(\mathbf{r})n(\mathbf{r'})W[|\mathbf{r}-\mathbf{r'}|,\bar{n}(\mathbf{r})]d\mathbf{r}d\mathbf{r'}, \qquad (5)$$

we express $K_{xc}$ in terms of functions $d$ and $f$, and find these functions using the normalization condition.

$$\int d\mathbf{r'} n(\mathbf{r'})G[|\mathbf{r}-\mathbf{r'}|,\bar{n}(\mathbf{r})] = -1. \qquad (6)$$

We proceed then in reciprocal space, which corresponds to using density perturbation of the form $\delta n(\mathbf{r}) = n_q e^{i\mathbf{qr}}$. Let $W_q$, $G_q$, $d_q$ and $f_{p,q}$ be the Fourier transforms of the corresponding functions. Then

$$d_q = -G_q/ng'_0. \qquad (7)$$

Since at $q \to 0$ the LDA should be restored,

$$\int d\mathbf{r'} W[|\mathbf{r}-\mathbf{r'}|,n] = 2\epsilon_{xc}/n. \qquad (8)$$

From this it immediately follows that

$$G_0 = -1/n, \qquad W_0 = 2\epsilon_{xc}/n. \qquad (9)$$

Thus $d_q = -nG_q$. Next variation of Eq.(6) gives us $f_{p,q}$. In fact, we need only diagonal elements, $f_{q,-q}$, for which we find $f_{q,-q} = 2nG_q(nG'_q + G_q)$. The second variation of Eq.(5) in terms of $d$ and $f$ is

$$K_{xc}(q) = W_q + nd_q W'_q + nd_q W'_0 + \frac{n^2}{2}(d_q^2 W''_0 + n^2 f_{q,-q} W'_0).$$

resulting in

$$K_{xc}(q) = W_q - n^2 G_q(W'_q + W'_0) + n^2(n^2 G_q^2 W'_0)'/2. \qquad (10)$$

The original formulation of the WDA used the homogeneous electron gas function for $G$. Since then, three forms of $G$ have been used in calculations, all of which result in improvement over LDA (in the admittedly limited number of tests performed to date). These are: the function $G$ derived for the uniform gas by Perdew and Wang [28], the Gunnarsson-Jones function $G^{GJ}(r) = C_1(n)\{1 - \exp[-\left(\frac{r}{C_2(n)}\right)^{-5}]\}$, and the Gritsenko et al. [39] function $G^{GRBA}(r) = C_1(n)\exp[-\left(\frac{r}{C_2(n)}\right)^k]\}$, $k = 1.5$ (note that the uniform gas function [28] is approximately given by the same expression with $k = 2$). We tested these functions for the densities $r_s = 1, 2$, and 5 and obtained modest agreement with the Monte Carlo results [40] (Cf. the left panel of Fig.2, where we plot the calculated exchange-correlation local field factor $I_{xc}(q) = \frac{q^2}{4\pi}K_{xc}(q)$, and compare it with Monte Carlo data [40]). By construction, $K_{xc}(0)$ is correct (and is in fact the LDA value). At $q \approx 1.5 - 1.8k_F$ $K_{xc}$ falls below its LDA value and continues to decrease at large $q$'s. However, a closer look reveals some disagreements: first, $I_{xc}^{WDA}(q)$ is considerably larger than the Monte-Carlo data for the wave vectors between $\approx 0.5k_F$ and $1.5k_F$. Second, $I_{xc}(q)$ in WDA tends to a constant value, while in Monte Carlo calculations it is $K_{xc}(q)$ itself that has a finite limit at $q \to \infty$, and $I_{xc}(q) \to const \cdot q^2$ at $q \to \infty$.

Can one correct these two deficiencies without compromising the correct one-electron limit of WDA? As discussed above, there is no particular reason to use the homogeneous electron gas pair correlation function for $G$ (nor, as discussed above, the exact pair correlation function for the inhomogeneous system, even if it had been known). Since using $G_0$ in WDA does not guarantee any improvement in describing properties of the homogeneous gas itself, one may use the freedom in $G(r)$ to adjust the WDA so that the calculated local field factor (and thus linear response function) is as accurate as possible. Inversion of eq.(10) yields $G(q)$ for a given $K_{xc}(q)$. It does not guarantee, however, that the result will be physical. So, as a first step, let us analyze Eq. (10). Let us first mention that the real $K_{xc}(q)$ changes its behavior from the long range limit to the short range limit near $q = 2k_F$, which plays the role of the inverse length scale. What is the characteristic length scale in WDA? To find that, we write $G_q = -\varphi(p/Q)/n$, with the condition $\varphi(0) = 1$, where $Q$ is some constant (both the Gunnarsson-Jones and the Gritsenko et al. functions are of this form). Then

$$W_0 = \frac{2}{\pi}\int_0^\infty G_q dq = -\frac{2Q}{\pi n_0}\int_0^\infty \varphi(x)dx = \frac{2\epsilon_{xc}(n_0)}{n_0}. \quad (11)$$

If we now define $Q(n) = -\pi\epsilon_{xc}(n)$, then the second condition on $\varphi(x)$ becomes $\int_0^\infty \varphi(x)dx = 1$. These two conditions reduce our freedom to adjust $G_q$: since the characteristic size of $\varphi(x)$ is of order of 1, the wave vector dependence of $G_q$ is defined by the ratio $q/Q = -q/\pi\epsilon_{xc}$. Apparently, varying the shape of the function $\varphi$ will not significatntly change the lentgh scale of the resulting $K_{xc}(q)$. Explicitly density-dependent functions $\varphi(x,n)$ may be needed to shift the hump from its position of $q \approx 1.5k_F$ to $q \approx 2k_F$. It is still an open question whether or not a physically sound function can be found with this property.

However, even if the "$2k_F$" problem is fixed, another, probably even more important problem remains: the short wave length behavior of $K_{xc}$. It is easy to see that if $G(q) \to 0$ at $q \to \infty$, then $W_p \to const/p^2$ at $p \to \infty$, and so does, according to Eq.(10), $K_{xc}$. On the other hand, as mentioned above, the correct $K_{xc}(q)$ goes to a constant at $q \to \infty$ as $q^2$, although the constant is smaller than $K_{xc}^{LDA}$. This result was predicted by Holas [41] and is physically important: it comes from the exchange-correlation contribution to kinetic energy (which is essentially local and decays slower with $q$ than the interaction part of $E_{xc}$). The present WDA misses

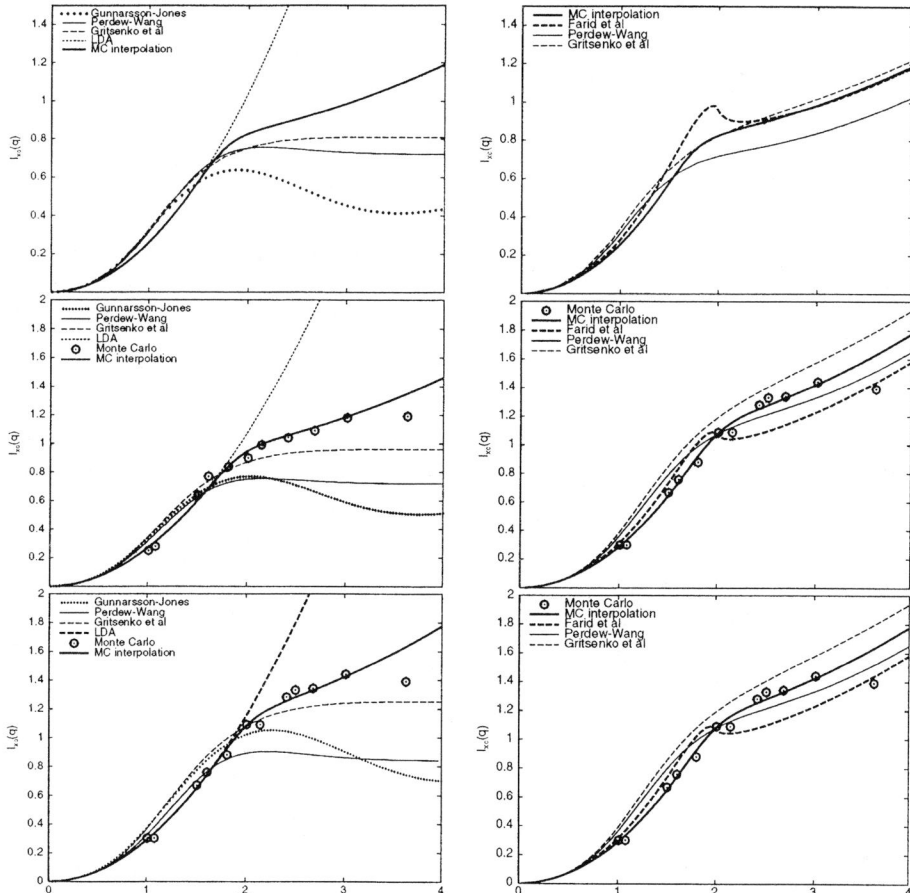

**FIGURE 2.** Left panel: exchange-correlation local field factor in the WDA of Ref. [30] (Gunnarsson-Jones), Ref. [39] (Gritsenko et al.), and derived from the homogeneous electron gas pair correlation function (Perdew-Wang), as compared with the Monte Carlo results (Monte Carlo) and the interpolating formula thereof (MC interpolation), as given in Ref. [40]. Densities, from top to bottom, correspond to $r_s = 1, 2, 5$. Right panel: the same for the modified WDA of Eq.16. Also the analytical formula of Farid et al. Ref. [42] is shown.

the corresponding physics. Fortunately, this is easy to correct. Farid et al. [42] tabulated the coefficient $\gamma$ that defines the asymptotic behavior of $K_{xc}(q \to \infty)$ as $K_{xc}(q \to \infty) = -\frac{4\pi}{q^2}\gamma(n)\frac{q^2}{k_F^2}$, where $\gamma(n)$ is a universal function, parametrized in Ref. [42]. Let us now modify the function $G(r)$

$$G(r) = G_1(r) + G_2(r) = A\delta(r)/4\pi r + G_2(r), \qquad (12)$$

Since $\int G_1(r)r^2 dr = 0$, the normalization condition for $G_2$ is the same as for $G$ itself. Since $4\pi \int G_1(r) r dr = A$, the LDA limit condition for $G_2$ becomes

$$4\pi \int G_2(r) r dr = 2\tilde{\epsilon}_{xc}(n)/n,$$
$$\tilde{\epsilon}_{xc}(n) = \epsilon_{xc}(n) - An/2. \qquad (13)$$

Thus

$$A = -4\pi\gamma(n)/k_F^2,$$
$$\tilde{\epsilon}_{xc}(n) = \epsilon_{xc}(n) + 2\pi n\gamma(n)/k_F^2 \qquad (14)$$

Now $G_p = G_{2p}$, $W_p = A + W_{2p}$, and $W'_p = A' + W'_{2p}$,

$$K_{xc}(q) = A + W_{2q} - n^2 G_{2q}(A' + W'_{2q} + W'_{2,0}) + n^2(n^2 G_q^2 W'_{2,0})'$$
$$= A - n_0^2 G_{2p} A' + \tilde{K}_{xc}, \qquad (15)$$

where $\tilde{K}_{xc}$ is calculated from $\tilde{\epsilon}_{xc}$ in exactly the same way as $K_{xc}$ is calculated from $\epsilon_{xc}$. The corresponding functional for the exchange-correlation energy is

$$E_{xc}^{AWDA} = \frac{1}{2}\int \frac{n(\mathbf{r})n(\mathbf{r}')}{(|\mathbf{r}-\mathbf{r}'|)} G[|\mathbf{r}-\mathbf{r}'|, \bar{n}(\mathbf{r})] d\mathbf{r}d\mathbf{r}' + \int n(\mathbf{r})\tilde{\epsilon}_{xc}[n(\mathbf{r})]d\mathbf{r}. \qquad (16)$$

Here $G(r)$ is normalized to $\tilde{\epsilon}_{xc}(\bar{n})$ instead of $\epsilon_{xc}(\bar{n})$. In practice, the first term gives rise to the standard expression for the WDA potential [22,21], and the second yields two additional terms, one from the variation of $n(\mathbf{r})$, and the other arising from $\delta\bar{n}(\mathbf{r}')/\delta n(\mathbf{r})$. Since we do *not* require that $G(r) = \int_0^1 [g(r;\lambda,\bar{n}) - 1]d\lambda$, where $g$ corresponds to the uniform gas, but rather consider it to be a flexible function satisfying two normalization conditions, further improvement of the method should be possible along the line described in the previous paragraph, namely the freedom in choosing $G(r)$ can be used to yield $K_{xc}$ according to Eq. (15) close to the linear response of the homogeneous electron gas, including correct behavior near $q = 2k_F$. In the right panel of Fig. 2, we show $I_{xc}$ calculated according to Eq. 16 with the different functional form of $G(r)$. Clearly, the results are much better than either the LDA or "conventional" WDA.

In short, we have calculated the exchange-correlation local field function $K_{xc}$ in the WDA, and found that besides the expected improvement over the LDA it has two major deficiencies: (1) it does not have correct asymptotic behavior at $q \to \infty$, and (2) the characteristic feature at $q \approx 2k_F$ is displaced towards smaller $q$'s. The former can be easily corrected by adding a delta-function component to $G(r)$, which results in Eq. (16). The latter is harder to fix, but there are still unused degrees of freedom in the formalism which may be used to tune the behavior near $2k_F$. Intuitively (cf. Ref. [38]), a method which retains exact one-electron limit of WDA, and at the same time is accurate in the opposite limit of the nearly uniform electron gas, seems promising for practical applications. However, tests on real materials will be needed to determine whether or not this modification of the WDA is advantageous in practice.

## V  CONCLUSIONS

WDA calculations of the equilibrium volume of several oxides show that with the electron gas form of $G$ and shell partitioning, the WDA yields much improved volumes over the LDA and GGA. No case was found where the WDA degrades the LDA results. Phonon frequencies and the ferroelectric distortion in $KNbO_3$ are in good accord with LDA predictions provided that the LDA calculations are performed at the experimental volume, which is effectively the same as the WDA volume. The direction of the ferroelectric soft mode is in good agreement with experiment, but all exchange correlation functionals tested yield a distortion that is 20% smaller than that reported in the powder neutron experiment of Hewat. [37] The implication is that this displacement should be re-examined experimentally. The above results combined with the fact that the WDA linear response of the electron gas is in substantially better agreement with Monte Carlo data than the LDA, and that, if desired, the approach has the flexibility to further improve this property, is suggestive that the WDA may be a generally more reliable method than the LDA.

We are thankful for helpful discussions with L.L. Boyer, H. Krakauer, O. Gunnarsson and R. Resta. This work is supported by the Office of Naval Research. Calculations were performed using DoD HPCMO facilities at NAVO and ASC. The CHSSI DoD Planewave code was employed for some of the calculations.

## REFERENCES

1. See for example, *Proceedings of The Fourth Williamsburg Workshop on First Principles Calculations for Ferroelectrics*, R.E. Cohen, ed., Ferroelectrics **194**, 1 (1997).
2. R.E. Cohen, Nature (London) **358**, 136 (1992).
3. R.E. Cohen and H. Krakauer, Ferroelectrics **136**, 65 (1992).
4. D.J. Singh and L.L. Boyer, Ferroelectrics **136**, 95 (1992).
5. A.V. Postnikov, T. Neumann, G. Borstel and M. Methfessel, Phys. Rev. B **48**, 5910 (1993)
6. A.V. Postnikov, T. Neumann and G. Borstel, Ferroelectrics **164**, 101 (1995).
7. R. Yu and H. Krakauer, Phys. Rev. Lett. **74**, 4067 (1995).

8. C.-Z. Wang, R. Yu and H. Krakauer, Ferroelectrics **194**, 97 (1997).
9. D.J. Singh, Ferroelectrics **164**, 143 (1995).
10. D.J. Singh, Ferroelectrics **194**, 299 (1997).
11. R.E. Cohen and H. Krakauer, Phys. Rev. B **42**, 6416 (1990).
12. D.J. Singh, Phys. Rev. B **52**, 12559 (1995).
13. S. Teslic, T. Egami and D. Viehland, Ferroelectrics **194**, 271 (1997).
14. H. Fujishita and S. Katano, J. Phys. Soc. Jpn. **66**, 3484 (1997).
15. A. Dal Corso, S. Baroni and R. Resta, Phys. Rev. B **49**, 5323 (1994).
16. R. Resta, Ferroelectrics **194**, 1 (1997).
17. S. DallOlio, R. Dovesi and R. Resta, Phys. Rev. B **56**, 10105 (1997).
18. G. Saghi-Szabo, R.E. Cohen and H. Krakauer, unpublished.
19. O. Gunnarsson, M. Jonson, and B.I. Lundquist, Phys. Lett. **59A**, 177 (1976).
20. O. Gunnarsson, M. Jonson, and B.I. Lundquist, Solid State Commun. **24**, 765 (1977).
21. J.A. Alonso and L.A. Girifalco, Phys. Rev. B, **17**, 3735 (1978).
22. O. Gunnarsson, M. Jonson, and B.I. Lundquist, Phys. Rev. B, **20**, 3136 (1979).
23. D.J. Singh, Phys. Rev. B, **48**, 14099 (1993)
24. M. Sadd and M.P. Teter, Phys. Rev. B, **54** 13643 (1996).
25. J.P.A. Charlesworth, Phys. Rev. B, **53**, 12666 (1996).
26. Ref. [25] is an exception, but note that the LDA results of this work differ from most previous calculations.
27. *Theory of the inhomogeneous electron gas*, Ed. by S. Lundqvist and N.H. March (Plenum, New York, 1983).
28. J.P. Perdew and Y. Wang, Phys. Rev. B **46**, 12947 (1993).
29. See for example, S. Goedecker and C.J. Umrigar, Phys. Rev. A **55**, 1765 (1997), and references therein.
30. O. Gunnarsson and R.O. Jones, Physica Scripta, **21**, 394 (1980).
31. D.J. Singh, S.A. Kajihara, S.G. Kim, C. Woodward, *DoD Planewave: A General Scalable Density Functional Code*, cst-www.nrl.navy.mil/people/singh/planewave.
32. D.J. Singh, H. Krakauer, C. Haas and A.Y. Liu, Phys. Rev. B **46**, 13065 (1992).
33. N. Troullier and J.L. Martins, Phys. Rev. B **43**, 1993 (1991).
34. D.J. Singh *Planewaves Pseudopotentials and the LAPW Method* (Kluwer, Boston, 1994).
35. W. Zhong, R.D. King-Smith and D. Vanderbilt, Phys. Rev. Lett. **72**, 3618 (1994).
36. M.D. Fontana, G. Metrat, J.L. Servoin and F. Gervais, J. Phys. C **17**, 483 (1984).
37. A.W. Hewat, J. Phys. C **6**, 2559 (1973).
38. A.R. Williams and U. von Barth, in Ref. [27].
39. O.V. Gritsenko, A. Rubio, L.C. Balbás, and J.A. Alonso, Chem. Phys. Lett **205**, 348 (1993)
40. S. Moroni, D.M. Ceperley, and G. Senatore, Phys. Rev. Lett **75**, 689 (1995).
41. A. Holas, in *Strongly Coupled Plasma Physics*, ed. by F.J. Rogers and H.E. De Witt (Plenum, NY, 1987)
42. B. Farid, V. Heine, G.E. Engel, and I.J. Robertson, Phys. Rev. B **48,** 11 602 (1993).

# Structural and vibronic properties of perovskites studied by using the Perdew - Burke - Ernzerhof GGA.

C.O Rodriguez*, D.L Novikov[†], M.G. Stachiotti** and N.E. Christensen[‡]

* IFLYSIB, Grupo de Física del Sólido,
C.C.565, 1900 La Plata, Argentina
[†] Science and Technology Center for Superconductivity, Department of Physics and Astronomy,
Northwestern University, Evanston, IL, 60208. USA
** Instituto de Física Rosario, Universidad Nacional de Rosario,
27 de Febrero 210 Bis, 2000 Rosario, Argentina
[‡] Institute of Physics and Astronomy, Aarhus University,
DK-8000 Aarhus C, Denmark.

**Abstract.**

The Perdew - Burke - Ernzerhof (PBE) [1] proposal of the Generalized Gradient Approximation (GGA) to Density Functional theory (DFT) was tested against its intrinsic uncertainty in choosing the coefficient $\kappa$, which is directly associated to the degree of localization of the exchange - correlation hole. A study of structural instabilities and zone center phonon eigenvalues and eigenvectors is presented. The originally proposed value of 0.804 (best suited for atoms and molecules) works well for some solids, whereas for others it must be decreased in order to predict equilibrium lattice parameters in good agreement with experiments. The effects on the structural instabilities and zone center phonon modes of varying $\kappa$ are examined for selected perovskites. The need of varying $\kappa$ from one one system to another reflects the fact that the localisation of the exchange-correlation hole is system dependent, and the sensitivity of the physical properties to its actual value illustrates the necessity of finding a universal function for $\kappa = g(\nabla^2 n/(2k_F)^2 n)$.

## INTRODUCTION

Perovskites have the general formula $ABO_3$. A variety of interesting properties depend on which elements, A and B, actually are commbined with oxygen to form the perovskite. Perovskites exhibit rich phase diagrams. The oxygen atoms are arranged in octahedra in the cubic symmetry surrounding B atoms, or distorted octahedra in crystals with lower symmetries. It is basically this B-metal-oxygen octahedral complex that determines the different characteristics of these compounds resulting in distinct electron distributions in the B-O covalent bonds. These perovskites have been extensively studied because of their possible technological applications, some of which are related to their ferroelectric (FE) properties. But a basic understanding of the mechanisms of their various transformations of phase and structure is also of great interest.

In this work we study the cubic phases of $SrTiO_3$, $KNbO_3$ and $BaTiO_3$, since they exemplify the rich variety of structure related and dynamical effects of the perovkite materials. The two compounds, $KNbO_3$ and $BaTiO_3$, both experience a similar sequence of phase transitions when temperature is lowered, with transition temperatures which have an approximately constant ratio : $T(BaTiO_3)/T(KNbO_3) = 5/3$. At high temperatures they are in the cubic paraelectric perovskite structure. The sequences to the ferroelectric phases are (1) from cubic to tetragonal, (2) tetragonal to orthorhombic and, finally (3) orthorhombic to rhombohedral. Thus the dynamics of these materials should be similar but with different energy scales.

$SrTiO_3$ has the simple cubic perovskite structure at high temperature, suffers an antiferrodistortive transition at 105 K and goes to a tetragonal phase in which the oxygen octahedra have rotated in opposite senses in neighboring unit cells. In the range 50-100 K there is a softening of the ferroelectric polar phonons that appears to extrapolate to a ferroelectric transition close to 20 K. It is believed that quantum fluctuations suppress long-range ferroelectric order at low temperatures since no FE transition is observed [2]. $SrTiO_3$ is therefore referred to as an incipient ferroelectric.

These ferroelectric perovskites, in particular their structural phase transitions, have been the subject of intense theoretical work, but only recently has first principles density functional theory been applied with success to analyze their properties at a microscopic level. The local spin-density approximation (LSDA) has provided a useful framework, although it systematically underestimates equilibrium volumes, and several properties are poorly reproduced at this volume. So, in such studies an ad-hoc correction has been made: the experimental rather than the calculated equilibrium volumes are employed. This has motivated a number of attempts to go beyond the LSDA in the search of a better framework. Specifically in the case of ferroelectric perovskites Generalized Gradient Aprroximation (GGA) and Weighted Density Approximation (WDA) have been applied in the later years [3].

In the Kohn-Sham density functional theory only the exchange-correlation energy $E_{XC} = E_X + E_C$ which is a functional of the electron spin densities must be approximated and can be expressed for slowly varying densities as the volume integrals of $n$ times $\epsilon_{xc}^{unif}$ in the LSDA case and $f(n\uparrow, n\downarrow, \nabla n\uparrow, \nabla n\downarrow)$ for the GGA case. For practical calculations the exchange-correlation energy density of a uniform electron gas, $\epsilon_{xc}^{unif}(n\uparrow, n\downarrow)$, and $f$ must be parametrized. The form of $\epsilon_{xc}^{unif}$ is now well established but which is the best choice of $f$ is still under debate.

In the recent work of PBE a simple derivation of a simple GGA-$f$ was presented which improves over the previous PW91 [4] in an accurate description of the linear response of the uniform gas, behaving correctly under uniform scaling and giving a smoother potential. The new PBE retains the correct features of LSDA and combines them with the most energetically important features of gradient corrected nonlocality. Studies of atomization energies for small molecules gave essentially the same results as PW91.

However, a remainig uncertainty relates to the value of the coefficient $\kappa$, in the spin-polarized enhancement factor $F_X(s)$ which is directly associated to the degree of localization of the exchange-correlation hole. In their original work PBE proposed $F_X(s) = 1 + \kappa - \kappa/(1 + \mu s^2/\kappa)$ which satisfied the inequality $F_X \leq 1.804$ with $\kappa = 0.804$ and with the value of $\mu \cong 0.21951$. A reduced value of $\kappa$ to the originally proposed 0.804 would worsen most atomic and molecular results, but might improve results in solids. The parameter $\kappa$ might be a weak function of the reduced Laplacian, $\kappa = g(\nabla^2 n/(2k_F)^2 n)$, which would also satisfy the condition 'D' in the PBE paper of uniform density scaling, i.e. that $E_X$ scales like $\lambda$.

In the present work we thus take $SrTiO_3$, $BaTiO_3$ and $KNbO_3$ as representative examples of the ferroelectric perovskites in their cubic phase. The PBE proposal of the GGA for the exchange correlation energy within DFT is tested against its intrinsic uncertainty in choosing the coefficient $\kappa$ and also compared to the LSDA results. For this we have selected three structure related properties which sense the energetics which are involved in the physics of the ferroelectric instabilities and phase transitions: (1) the energy volume curves from which lattice parameters and bulk moduli are derived (2) $\Gamma_{15}$ modes for the cubic lattice and the influence of volume and (3) the tendency to suffer an FE transition when the B atom is displaced in the (001), (110) and (111) directions. This distortion pattern does not correspond to the actual ferroelectric mode which will most probably involve not only z-axis displacements of the B cation, but also motions of the the oxygens and the A cations. Further, the distortion modes are different for the different materials. Our choice of a unique pattern is convenient for a comparison between materials, and it allows an analysis of possible instabilities for different displacements.

A study of LSDA and PBE-GGA ($\kappa$=0.804) has recently been made for $s - p$ materials [5], and another, similar to the present, also including magnetic effects, has been made for the 3d, 4d and 5d transition metal series [6].

# MODEL AND COMPUTATIONAL DETAILS

The calculations presented in this work were performed within the LSDA and the PBE-GGA to density functional theory, using the full-potential LAPW method. In this method no shape approximation on either the potential or the electronic charge density is made. We use the WIEN97 implementation of the method [7] which allows the inclusion of local orbitals (LO) in the basis, improving upon linearization and making possible a consistent treatment of semicore and valence states in one energy window hence insuring proper orthogonality [8]. The Ceperley Alder parametrization for the LSDA exchange-correlation potential is used [9].

The muffin-tin sphere radii ($R_i$)= 2.0, 1.95, 1.50, 2.0, 2.0, and 1.90 a.u. were used for Sr, Ti, O, Ba, K, and Nb, respectively. The value of the parameter $RK_{max}$, which controls the size of the basis sets in these calculations, was chosen to be 8 for all systems studied. This gives well converged basis sets consisting of approximately 1000, 1150 and 1200 LAPW functions plus local orbitals for $SrTiO_3$, $BaTiO_3$ and $KNbO_3$ respectively. In order to obtain sufficient accuracy with inclusion of the GGA functionals a relatively high plane wave cut-off energy (256 Ry) was necessary in order to obtain a converged interstitial representation of the potential. This choice of parameters was justified by performing calculations for other $R_i$ values and by increasing $RK_{max}$. Although an increase of $RK_{max}$ from 8 to 9 had no significant effects on most of physical properties that we studied, convergence of the cohesive energy to a precision of 1 mRy per formula unit required $RK_{max}$=10 with the choice of $R_i$ given above.

We introduced LO to include the following orbitals in the basis set: Ti-$3s$ and $3p$, Ba-$5s$, $5p$ and $4d$, Sr-$4s$ and $4p$, K-$3s$ and $3p$, Nb-$4s$ and $4p$ and O-$2s$.

Integrations in reciprocal space were performed using the special points method. We used 6×6×6 meshes which represent 250 k-points in the first Brillouin zone. This corresponds to 28 special k-points in the irreducible wedge for the rhombohedral structure, 36 for the orthorhombic, 18 for the tetragonal and 10 in the cubic. Convergence tests indicates that only small changes result from increasing to a denser k-mesh.

Of the two allowed symmetries for zero wave-vector ($\Gamma$ point) we only studied the triply degenerate $\Gamma_{15}$ modes. There are three of these, not counting the zero frequency acoustic mode. These, are infrared active and include the "soft- ferroelectric" mode. We determined the phonon frequencies and polarizations for this particular symmetry by calculating atomic forces for several small displacements ( $\sim 0.001 \mathring{A}$ ) consistent with the symmetry and small enough to be in the linear regime. From the force as a function of displacement the dynamical matrix was obtained.

## RESULTS

In Table 1 the lattice parameters and bulk moduli as derived from energy-volume curves are shown when using the LSDA and the PBE-GGA (with $\kappa = 0.804$ as originally proposed by PBE and which will give very similar results as the PW91), together with the experimental values. The equilibrium volumes obtained from the LSDA calculations are 3.3 %, 5.3 % and 4.3 % smaller than experiment for $SrTiO_3$, $BaTiO_3$ and $KNbO_3$ respectively. This agrees with the generally observed tendency to overbinding in this approximation. The LSDA bulk moduli are overestimated when evaluated at these too small volumes. The values calculated at the experimental volumes (see the values in brackets) agree better with the experiments.

As is the general trend in the application of GGA functionals to solids, it expands and softens bonds, an effect that sometimes corrects and sometimes overcorrects the LSDA prediction. All GGA calculations with $\kappa$=0.804 overestimate the equilibrium volumes, namely by 3.3 %, 1.6 % and 1.6 % for $SrTiO_3$, $BaTiO_3$ and $KNbO_3$, respectively.

**TABLE 1.** Lattice parameter (in au) and Bulk modulus (in GPa) obtained in the present study for LSDA and PBE-GGA($\kappa = 0.804$) and compared with experimental values. In brackets: bulk modulus at the experimental equilibrium volume.

|  |  | LSDA | PBE-GGA ($\kappa = 0.804$) | Experiment |
|---|---|---|---|---|
| $SrTiO_3$ | a | 7.30 | 7.46 | 7.38 [a] |
|  | B | 204(176) | 167(194) | 174-183[a] |
| $BaTiO_3$ | a | 7.44 | 7.61 | 7.57[a] |
|  | B | 195(155) | 160(173) |  |
| $KNbO_3$ | a | 7.48 | 7.63 | 7.59 [a] |
|  | B | 206(155) | 171(186) | 138 [a] |

[a] Ref. [10]

We agree with previous calculations by Singh [11] for $KNbO_3$, but our results for $BaTiO_3$ are different. We have not found any GGA calculations to compare with in the case of $SrTiO_3$. In Fig. 1 the values of $V/V_o$ ($V_o$, the corresponding experimental value) are given for $SrTiO_3$, $BaTiO_3$ and $KNbO_3$ when the value of $\kappa$ in the PBE-GGA is decreased from 0.804 to 0.5. As can be seen both $BaTiO_3$ and $KNbO_3$ would give perfect determination of the lattice parameter ($V/V_o=1.$) for $\kappa \sim 0.6$. In the case of $SrTiO_3$ the value should be further reduced below 0.5.

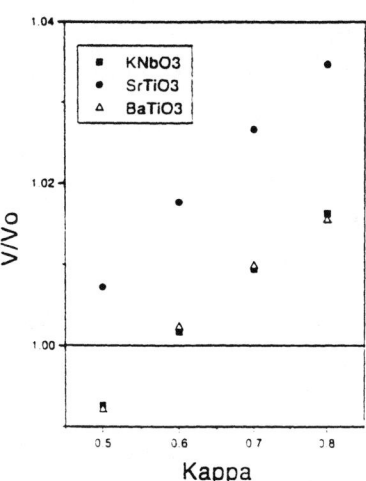

**FIGURE 1.** Equilibrium volumes relative to the experimental ones of $SrTiO_3$, $BaTiO_3$ and $KNbO_3$ as a function of the PBE-GGA intrinsic parameter $\kappa$.

In Table 2. the frequencies and eigenvectors of the $\Gamma_{15}$ triply degenerate mode are shown. For all three compounds the LSDA answer calculated at the experimental lattice constant is detailed. An unstable mode appears that corresponds to the vibration of the B and A atoms against the O atoms. For the case of $SrTiO_3$ we detail the same calculation at a smaller lattice parameter, and see how frequencies change when the volume is reduced : the instabilities even dissapear if compression is carried on to the LSDA optimized volume. Our results for frequencies and eigenvectors for $SrTiO_3$ and $BaTiO_3$ are nearly identical to those of Ghosez [13] but differ (mostly in the soft mode frequency, 183i vs 115i) with the results of Singh for $KNbO_3$ [12]. A comparisson with experiment has already been done in the work of Ghosez [13] and Singh [12] already cited, showing an overall agreement. For the case of $SrTiO_3$ the instability dissapears when PBE-GGA is used. Two different values of $\kappa$ (0.804 and 0.6) were studied. The soft-mode frequency reduces somewhat when $\kappa$ reduces. Note that the eigenvector is practically unchanged.

**TABLE 2.** Frequencies and eigenvectors for $\Gamma_{15}$ modes in $SrTiO_3$, $BaTiO_3$ and $KNbO_3$. It exemplifies the modifications introduced by differerent lattice constants, LSDA, PBE-GGA and use of different values of $\kappa$.

| freq. (cm$^{-1}$) | | eigenvector | | | |
|---|---|---|---|---|---|
| | | Sr | Ti | O1 | O2 and O3 |
| $SrTiO_3$ | 87i | 0.30 | 0.43 | -0.47 | -0.70 |
| LSDA, a=7.38 au. | 149 | 0.65 | -0.73 | -0.041 | -0.15 |
| | 522 | 0.041 | -0.031 | -0.82 | 0.55 |
| $SrTiO_3$ | 59i | 0.34 | 0.39 | -0.45 | -0.72 |
| LSDA, a=7.358au. | 156 | 0.63 | -0.76 | -0.025 | -0.09 |
| | 532 | 0.038 | -0.009 | -0.83 | 0.54 |
| $SrTiO_3$ | 69 | 0.34 | 0.39 | -0.44 | -0.73 |
| GGA, a=7.38au. | 166 | 0.63 | -0.76 | -0.05 | -0.08 |
| $\kappa = 0.804$ | 546 | 0.027 | 0.018 | -0.85 | 0.53 |
| $SrTiO_3$ | 58 | 0.35 | 0.38 | -0.44 | -0.73 |
| GGA, a=7.38au. | 165 | 0.63 | -0.77 | -0.04 | -0.075 |
| $\kappa = 0.6$ | 543 | 0.029 | 0.014 | -0.85 | 0.53 |
| | | Ba | Ti | O1 | O2 and O3 |
| $BaTiO_3$ | 242i | 0.016 | 0.65 | -0.64 | -0.39 |
| LSDA, a=7.57 au. | 158 | 0.64 | -0.55 | -0.26 | -0.47 |
| | 449 | 0.03 | -0.24 | -0.67 | 0.70 |
| | | K | Nb | O1 | O2 and O3 |
| $KNbO_3$ | 183i | 0.018 | -0.59 | 0.55 | 0.59 |
| LSDA, a=7.50 au. | 171 | 0.88 | -0.36 | -0.15 | -0.25 |
| | 480 | 0.022 | -0.075 | -0.76 | 0.64 |

The ferroelectric instabilty involves displacement of the A and B atoms against the O atoms. One can study these by making these displacements proportional to the experimental measured ones. For example as in the studies of Ghosez [13] and Singh. [11].

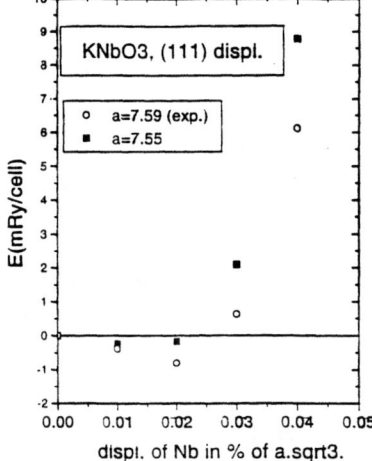

**FIGURE 2.** Total energy versus Nb displacements in the (111) direction for $KNbO_3$, obtained at the experimental lattice constant and a compressed one. The LSDA exchange-correlation energy was used.

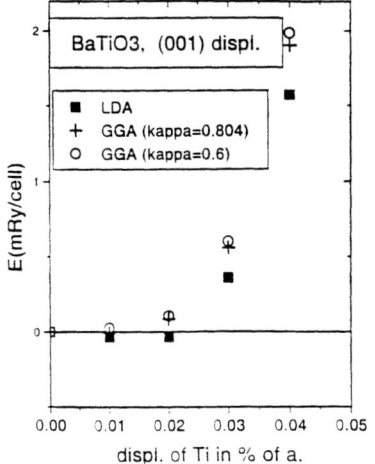

**FIGURE 3.** Total energy versus Ti displacements in the (001) direction for $BaTiO_3$, obtained at the experimental lattice constant for different exchange-correlation energies: LSDA and PBE-GGA ($\kappa = 0.804$ and $0.6$).

We have performed total energy calculations as a function of displacement of the B atom alone in the (111), (110) and (001) directions. Although this is not the actual pattern it still allows us to sense the energetics when different exchange-correlations are used and even sense the tendency to suffer a ferroelectric transition. In Fig. 2 the LSDA energy as a function of displacement of the B atom in the (111) direction in KNbO$_3$ for the experimental lattice parameter is shown. An instability with an energy lowering of 1 mRy/cell which becomes smaller for K moving in the (110) direction and even smaller for a (001) displacement, describing the correct sequence of reduction of the potential wells. When the lattice parameter is reduced to 7.55 a.u. the double well nearly dissapears in resemblance to what happens when the actual ferroelectric distortion pattern is used [11]. In Figs. 3 and 4 we will concentrate on results for BaTiO$_3$. They show the energy changes for displacements of the Ti atom along (001) and (111). The calculations were made for LSDA, and PBE-GGA with $\kappa$=0.804 and 0.6. In the LSDA case an instability occurs with a much larger well in the (111) case. This pattern is thus consistent with the experimental cubic → tetragonal → orthorhombic → rhombohedral, phase changes with temperature and with the calculations of Ghosez [13]. When PBE-GGA is used the instabilities are reduced or nearly disappear and their energetics are very little dependent on the $\kappa$ value used.

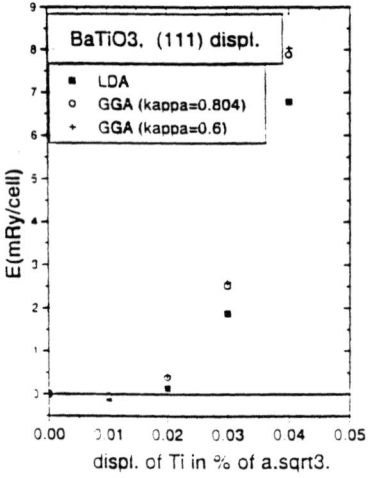

**FIGURE 4.** Total energy versus Ti displacements in the (111) direction for BaTiO$_3$, obtained at the experimental lattice constant for different exchange-correlation energies: LSDA and PBE-GGA ($\kappa = 0.804$ and 0.6).

## CONCLUSIONS

Results of LSDA and PBE-GGA calculations have been reported for structural properties relevant for the physics of the ferroelectric instabilities of perovskites. In the case of PBE-GGA calculations were made with different values of the coefficient $\kappa$ which is related to how the localization of the exchange correlation hole is treated. SrTiO$_3$, BaTiO$_3$ and KNbO$_3$ were taken as representative example materials. The usual underestimation of LSDA for the equilibrium volume is obtained (an average of approximately 4% for the three materials )in complete accord with all other theoretical studies. GGA, on the other hand overestimates the equilibrium volumes when $\kappa=0.804$ (the originally proposed value) is used. For values of 0.6 in the cases of $BaTiO_3$ and $KNbO_3$ and around 0.5 in $SrTiO_3$ a correct equilibrium volume ($V/V_o=1.0$) can be obtained. A strong volume dependence of the ferrolelectric mode frequency and instability energy appears and are qualitatively similar in both LSDA and PBE-GGA. For a given lattice parameter and in particular for the cases where the potential wells where atoms move are very flat the small differences in energetics between LSDA and PBE-GGA might give different predictions which would modify the underlying physics. The use of different $\kappa$ values gives quite similar energetics.

Acknowledgments
This work was supported by Consejo Nacional de Investigaciones Científicas y Técnicas de la República Argentina, the Danish Natural Science Research Council (Grant No. 9600998) and the Commission of the European Communities, Contract No. CI1*-CT92-0086. M.S. also thanks Consejo de Investigaciones de la Universidad Nacional de Rosario for support.

# REFERENCES

1. J.P. Perdew, K. Burke and M. Ernzerhof, Phys.Rev.Lett.**74**, 4067 (1995).
2. K.A. Müller and H. Burkard, Phys. Rev. B **19**, 3593 (1979); R. Viana, P. Lunkenheimer, J. Hemberger, R. Böhmer and A. Loidl, Phys. Rev. B **50**, 601 (1994);
3. D.Singh, Ferroelectrics **194**, 299 (1997).
4. J.P. Perdew, J.A. Chevary, S.H. Vosko, K.A. Jackson, M.R. Pederson, D.J. Singh and C. Fiolhais, Phys.Rev.B **46**, 6671 (1992).
5. In-Ho Lee and R.M. Matin, Phys.Rev.B **56**, 7197 (1997).
6. D.L. Novikov and A.J. Freeman , Bull. Am. Phys. Soc. **42**, 149, 1997. ; E.L. Peltzer y Blanca, C.O. Rodriguez, J. Shitu, and D.L. Novikov, (to be published)
7. P. Blaha, K. Schwarz, P. Dufek and R. Augustyn, WIEN97, Technical University of Vienna 1997. (Improved and updated Unix version of the original copyrighted WIEN-code, which was published by P. Blaha, K. Schwarz, P. Sorantin and S.B. Trickey, Comput. Phys. Commun. **59**, 399 1990).
8. D. Singh, Phys. Rev. B **43**, 6388 (1991)
9. J.P Perdew and Y. Wang , Phys. Rev. B **45**, 13244 (1992).
10. T. Mitsui et al., *Ferro- and Antiferroelectric Substances*, Landolt-Börnstein, Group III, Vol. 3, edited by K.-H. Hellwege and A. M. Hellwege (Springer-Verlag, Berlin, 1969).
11. D.Singh, Ferroelectrics **164**, 143 (1995).
12. D.Singh and L.Boyer, Ferroelectrics **136**, 95 (1992).
13. P.H.Ghosez, PhD thesis, Universite Catholique de Louvain, 1997.

# Modelling and microscopic dynamics of KNbO$_3$ from first-principles

M.G.Stachiotti*, M.Sepliarsky*, R.L.Migoni* and C.O.Rodriguez[†]

* Instituto de Física Rosario, Universidad Nacional de Rosario,
27 de Febrero 210 Bis, 2000 Rosario, Argentina
† IFLYSIB, Grupo de Física del Sólido, C.C.565, 1900 La Plata, Argentina

**Abstract.** First-principles methods for computer simulations are rapidly improving in speed and accuracy. However, the highly demanding nature of these computations has the consequence that direct simulations of temperature driven structural transitions in perovskites are not computationally feasible at present.

In this work, we present an ab-initio supported determination of interatomic potentials, in the framework of a shell model, in order to study the structural behavior of KNbO$_3$ at finite temperature. To this purpose, the adiabatic potential is evaluated for different ferroelectric distortions and the model potential parameters are fitted to first-principles total energy calculations. The phase diagram as a function of temperature is obtained through a constant-pressure molecular dynamics simulation. The analysis of the dynamical structure factor in the different phases provides a deeper insight into the nature of the dynamics associated with the structural transitions.

## INTRODUCTION

Due to the wide variety of exhibited physical properties, the ABO$_3$ perovskites constitute a particularly interesting family for the study of the relations between the structural phase transitions, their mechamism and sequency. Among the ferroelectric ones, KNbO$_3$ has received a great theoretical and experimental attention. This compound undergoes a sequence of three phase transitions. With decreasing temperature it transforms from cubic paraelectric to tetragonal ferroelectric at 701 K, becomes orthorhombic at 488 K, and finally rhombohedral at 210 K.

A still open question concerns the dynamical nature of these structural transitions. Indeed, numerous studies have been carried out in order to elucidate whether these transitions have a displacive or an order-disorder character. A displacive mechanism driven by a soft TO mode which becomes unstable as the temperature is lowered was initially proposed; however, many results are inconsistent with this soft mode picture. If hyper-Raman and infrared reflectivity spectra in the cubic and tetragonal phases of KNbO$_3$ are interpreted in terms of a soft mode, this has to be assigned such a high damping that it can hardly be distinguished from a relaxator [1,2]. In fact, an unified interpretation of the experimental data has been given by assuming the coexistence of a relaxational mode and a soft phonon [1,3].

An alternate picture for the occurence of the structural transitions was provided by the eight-site model. According to this order-disorder model, the potential energy surface has a maximum for the cubic perovskite structure and eight degenerate minima for the [111] displacements of the transition metal ion that correspond to the low-temperature rhombohedral structure. In the cubic phase, the eight sites are occupied with equal probability, and this symmetry is broken as the temperature is lowered: four sites are occupied in the tetragonal phase, two sites in the orthorhombic phase, and finally, only one site is occupied in the rhombohedral structure. However, in the orthorhombic phase of $KNbO_3$ a broad soft mode peak is clearly observed in Raman experiments near the orthorhombic-rhombohedral transition, without any quasielastic component [3], suggesting a more displacive-like dynamics for this transition. Indeed, a more recent femtosecond time-resolved spectroscopy study for the orthorhombic phase of $KNbO_3$ rule out relaxational contributions of the same symmetry as the soft mode [4].

First principles calculations have contributed greatly to understand the origins of the transitions in $KNbO_3$ [5–8] and related perovskites [9–13]. These are accurate calculations which provide detailed information about electronic charge distributions and the character of electronic bonding. While the two more widely used approximations to Density Functional Theory, LDA and GGA, under and overestimates the equilibrium volume for $KNbO_3$ and $BaTiO_3$ [8]; they both yield correct predictions of the ground state structure and the occurrence of ferroelectric phases at the experimental volume, providing considerable insight into the nature of the soft-mode total energy surface.

First-principles techniques are more powerful than any calculation based on empirical models. Nevertheless they are quite computer demanding and simulations of temperature driven structural transitions in perovskites are not computationally feasible at present. A successful approach to study finite temperature properties has been developed on the grounds of effective Hamiltonians whose parameters were fully determined from ab-initio calculations. This scheme has been applied to many ferroelectric perovskites [14–16]. However, the restricted dynamics considered and the lack of an atomistic description make this approach inappropriate to investigate many interesting properties, as the anharmonic lattice dynamics, the influence of point defects, the structural behavior of mixed compounds, etc.

The goal of this work is to obtain an atomistic model which describes accurately the dynamical properties and phase transitions sequence of $KNbO_3$. To this purpose, interatomic potentials are determined, in the framework of a shell model, by fitting ab-initio total energies calculations for different ferroelectric distortions. The temperature behavior of the system and the dynamical nature of its structural transitions are then investigated through molecular dynamics simulations.

# MODEL AND COMPUTATIONAL DETAILS

The lattice dynamics in the cubic phase of $KNbO_3$ [17] has been carried out in the framework of a shell model whith an anisotropic nonlinear core-shell interaction at the oxygen ions [18,19]. The measured phonon dispersion curves and their temperature dependence have been well reproduced in the soft-mode regime within the self-consistent phonon approximation. In this model, the $A$ and $B$ ions are considered isotropically polarizable. However, an anisotropic core-shell interaction is considered for the $O^{-2}$ ion, reflecting its site symmetry. In fact, the oxygen polarizability depends strongly on the crystal environment and additional polarization effects are expected from the $O_p$-$Nb_d$ hybridization. The above anisotropy is described by two different linear core-shell coupling constants $k_{OA}$ (in the directions of the $A$ ions) and $k_{OB}$ (in the direction of the $B$ ion). An additional fourth order contribution to the core-shell interaction along the direction of the B ion is taken into account by a coupling constant $k_{OB,B}$.

However, to obtain a model which describes the phase transition sequence of $KNbO_3$, it is necessary to allow for the homogenous strains involved in the phases. This can be achieved by replacing harmonic force constants by interatomic potentials. So, we first calculate potential parameters from the force constants by assuming pairwise $A - O$, $B - O$ and $O - O$ interactions. We choose to represent these by Buckingham potentials: $V(r) = ae^{(-br)} - \frac{c}{r^6}$. Actually, the Van der Waals term is included only for the $O - O$ interaction because it is attractive, as it turns out from the force constants. In a further step, the model parameters are improved by fitting ab-initio total energy calculations for different ferroelectric distortions, with and without lattice strain.

To evaluate the adiabatic energy surface for ferroelectric distortions, we have calculated the potential energy of the model for several atomic displacements of the transition metal sublattice. To this purpose, the shell coordinates (which represent the electronic degrees of freedom) are evaluated, for a given core configuration, by solving the adiabatic condition. Once the equilibrium solution for the shell coordinates is obtained iteratively by a steepest descent procedure, the potential energy is computed.

For the total energy calculations we used the full-potential linear muffin-tin orbital (LMTO) method, within the LDA and the Ceperly-Alder exchange-correlation potential. In the LMTO, no shape approximations are made for either the charge density or the potential. Details of the method are given elsewhere [20,21]. The calculations were performed employing the same basis set used by Postnikov et al. [5] and the Brillouin zone integrations have been well converged in the number of k-points.

For the investigation of the temperature driven structural transitions we use a Parrinello-Rahman molecular dynamics simulation. The study is carried out using the DL-POLY package [22], where the adiabatic dynamics of the electronic shells is approximated by assigning small masses to them. The runs were performed employing a Hoover constant-(T,P) algorithm with external pressure set to zero;

all cell lengths and cell angles were allowed to fluctuate. Periodic boundary conditions over 4x4x4 primitive cells were considered; the basic molecular dynamics cell therefore contained 320 ions (plus 320 shells which are additional degrees of freedom). The time step was 0.004 ps, which provided enough accuracy for the integration of the shell coordinates. The total time of each simulation, after 5 ps of thermalization, was 45 ps.

After equilibration, the dynamical structure factor $S(\mathbf{q},\omega)$ is calculated with no thermostat and barostat, i.e. using conventional energy conserving dynamics, as the space-time Fourier transform of the core-core displacement correlation function:

$$S(\mathbf{Q},\omega) = \int_{-\infty}^{+\infty} dt e^{i\omega t} \sum_{l\kappa} \sum_{l'\kappa'} e^{i\mathbf{Q}\cdot\left(\mathbf{R}_\kappa^l - \mathbf{R}_{\kappa'}^{l'}\right)} \langle \mathbf{Q}\cdot\mathbf{u}_\kappa^l(t) \mathbf{Q}\cdot\mathbf{u}_{\kappa'}^{l'}(0) \rangle \qquad (1)$$

where $\mathbf{Q}$ is an arbitrary wave vector, which can be decomposed in a reciprocal lattice vector $\mathbf{G}$ and a wave vector $\mathbf{q}$ within the first Brillouin zone, $\mathbf{Q} = \mathbf{G} + \mathbf{q}$. A gaussian smoothing procedure is applied before the time Fourier transform is performed.

## RESULTS

The total energy as a function of the transition metal displacements along [001], [011] and [111] directions is shown in Figure 1, where the results of the model are compared with the LMTO calculations.

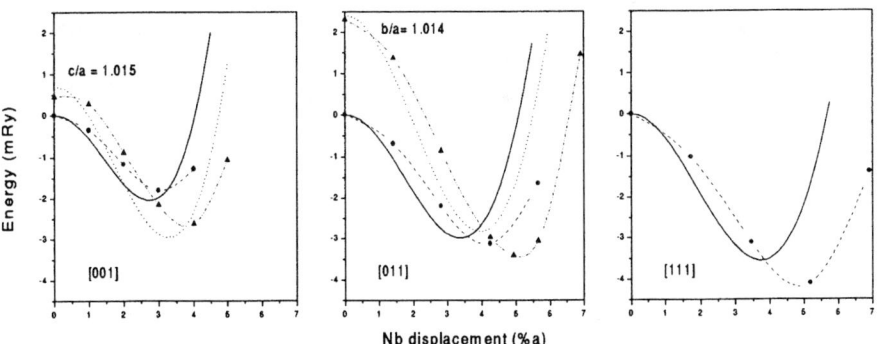

**FIGURE 1.** Total energy versus Nb displacements, obtained at the experimental lattice constant, without strain and with tetragonal and orthorrombic strains. The energies are referred to the cubic structure. The unstrained crystal results are represented with full lines for the model and circle points for the LMTO calculations. The results for the strained lattices are shown by dotted lines for the model and triangule points for the LMTO.

A reasonable agreement is achieved although the model leads to smaller values for the off-center transition metal shifts. In order to study the relevance of the lattice strain on the energetics, we performed displacements along the three mentioned directions for strained structures at the experimental values of $\frac{c}{a}$, $\frac{b}{a}$ and $\alpha$. The results are also shown in Figure 1. The model reproduces satisfactorily the strong dependence of the energy on the tetragonal and orthorhombic strains. We find a negligible effect on the total energies for the rhombohedral strain, so it is not plotted in the figure. The relative values for the energy wells of the three strained structures is consistent with the experimentally observed phase transitions sequence.

An interesting test of our modelling concerns its phonon dispersion relations. Recently, Yu and Krakauer [23] have performed first-principles phonon calculations for the ideal cubic perovskite structure of $KNbO_3$, using a linear response approach within the framework of the LAPW method. This calculation reveals structural instabilities with pronounced two-dimensional character in the Brillouin zone, corresponding to chains of displaced Nb ions oriented along the [001] directions. To check if our model is able to reproduce such kind of instabilities, we compute the phonon dispersion curves within the harmonic approximation for the Buckingham potentials and retaining only the harmonic core-shell couplings at the oxygen ions. The result is shown in Fig.2. The imaginary phonon frequency for the ferroelectric mode at $\Gamma$ turns out to be almost twice as large in our model calculation as compared with the ab-initio result. Nevertheless, it is remarkable that the model reproduces the wave vector dependence of the instabilities, as obtained by the ab-initio linear response approach (compare Fig.2 with Fig.1 of Ref [23]).

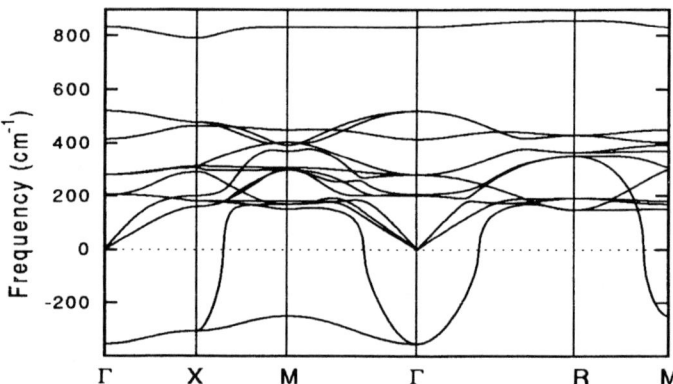

**FIGURE 2.** Calculated phonon dispersions in the cubic structure at the experimental lattice constant. Imaginary phonon frequencies are represented as negative values.

We have used the above described model to perform constant-pressure MD simulations at several temperatures. In Fig. 3(a) we plot the order parameters (the three components of the mean polarization) as a function of temperature. The corresponding cell parameters are displayed in Fig. 3(b).

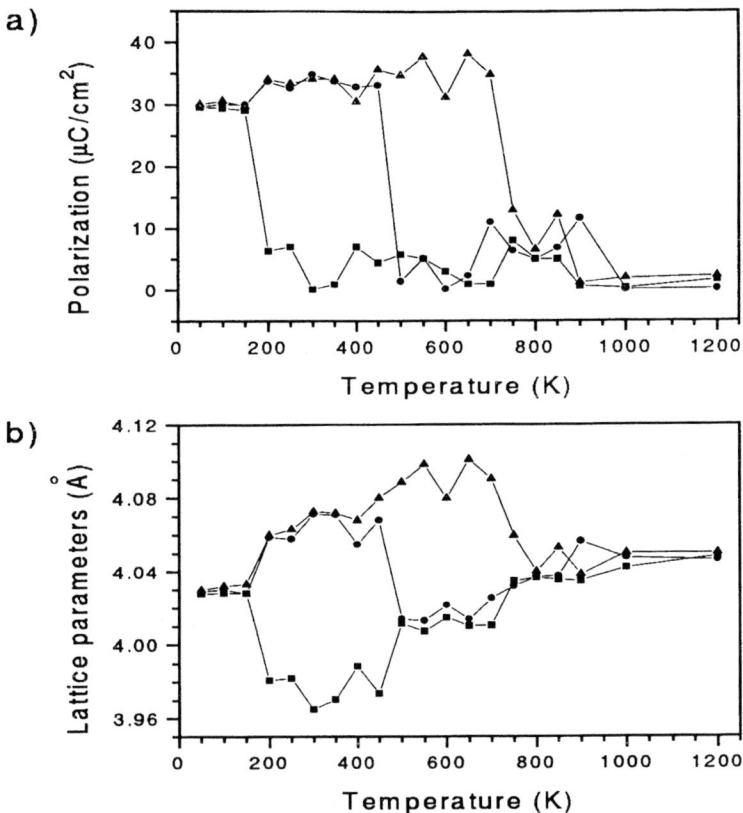

**FIGURE 3.** Phase diagram of KNbO$_3$: a) the three components of the average polarization as a function of temperature. b) the corresponding cell parameters.

At high temperatures, the averaged polarizations $p_x$, $p_y$ and $p_z$ are all very close to zero and the three lattice constants have almost identical values. As the system is cooled down below 725 K, $p_x$ acquires a value clearly different from zero, while $p_y \simeq p_z \simeq 0$, and the structure presents a considerable tetragonal strain (see Fig. 3(b)). This indicates the transition from the paraelectric cubic to the ferroelectric tetragonal phase. When the temperature is further reduced, the two lower ferroelectric phases appear: the orthorhombic one below $\sim$ 475 K, with clearly finite $p_x \simeq p_y$ and still $p_z \simeq 0$, and finally the rhombohedral phase below $\sim$ 175 K, with approximately equal values of the three polarization components. Although the model gives slighlty large cell parameters and distortions compared with experimental data [1], it shows transitions in good accord with the experimental transition temperatures. This is an indication that the relative values of the instability energies in different directions yield the correct sequence of phase transitions in $KNbO_3$.

To gain insight into the dynamical excitations relevant to the ferroelectric phase transitions, we calculate the dynamical structure factor $S(\mathbf{q}, \omega)$. On the left hand side of Figure 4 we show the low-frequency range of $S(\mathbf{q} = 0, \omega)$ at 800 K (cubic phase), 525 K (tetragonal phase) and 250 K (orthorhombic phase).

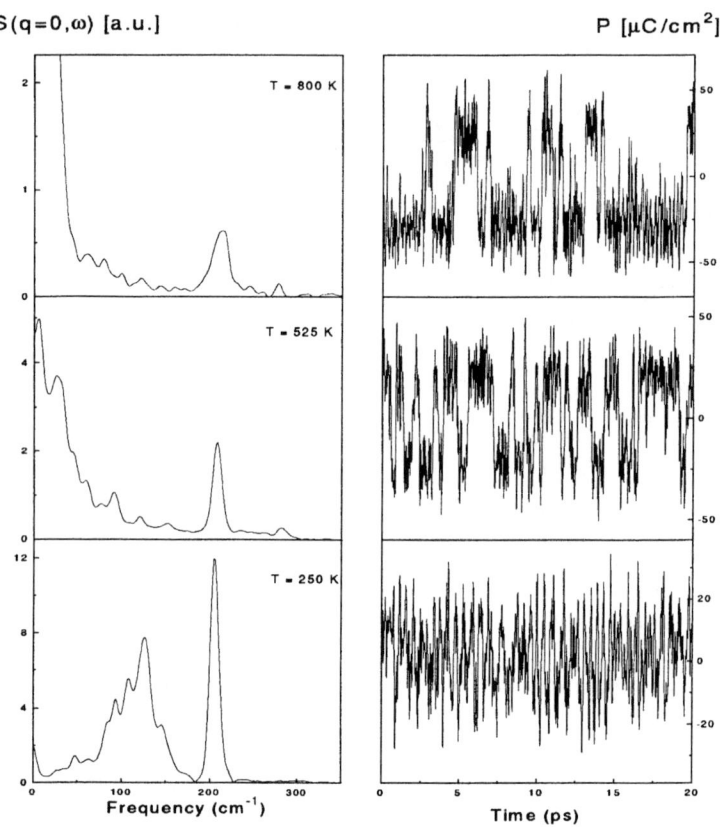

**FIGURE 4.** $S(\mathbf{q} = 0, \omega)$ and time evolution of the z-component polarization in a single cell at several temperatures.

In both the cubic and tetragonal phases a quasielastic peak is observed, while no peak appears corresponding to the ferroelectric (lowest TO) mode. The peaks observed at $\approx 200$ cm$^{-1}$ and $\approx 280$ cm$^{-1}$ correspond in the cubic phase to the second lowest infrared active TO($\Gamma_{15}$) mode and the silent $\Gamma_{25}$ mode, respectively. On the other hand, in the orthorhombic phase a broad peak centered at $\approx 100$ cm$^{-1}$

is observed, which is assigned to the ferroelectric mode, while no central component appears. These results are in good agreement with the experimental observations mencioned in the Introduction.

To clarify the nature of the dynamics leading to the above remarqued features of $S(\mathbf{q} = 0, \omega)$, we show on the right hand side of Figure 4 the time evolution of a single cell polarization component (a vanishing one in average) at the same temperatures. We observe for both the cubic and tetragonal phases that fast oscillations around finite polarization values coexist with much slower polarization reversals. So, the dynamics posseses two components with different time scales. While one component is associated with quasi-harmonic oscillations around an off-center position, the other refers to a relaxational motion between equilibrium sites.

To further clarify this dynamical behavior, we plot in Fig. 5 the time evolution of two polarization components (y and z) for a single cell in the cubic phase. A similar figure is obtained if we consider the third component (x).

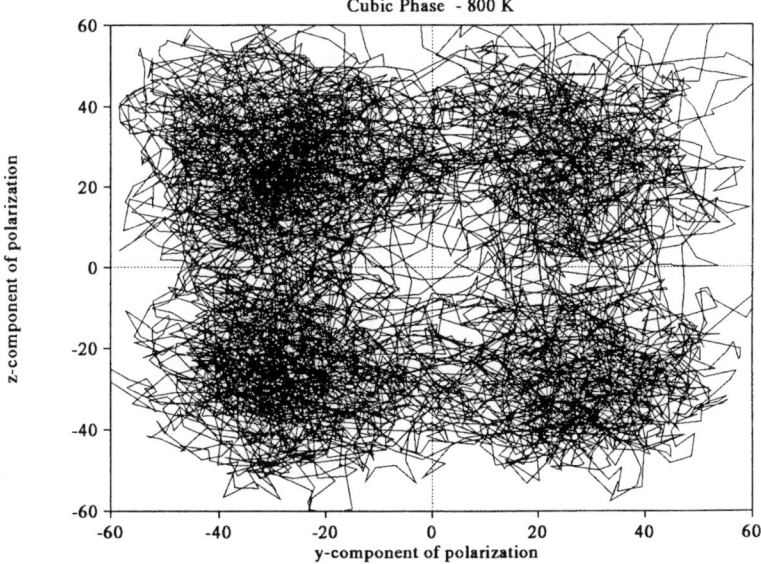

**FIGURE 5.** Time evolution of two polarization components for a single cell in the cubic phase.

It is clear that the dynamics associated with the appeerence of the quasi-elastic component is a relaxational motion of the local polarization between the eight energy minima along [111] directions. In the tetragonal phase, one component $(p_x)$ remains oscillating around a non-zero value while the other two show a similar relaxational dynamics between four energy minima.

This dynamical behavior is in agreement with the picture provided by the eight site model. The relaxation proccess is therefore mainly responsible for the C-T and

T-O phase transitions. Such a relaxational motion is absent in the low temperature range of the orthorhombic phase, where we observe only fast oscillations around zero polarization (see Fig.4) indicating a more displacive-like dynamics for the O-R transition.

## CONCLUSION AND PERSPECTIVE

We have obtained an atomistic model for $KNbO_3$ which describes its structural instabilities in good agreement with LDA total energy calculations. A further molecular dynamics simulation allowed us to evaluate the phase diagram and provided a deeper insigh into the dynamical nature of the structural transitions. We find that the C-T and T-O transitions have an order-disorder character, while the O-R is more displacive-like. A study of dynamical correlations between local polarizations forming precursor clusters in the cubic phase is in progress.

This modelling approach can be extended to other members of the ferroelectric perovskite family. We are now modelling $BaTiO_3$, which exhibits the same phase transition sequence of $KNbO_3$ but with much smaller instability energies associated with the occurrence of the ferroelectric phases.

Acknowledgments
This work was supported by Consejo Nacional de Investigaciones Científicas y Técnicas de la República Argentina. M.S. thanks also support from Consejo de Investigaciones de la Universidad Nacional de Rosario.

## REFERENCES

1. M.D.Fontana, G.Métrat, J.Servoin and F.Gervais, J.Phys.C **16**, 483 (1984).
2. H.Vogt, M.D.Fontana, G.E.Kugel and P.Günter, Phys. Rev. B **34**, 410 (1986).
3. M.D.Fontana, G.Ridah, G.E.Kugel and C.Carabatos-Nedelec, J. Phys. C **21**, 5853 (1988).
4. T.Dougherty, G.Wiederrecht, K.Nelson, M.Garret, H.Jenssen and C.Warde, Phys. Rev. B **50**, 8996 (1994).
5. A.Postnikov, T.Neumann, G.Borstel and M.Methfessel, Phys.Rev.B **48**, 5910 (1993).
6. M.Posternak, R.Resta and A.Baldereschi, Phys. Rev. B **50** 8911 (1994).
7. D.Singh and L.Boyer, Ferroelectrics **136**, 95 (1992).
8. D.Singh, Ferroelectrics **164**, 143 (1995).
9. R.Cohen and H.Krakauer, Phys.Rev.B **42**, 6416 (1990).
10. R.Cohen, Nature **358**, 136 (1992).
11. R.D.King-Smith and D.Vanderbilt, Phys. Rev. B **49**, 5828 (1994).
12. PH.Ghosez, X.Gonze and J.-P. Michenaud, Ferroelectrics **164**, 113 (1995).

13. D.Singh, Phys. Rev. B **53**, 176 (1996).
14. W.Zhong, D.Vanderbilt and K.Rabe, Phys.Rev.Lett. **73**, 1861 (1994).
15. W.Zhong, D.Vanderbilt and K.Rabe, Phys.Rev.B **52**, 6301 (1995).
16. K.Rabe and U.Waghmare, Ferroelectrics **194**, 119 (1997).
17. G.E.Kugel, M.D.Fontana and W.Kress, Phys.Rev.B **35**, 813 (1987).
18. R. Migoni and H. Bilz and D. Bäuerle, Phys. Rev. Lett. **37**, 1155 (1976)
19. C.Perry, R.Currat, H.Buhay, R.Migoni, W.Stirling and J.Axe, Phys. Rev. B **39**, 8666 (1989)
20. M. Methfessel, Phys. Rev. B **38**, 1537 (1988).
21. M. Methfessel, C.O. Rodriguez, and O.K. Andersen, Phys. Rev. B **40**, 2009 (1989).
22. DL-POLY is a package of molecular simulation routines written by W.Smith and T.R.Forester, Daresbury and Rutherford Appleton Laboratory, Daresbury, UK.
23. R.Yu and H.Krakauer, Phys.Rev.Lett.**74**, 4067 (1995).

# Mechanisms and kinetics of the initial stages of ferroelectrics island film growth under non-uniform conditions

D.A.Grigoriev and S.A.Kukushkin

*Institute of Mechanical Engineering Problems, Russian Academy of Sciences, Bolshoj 61, V.O., St. Petersburg 199178, Russia, e-mail : gr@ipme.ru*

**Abstract.**
The kinetics of the initial stages of ferroelectrics films growth under spatial non-uniform conditions is investigated. The complete closed system of equations relating the characteristics of growing island film with the process parameters is developed. The solution of that system allowed an assessment of all basic features of island film such as island size distribution function, mean radius of island and ratio of substrate occupation as the time and spatial coordinate dependence. The causes of the phenomenon why films produced by CVD-method show non-uniformity of characteristic over the substrate surface are revealed and mathematical description is given. Some recommendations on the development of films with prearranged properties are provided.

## I INTRODUCTION

Ferroelectric films are recently gaining in importance for technology as well as for physics of phase transitions. Such films can be produced by different methods. They are laser evaporation, CVD, ion-plasma deposition, sol-gel technology etc. [1,2]. One of the essential tasks which experimenters and technologists face is the production of ferroelectric films with uniform structural and hence physical properties over the substrate area. Meanwhile, the experiments on deposition of $SrTiO_3$ films onto quite large substrate by ion-plasma deposition and $PbTiO_3$ films by CVD have shown that the film thickness and physical properties vary in different points of the surface. The experimenters associate this phenomenon with the fact that due to some features of mass transfer on the substrate in its different points the flux of the matter from which the film originates is disparate.

There are number of theoretical works describing the mass transfer on the substrate. For instance, the paper dedicated to Monte-Carlo modeling of matter delivery when at the film is grown by ion-plasma deposition method [3].

There are a whole scope of works on CVD mentioned above which boil down to the solution of convection diffusion equation to finding a power of flux on the surface. Nevertheless, the data acquired on flux of the components don't allow a comprehensive understanding of the growing film structure. Moreover, a comparison of experiments where a mass of $SrTiO_3$ grown were measured with the flux value simulated has shown that only a minor portion of the matter entered the substrate passes into the film.

So, it seems to be interesting to associate the transport mass theories existing with a modern theory of island film evaluation.

The basic contemporary theories of the film growth are given in the detailed review [1]. However, most theories of different stages of FOPT existing consider the film growth when the matter supply is uniform over the substrate area (Fig.1). Example, of method involving a spatial non-uniformity of matter supply is found in CVD which is actively used in the production of $BaTiO_3$ films [2] .

## II  FORMULATION OF THE PROBLEM AND PHYSICAL NATURE OF THE PROCESS

The body of the method is blowing on the substrates with carrier gas containing one or more admixed components which serves as the matter source for the film growth. The film can grow either directly from admixed components or products of their reaction or from their breakdown products ( in case metal organic compounds are used as admixture [2]). In this respect CVD-process can be divided into two stages for the sake of simplicity. At the first stage occurs delivery of admixed component to the substrate surface and chemical reaction resulting in evolving of substances from which the film originates. For example, breakdown of metal organic compounds with liberation of oxide [2] etc. This process is given by the transport equation with appropriate boundary conditions which are considered bellow.

At the beginning of the second stage the matter originator of film growth will be accumulated on the substrate in the amount exceeding the equilibrium one. Accumulation of the substance proposed for the film growth can proceed directly through the components delivered to the surface as well as chemical reaction between them. Then first - order phase transition with the formation of new phase solid islands takes place.

First order phase transition on the solid surface is usually divided into several stages [1]. First fluctuation formation of new phase islands on the substrate surface takes place, then the islands grow independently without change in their number and after all the coarsening process or Ostwald ripening begins in the island film. This process features integrated thermal and diffusion fields formation where the islands with the size less than a critical value dissolve and ones with bigger size grown. It is followed by reduction of a total number of islands and definite distribution of islands with respect to their size and phase constitution (in case of multicomponent system) is established. It is the most protracted stage and as a rule, just at this stage the film structure is finally completed, as it is reported in

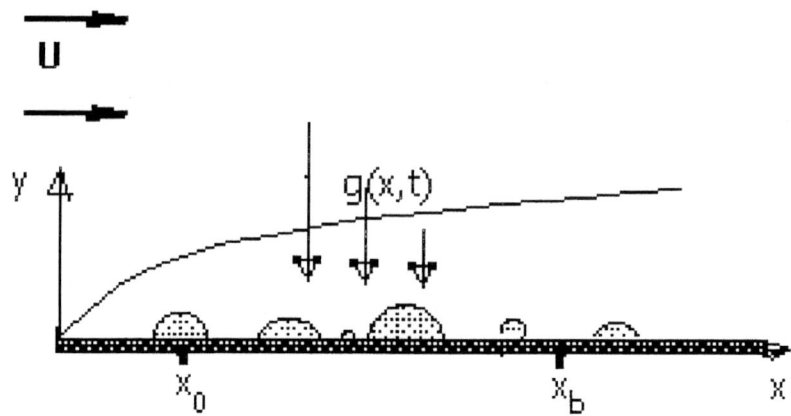

**FIGURE 1.** Schematic diagram of the process. Arrows show the direction of carrier gas flow. 1 — the substrate, 2 — the islands, 3 — the hydrodynamic boundary layer.

a number of experimental and theoretical works [1]. Owing to hydrodynamics of the blowing the substrate by carrier gas the quantity of reactants entering and reacting on its surface varies on different areas. In consequence, the concentration of substance from which the film will grow varies in different parts of the substrate, i.e. film growth and respectively its structure will be non-homogeneous over the surface area.

Let us consider as an example a variation of CVD-method when the surface is located along the carrier gas flow ( Fig.1 ). In order to make clear a physical nature of the processes occurring we here investigate the case when islands of like composition are formed on the surface. Those islands are built up from two components at least. The substrate surface can formally serve as the second component. Let the process occur under isothermal conditions i.e. the temperature of the substrate is a constant. Fig.1 shows schematic diagram of the process. As stated in [1] the island ensemble must be situated within the boundary gas layer that is critical condition for Ostwald ripening. In case of boundary layer stall Ostwald ripening is disrupted, the islands begin to grow independently, and integrated diffusion field is disturbed with the result that the film does not take predicted structural and phase constitution. This phenomena may probably explain the fact that CVD-method is effective for application of high quality films and coats generally on hydrodinamicaly streamlined items.

# III THE EXAMPLE OF CALCULATION OF REACTANT FLOW FROM CARRIER GAS ONTO THE SUBSTRATE

Let us assume that length and width of the plate we have chosen as a substrate are much greater than its thickness (Fig. 1). In this case it can be refered as an infinitely thin one . The substrate is blown by carrier gas at the pressure lower than atmospheric one and at constant temperature. The gas flow rate in the distance from the substrate is equal to $U$. The reactants with concentrations $C_i$ are admixed to the carrier gas. We now write down two-dimensional stationary transport equation (1) for each admixed component in incompressible carrier gas which moves straightforward and laminar with boundary conditions on the reaction surface of mixed kinetics type.

$$V_x \frac{\partial C_i}{\partial x} + V_y \frac{\partial C_i}{\partial y} = D_i \frac{\partial^2 C_i}{\partial y^2}, \tag{1}$$

$$D_{C_i} \left( \frac{\partial C_i}{\partial y} \right) \bigg|_{y=0} = k^* C_i^j,$$

where $C_i$ is concentration of admixed component $i$ in carrier gas $V_x$, $V_y$ are gas velocity components, $D_{C_i}$ is coefficient of diffusion of component $i$ in the gas; $k^*$ is constant of chemical reaction of formation of the matter from which the film will originate (e.g. for example the decomposition reaction of metallorganic compound [2]). The solution of the equation (1) presents no problem in this case so we can put down at once the value of flow power of the component which falls onto the substrate at a point $y = 0$ depending on coordinate $x$ that is important for further considerations.

$$g_{gi}(x) = D_{C_i} \left( \frac{\partial C_i}{\partial y} \right) \bigg|_{y=0} \approx \frac{3^{2/3}}{2} \frac{C_{0i} D_{C_{0i}} \sqrt{0,665}\, U^{3/4}}{\eta^{1/4} \Gamma(1/3) \sqrt{x + x_0}}, \tag{2}$$

where $\eta$ is dynamic viscosity of gas, $U$ is flow rate of carrier gas in the distance from the substrate; $x$ is a distance from the substrate edge, $x_0$ is the left boundary of area where the Ostwald ripening proceeds (Fig.1); $\Gamma(z)$ is gamma function, whose value can be found in mathematical tables, $z$ is an argument.

It should be noted — that in order to consider the film growth it was essential for us to get only the type of each component flux in the form of $g_g(x)t^{n-1}$. In our case $n = 1$, i.e. — the flux is stationary. We now turn to the discussion of the processes on the substrate surface during the film growth by CVD.

# IV ISLAND FILMS EVOLUTION BY CVD-METHOD

Let us assume that there is an ensemble of islands of like composition i.e. the same phase[1] on the substrate surface (see Fig. 1). The islands have equilibrium shape, for instance shape of sphere with radius $R$. Cylindrical islands can be described by analogy. The species arrived from the carrier gas react between themselves on the substrate surface according to the equation

$$\nu\mu = k, \qquad (3)$$

where $\mu$ and $\nu$ are concentration of components entering the substrate, $k$ is a chemical reaction constant. Stoichiometic ratio of the reaction is assumed to be one.

Due to non-uniformity of the matter supply from the gas in terms of the substrate length there are diffusion flows originating along the substrate. In this case mater distribution over the substrate should be described by diffusion equation for each component. Such equations should include the power of the mater sink into the islands of a new phase $\frac{d}{dt}I(x,t)$, where $I(x,t) = \frac{1}{2}\chi \int_0^\infty f(R,x,t)R^3(x,t)\,dR$ is volume of the matter in the islands; $f(R,x,t)$ is a size distribution function, $\chi = \frac{1/3\pi(2-3\cos\Theta+\cos^3\Theta)}{V_m N_n Q_0}$; $\Theta$ is boundary angle, $V_m$ is volume per atom (molecule) in a new phase; $N_n$ is a number of adsorbtion sites per unit surface, $Q_0$ is quantity of the matter of a new phase by onset of Ostwald ripening. Multiplier $\frac{1}{2}$ in the expression for $I(x,t)$ is used in order not to allow for sink into islands twice. If stoichiometric coefficients in (3) are not equal to one, this multiplier should be replaced with the ratio of respectful stoichiometric coefficient. The source of the phase determined by a minimum component source $g_g^s(x)$, should be included in diffusion equation also with multiplier $\frac{1}{2}$. Thus, the equation of component diffusion along the surface with appropriate boundary conditions has the form:

$$\frac{\partial \nu}{\partial t} = D_\nu \frac{\partial^2 \nu}{\partial x^2} + \frac{1}{2}g_g^s(x) - \frac{d}{dt}I(x,t); \qquad (4)$$

$$\frac{\partial \mu}{\partial t} = D_\mu \frac{\partial^2 \mu}{\partial x^2} + \frac{1}{2}g_g^s(x) - \frac{d}{dt}I(x,t); \qquad (5)$$

$$mu|_{x=0} = \mu_0; \quad \nu|_{x=0} = \nu_0 \quad \mu_{x=\infty} = \mu_\infty \quad \nu_{x=\infty} = \nu_\infty, \qquad (6)$$

According to [1] the continuity equation for the distribution function of islands in terms [1] of dimensions should also be included in the equation system describing Ostwald ripening of island ensemble:

---

[1] Polyphase film evolution is not observed in this paper in order not to shade physical sense of the processes

$$\frac{\partial f(R,t,x)}{\partial t} + \frac{\partial}{\partial R}\left(f(R,t,x)V_R(R)\right) = 0, \quad f(R,t,x)|_{t=0} = f_0, \tag{7}$$

where $f_0$ is the initial distribution function; $V_R(R)$ are the expressions for island grow velocity. It is shown in [1] that $V_R(R)$ is determined from the solution of diffusion equation for a singled out island in integrated field governed by other islands and depends on mean concentration of system components. In our case if mean concentration varies slightly along the length of the substrate, it can be regarded as average one and the data from [1] may be used. Particularly, when the island growth rate is limited by the rate of chemical bonds formation on the particle surface, i.e. speed of building the components into crystal lattice of the island :

$$V_R = \frac{2\sigma\beta V_m^2 \psi_1(\theta)\alpha(\theta)}{K_B T R}\left(\frac{R}{R_{cr}} - 1\right); \tag{8}$$

where $\beta$ is specific boundary flow per island , $\sigma$ is surface tension, $K_B$ is Boltzmann constant, $R_{cr}$ is island critical radius, $T$ is the process temperature.

The system of equations (4)— (7) is the complete closed nonlinear system of the equation describing Ostwald ripening, in the island film under conditions typical for CVD-process. First simultaneous solution of the equations (3) and (4)- (5), will be found. Let us subtract (4) from (5) and denote the concentration gradient by $M = \mu - \nu$. For simplicity let component diffusion coefficients be equal to each other $D_\mu = D_\nu = D$. Then we introduce a new independent variable $m = \frac{x+x_0}{\sqrt{D(t+t_0)}}$. Having solved (4)— (7) and (3) simultaneously we obtain the expression for total flux on the surface $g_\Sigma(x,t) = \frac{1}{2}g_g(x) - \frac{d}{dt}I(x,t)$ at large times $t$:

$$g_\Sigma(x,t) = \frac{2k(M_0 - M_\infty)^2 \exp\left(-\frac{1}{2}\frac{(x+x_0)^2}{D(t+t_0)}\right)}{\pi D(t+t_0)(M_0+4k)^{3/2}}. \tag{9}$$

where $M = M_0 - (M_0 - M_\infty)\Phi(\frac{x+x_0}{\sqrt{D(t+t_0)}})$; $\Phi(z)$ — is a error function.

We now write down the equation of mass balance on the substrate at large times $t$ [1] :

$$\int_0^t g_\Sigma(x,t) = \chi \int_0^\infty f(R,x,t) R^3(x,t)\, dR, \tag{10}$$

substituting the flow value (9) into (10) and integrating the left-hand side of eq. (10) gives:

$$\frac{2k(M_0 - M_\infty)^2}{\pi(M_0^2+4k)^{3/2}}\left\{\text{Ei}\left(-\frac{1}{2}\frac{(x+x_0)^2}{D(t+t_0)}\right) - \text{Ei}\left(-\frac{(x+x_0)^2}{D(t_0)}\right)\right\} + \tag{11}$$

$$+\frac{3^{2/3}}{2}\frac{D_{C_i}\sqrt{0,665}U^{3/4}}{\eta^{1/4}\Gamma(1/3)\sqrt{x+x_0}}t = \chi\int_0^\infty f(R,x,t) R^3(x,t)\, dR,$$

where $Ei(z)$ is an integral exponential function. We now consider two limiting case that are of physical importance.

**1.** Let us assume that the following relation prevails between the terms of the left-hand side of eq. (12) in the given time space interval:

$$\frac{2k(M_0 - M_\infty)^2}{\pi(M_0^2 + 4k)^{3/2}} \left\{ Ei\left(-\frac{1}{2}\frac{(x+x_0)^2}{D(t+t_0)}\right) - Ei\left(-\frac{(x+x_0)^2}{D(t_0)}\right) \right\} \ll \qquad (12)$$

$$\ll \frac{3^{2/3}}{2}\frac{D_{C_i}\sqrt{0,665}U^{3/4}}{\eta^{1/4}\Gamma(1/3)\sqrt{x+x_0}}t,$$

From physical point of view it indicates that matter flux from carrier gas onto the surface is intensive enough and the first term of eq. (12) can be neglected. Thus, eq. (12) takes the form:

$$\frac{3^{2/3}}{2}\frac{D_{C_i}\sqrt{0,665}U^{3/4}}{\eta^{1/4}\Gamma(1/3)\sqrt{x+x_0}}t = \chi \int_0^\infty f(R,x,t) R^3(x,t)\, dR. \qquad (13)$$

Then we introduce new variables $u = R/R_{cr}$ and $X = \Delta_0/\Delta(x,t)$ where $\Delta(x,t) = \frac{\nu - \nu_\infty}{\nu_\infty} + \frac{u - u_\infty}{u_\infty}$ is relative saturation in terms of phase $\Delta_0$ is initial saturation. The parameter $\tau = \ln(X^2)$ which unambiguously associated with time at large $t$ should be chosen as "time". Converting (20) with new variables we have (13) into the form that agrees with the equation solved in [1], with only difference that $\chi^* = \frac{3^{2/3}}{2}\frac{\chi \eta^{1/4}\Gamma(1/3)\sqrt{x+x_0}}{D_{C_i}\sqrt{0,665}U^{3/4}}$ in our case depends on spatial coordinate $x$.

It should be emphasized that having transformed the expression (2) into $g_g\, t^{n-1}$ we see that it formally corresponds to $n = 1$. As shown in [1] the matter sources are damped at $n < 3/p$ and at $n \geq 3/p$ the matter sources are undamped. So under the stationary conditions being discussed (i.e. at $n = 1$) the source can be damped at transport mass mechanism conforming to $p = 2$ as well as continuous at $p = 3$ or swelling at $p = 4$. Thus, when applying different matters the similar conditions of the process can result in the formation of radically different systems from monodispersed (at the given point of the space) at $p \leq 3$, to polydisperse ones with the size distribution discussed below. We here study the special case of $p = 2$, the other variants can be obtained the same way in accordance with the results of [1] and present analysis.

Without exhibiting here any intermediate calculations we now write down at once the basic relationships characterizing the island film at the Ostwald ripening stage when it grows by CVD-method providing that component distribution is dictated mainly by hydrodynamic matter supply from gas phase.

Thus, in this case the critical radius of islands is $R_{cr}^p = R_{cr0}^p + A_p t$, where $R_{cr0}$ is a critical radius of islands before the onset of Ostwald repining and the value of the coefficient $A_p$ can be found in [1]. For example, at $p = 2$ the coefficient $A_2$ is $A_2 = \frac{\beta \sigma V_m^2 \psi_1(\theta)\alpha(\theta)}{K_B T}$. Mean radius of islands varies with time as:

$$\bar{R}(x,t) = C_{pn}(A_p t)^{1/p} \qquad (14)$$

where: $C_{pn} = (\int_0^{u_0} P_p^s(u)u\,du)/(\int_0^{u_0} P_p^s(u)\,du)$ ; $u_0$ is locking point; at $p=2$ parameter $u_0$ takes the value 2(See [1] ).

The number of islands varies in accordance with the law:

$$N_g(x,t) = \frac{G_{gp}}{(3/p-1)(A_p t)^{3/p-1}\sqrt{x+x_0}}. \qquad (15)$$

$$G_{gp} = \frac{3^{2/3}}{2}\frac{\gamma_0\; D_{C_i}\;\sqrt{0,665}\; U^{3/4}\, V_m\, R_{k0}^{p-2}\, N}{\eta^{1/4}\Gamma(1/3)\,\pi A_p} \int_0^{u_0} \frac{e^{-(3/p-n)\tau(u)}u^3 du}{(du^{p-1}/d\tau)^s},$$

where $\tau(u) = \int_0^{u_0} du/(du^{p-1}/d\tau)^s$ [1], $N$ is number of adsorbtion sites per unit area, $\gamma_0$ is the coefficient determined according to [1].

In the general case island size distribution function is of the form:

$$f(R,x,t) = \frac{N(x,t)}{R_{cr}(t)} P_p(u). \qquad (16)$$

The form of functions $P_p(u)$ is shown in [1]. E.g., at $p=2$

$$P_2(u) = \begin{cases} \dfrac{2e^{3-2n}\,u\,\exp\!\left(-\frac{3-2n}{1-u/2}\right)}{(2-u)^{2+2\,(3/2-n)}} & u < 2 \\ 0 & u \geq 2, \end{cases}$$

where $n$ is equal to one for the stationary case being discussed. When the process is non-stationary $n$ is not equal to one. Fig. 2 shows the form of function $f(R,x,t)$ at a specific moment of time $t$.

The ratio of the substrate occupation by islands in the case of semi-spherical islands is given by:

$$\xi(x,t) = \frac{4\pi\, G_{pg}\, Cpn^2(A_p t)^{1-1/p}}{(3/p-1)\sqrt{x+x_0}}. \qquad (17)$$

In so doing the rate of substrate occupation by islands increases and time of continuos film formation depending on coordinate can be estimated by the expression:

$$t_g = \frac{1}{A_p}\left(\frac{(3/p-1)\sqrt{x+x_0}}{4\pi\, G_{pg}\, C_{pn}^2}\right)^{\frac{p}{p-1}} \qquad (18)$$

**2.** Next we consider the case when matter flux from gas phase onto the substrate is small comparing with diffusion transport of components over the substrate surface (13). It may take place for instance, when the concentration of admixed components in the distance from the substrate or transport gas velocity is small (See eq. (2)). Then we have (12) in the form:

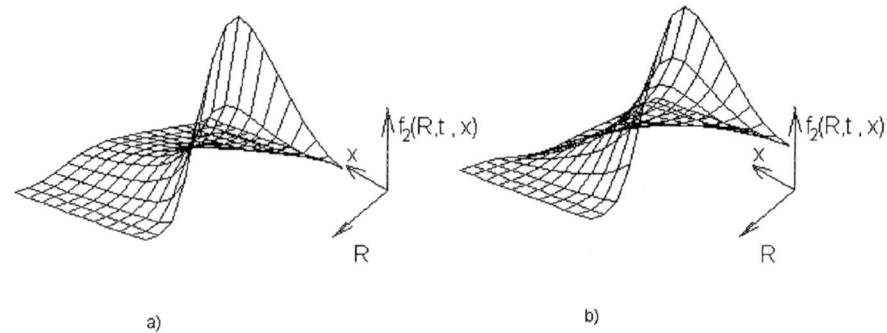

**FIGURE 2.** a) Dependence between island distribution function $f_p(R, x, t)$ and radius $R$ and coordinate $x$ providing that island ensemble evolution is controlled through matter supply from carrier gas at the given moment t. b) Dependence between island distribution function $f_p(R, x, t)$ and radius $R$ and coordinate $x$ providing that island ensemble evolution is controlled through difusion along the substrate at the given moment t.

$$\frac{2k(M_0 - M_\infty)^2}{\pi(M_0^2 + 4k)^{3/2}} \left\{ \text{Ei}\left(-\frac{1}{2}\frac{(x+x_0)^2}{D(t+t_0)}\right) - \text{Ei}\left(-\frac{(x+x_0)^2}{D(t_0)}\right) \right\} = \quad (19)$$

$$= \chi \int_0^\infty f(R, x, t) R^3(x, t) \, dR.$$

Now we consider the expression in the left-hand side at large $t$. By equating the left-hand side to zero and solving equation for $x$ we get the right boundary of Ostwald ripening region (Fig. 1):

$$x_b = \sqrt{\frac{2D(t+t_0)}{\gamma}} - x_0.$$

where $\gamma = 1.781\ldots$ is Euler constant. Thus, Ostwald ripening of island ensemble occurs in the region $x_0 < x < x_b$ bound which extends in the course of time. For this case we write down all main parameters of island film without demonstration of intermediate calculations.

The mean radius of islands in this approximation is determined by (14) and number of islands per unit surface is given by:

$$N_d(x, t) = \frac{G_{pd} \ln\left(\frac{2D}{\gamma(x+x_0)^2}\right)}{3/p(A_p t)^{3/p}}. \quad (20)$$

where: $G_{pd} = \frac{2\gamma_0 k(M_0 - M_\infty)^2 V_m R_{k0}^{p-2} N}{(M_0^2 + 4k)^{3/2} \pi} \int_0^{u_0} \frac{e^{-(3/p-n)\tau(u)} u^3 \, du}{(du^{p-1}/d\tau)^s}.$

The distribution function in this case will have the form (16) where $n = 0$ in the expression for $P_p(u)$. The form of this function is represented on Fig.2b. The ratio of occupation of the substrate by islands is time variant according to the law:

$$\xi(x,t) = \frac{4\pi G_{pd}}{3/p\,(A_p\,t)^{1/p}} \ln\left(\frac{2D}{\gamma\,(x+x_0)^2}\right) \tag{21}$$

## V CONCLUSIONS

1. A spatial non-uniformity of the matter flux on the substrate resulted in concentration gradients has substantial influence on coarsening kinetics of the island film.

2. It is revealed that evolution of an island ensemble at the Ostwald ripening stage can be controlled either by matter flow or when the flows are weak by diffusional redistribution of the mass. The mathematical expression for this criteria has been obtained.

3. We have established that is a tentative analytical relationships among the parameters of CVD-process and physical constants of the film material as well as such characteristics of an island film as distribution of island sizes, numbers, mean and critical radii in terms of time and spatial coordinate.

4. To initiate Ostwald ripening it is essential that the film be in the boundary layer of a carrier gas. Otherwise, there is no supersaturation drop and the film structure can't be predicted.

5. We have obtained the expression for ratio of a substrate occupation by islands and for time of continuous film formation at given point on the surface.

### Acknowledgments

Thanks are due to Organizing Committee of "Fifth Williamsburg Workshop on First-Principles Calculations for Ferroelectrics" for support my participation.

This work was performed partial with the support of the Russian Foundation for Basic Research (Project $N^0$ 96-03-32396) and project "Integration" $N^0$ 589.

## REFERENCES

1. S.A. Kukushkin, A.V. Osipov, New phase formation on Solid surfaces and thin film condansation. //Progress in Surface Science. 1996, V. 51, N 1, pp 1-107.
2. Srinivasan Sivaram. Chemical Vapor Deposition. Thermal and plasma deposition of electronic materials. New York. Chapman & Hall, 1995. 292 pp.
3. Petrov P.K., Volpyas V.A., and Chakalov R.A., 3D Monte Carlo simulation of sputtered atom transport in the process of ion-plasma sputter deposition of multicomponent thin films // Vacuum, 1998, In Press.

# Prospects for Gigabit Ferroelectric Nonvolatile Memories using Strontium Bismuth Tantalate Thin Films

J. F. Scott

School of Physics
University of New South Wales
Sydney 2052, Australia

**Abstract.** We discuss three aspects of the development of Gbit density nonvolatile random access memories (NV RAMs) utilizing ferroelectric thin films and, in particular, $SrBi_2Ta_2O_9$ (SBT) as the active storage medium.

## INTRODUCTION

Since the development of strontium bismuth tantalate as a fatigue-free thin-film memory material (1), the physics understanding has lagged the engineering developments. Already Matsushita and others (Fujitsu, NEC) have manufactured smart credit cards with this material (2), embedded memories up to 64 Kbit in Si microcontrollers and microprocessors (3,4), and free-standing NVRAMs of 256 Kbit (5,6) and 1 Mbit (7). In general, however, these products have been made without detailed knowledge of device models. In particular, the band structures and leakage charge mechanisms at the electrode-dielectric interfaces have been poorly understood.

## BAND-MATCH-UPS AT THE ELECTRODE-FERROELECTRIC INTERFACE

In 1997 our group determined the electron affinity of SBT (8), its bandgap (9,10), and the pinning level of the Pt electrode Fermi surface by surface states in the SBT films. The latter value was found to be 2.1 eV, almost precisely mid-gap, as shown in Fig.1. The band-bending of the SBT conduction band was thereby calculated to be 0.3 eV, and the Schottky barrier height to be 1 eV, a value in very good agreement with experimental measurements of 0.83-0.90 eV (11,12).

## Pinning by Surface States

Although the value of 2.1 eV for the surface state pinning energy in SBT seemed reliable, there was no direct confirmation of this until December 1997, when S. Thurgate at Murdock Univ. measured the surface state trap energies via surface photovoltaic techniques on our films. His preliminary data show that a large density of states exists at 2.18 eV, which we tentatively assign to bismuth states; these data need to be confirmed, however, to make certain that they are not artifacts produced by second-order effects from the bandgap at 4.2 eV. It is known that Bi is rather volatile in SBT and exists as both bismuth oxide ($Bi_2O_3$) and metallic Bi (13,14). As a result, we have undertaken a number of experiments in which excess Bi was intentionally put into the SBT.

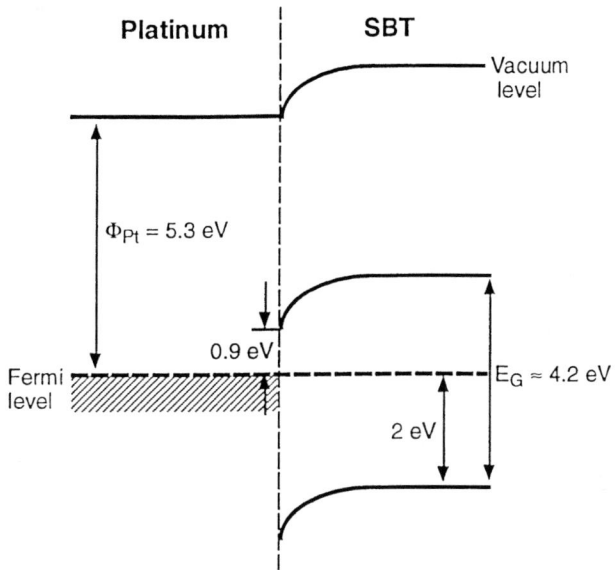

**FIGURE 1.** Band structure match-up (8) for SBT on Pt electrodes, showing mid-gap pinning at 2.1 eV.

# SPACE CHARGE LIMITED CURRENTS IN Bi-EXCESS SBT

## Extension of Rose's Theory to Ferroelectrics

### Experiment

In work to be published separately we find (15) that stoichiometric SBT of ca. 200 nm thickness is Schottky-limited in its conduction on Pt electrodes, but small amounts of excess Bi (>1%) produce ohmic conductivity up to 1.0 V (strictly speaking, current linear in voltage, which is not necessarily ohmic in that it can still be interface-controlled and not proportional to thickness (16)) and space-charge-limited currents from 1.0 V to breakdown. The same thing is true of bismuth titanate films with excess Bi (17).

### Theory

The theory of space-charge-limited currents (SCLCs) in dielectrics with conduction controlled by deep traps was developed in detail by Rose (18). Its temperature dependence has been compared with experimental data. His key result can be presented in the following form:

$$J(V,T) = \text{const.} \times (V/d)(AV/d^2)^{T^*/T} \quad (1)$$

where V is applied voltage; d, film thickness; and $T^*$, a parameter *related* to average trap energy. $T^*$ is of order 300K, so that the ratio $T^*/T$ is of order unity, and hence J(V) approximates a constant times $V^2/d^3$ for most situations, the familiar Mott-Gurney Law. (Note that $T^*$ is not *equal* to the average trap depth, which might be ca. 0.5 eV = 5000K.)

In Rose's model and its earlier applications, ferroelectrics were not considered. Typical application was to insulating or semiconducting dielectrics in which the constant A in Eq.1 was independent of T and V. The only T-dependence in J came from the trap-filling statistical weighting factors. (A is also proportional to the carrier mobility, but in a ferroelectric at room temperature or above this is either limited by grain boundaries -- in a ceramic film -- or phonon scattering in a crystal film -- and is rather T-independent.) These factors will make current density increase with increasing temperatures (note that as T increases in Eq.1, $T^*/T$ decreases and hence J increases, since the quantity raised to the $T^*/T$ power is less than unity for all feasible voltages V). What is neglected here quantitatively is that A is also proportional to the dielectric constant $\varepsilon$ (18,19), so that for ferroelectrics it is necessary to generalize Rose's theory to include explicitly the V and T dependence of $\varepsilon$. The T-dependence for the dielectric constant of a thin ferroelectric film is that of a Curie-Weiss behaviour modified by a surface layer. For PZT this has been modeled accurately by Scott et al.(20) as

$$\varepsilon = \varepsilon_s \varepsilon_b (2d_s + d_b)(2d_s\varepsilon_b + \varepsilon_s d_b) \quad (2)$$

where the subscript s,b means the surface layer of the film or the bulk interior of the film. For PZT Scott et al. found $d_s = 15$ nm; $d_b = 300$ nm; $\varepsilon_s = 200$ (independent of T -- a damaged surface layer); and

$$\varepsilon_b = 10^6 \text{ K}/(670\text{K}-T) \quad (3)$$

-- a bulk Curie-Weiss dependence with a Curie constant of a million degrees and a Curie temperature of 670K, both approximating known single-crystal values.

This T-dependence of coefficient A from Eqs.2,3 must be inserted in Eq.1. For SBT the Curie constant is $2 \times 10^5$ K and the Curie temperature is 463K. Since SBT is known to have surface layers very much thinner than in PZT, we can neglect $d_s$ in the equations above.

The voltage dependence of $\varepsilon$ is also very complicated. It can be determined independently from the C(V) data measured on each film. But the analytical form of dielectric constant is

$$\varepsilon(T,V) = 0.64 \, \varepsilon(T,0) \, (1+\alpha)^{-1/n} \quad (4)$$

where $\alpha = 5 \times 10^{11}$ (SI units) $\varepsilon^3(T,0) (V/d)^2$. n can vary from ca. 2 to 4 (Sze), depending upon the space-charge distribution assumed (quadratic, quartic) for charge versus depth from the interface. Based upon earlier (1961) work of Johnston, Outzourhit et al. (21) assume n = 3/2. The V-dependence of Eq.4 must be combined with Eqs.2 and 3 to give the final $\varepsilon(T,V)$ and hence $A(T,V)$ in Eq.1. This term can dominate and produce J values that increase with T rather than decrease, as in Rose's theory. However, in the present SBT study, voltages varied only from 2.8 to 4.2 V. This 50% variation gives only a 10-15% variation in Eq.4 and hence can be neglected.

As shown in Fig.2, Eq.1 with these modifications fits the SBT:Bi data extremely well. Attempts to fit these data at all T and V to a Schottky model gave unphysical results (negative barrier heights, etc.).

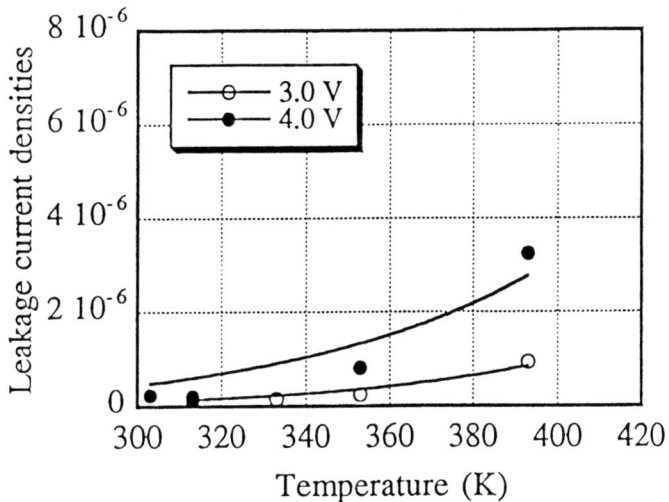

**FIGURE 2.** Temperature dependence of leakage current in SBT with bismuth excess. Curves are Eq.1 with T-dependence from Eqs.2,3 [K. Watanabe, A. J. Hartmann, and J. F. Scott, unpublished.] All curves use the same value of T*.

## PATTERN SELF-ASSEMBLY

When Bi reaches the surface of SBT films it beads up. In vacuum this produces submicron-diameter metallic Bi droplets; in air, it produces bismuth oxide ($\delta$-$Bi_2O_3$, as shown by TEM), or if hydrogen is present in the film, free metallic bismuth. Upon cooling, these droplets form nanoelectrodes ca. 0.1 x 0.1 x 0.05 microns in size, which can be contacted (e.g., via Hg dots) and used to switch submicron regions of the ferroelectric films. Note that the defect-fluorite structure of $\delta$-$Bi_2O_3$ is the highest conductivity (22) oxide known (resistivity ca. 1 W-cm), so that these mesa-like nano-structures can be used as electrodes whether oxidized or not.

Plan view and cross-sections of these self-patterning nano-electrodes are for bismuth titanate. Nearly identical structures have been observed in SBT by Watanabe, and slightly different SBT:Bi structures reported by Zafar et al. (23). These nanoelectrodes are 50x smaller than the smallest switched ferroelectric cells from NEC (24). Attempts to make 0.5 x 0.5 micron cells at Hitachi failed (25), due to the inability of the Ar ion etch to reach the bottom of each cell (aspect ratio problem in 120 nm thick films). These cells are about half the size of the 200 nm Pt post electrodes reported by Tegal (26), which were non-switching devices.

## CONCLUSIONS

The main problem remaining is complete registration of the nanoelectrodes on the films, which is only partially ordered in distance and angle. We think that this can be done via 10-nm beam-width ion implantation (27). In such a way, an electroded 1 Gbit chip could be made without conventional lithography, via seeded self-patterning.

## ACKNOWLEDGMENTS

Work supported by a Humboldt Forschungspreis in Materials Science 1997, and by a grant from the Australian ARC.

## REFERENCES

1. Paz de Araujo, C. A., et al., Nature **374**, 627 (1995).
2. Sumi, T., et al., Jpn. J. Appl. Phys. **35**, 1516 (1996).
3. Fukushima, T., et al., VLSI Circuit Digest of Technical Papers (1996), p.46.
4. Fujitsu Corp. press release (8 Dec 1997).
5. Sumi, T., et al., Proc. ISSCC, San Diego (16 Feb 1995).
6. Nakao, T., et al., FMA-1 **3** (Kyoto, 30 May 1996, p.23).
7. Koike, H., et al., Proc. ISSCC, San Diego (8 Feb 1996).
8. Hartmann, A. J., et al., Proc. Phys. Soc. Korea (1998, in press; proc. IMF-9).
9. Hartmann, A. J., et al., Integ. Ferroelec. **18**, 601 (1997).
10. Melnick, B. M., et al., Integ. Ferroelec. **15**, 221 (1997).
11. Watanabe, K., et al., Integ. Ferroelec. **14**, 95 (1997).
12. Lee, J. K., et al., Integ. Ferroelec. **15**, 389 (1997).
13. Scott, J. F., et al., J. Physique (Paris, in press 1998).
14. Gutleben, C. D., MRS Proc. **433**, 109 (1996).
15. Watanabe, K., Hartmann, A. J., and Scott, J. F., (Appl. Phys. Lett., submitted).
16. Dey, S. K., et al., Integ. Ferroelec. **7**, 341 (1995).
17. Alexe, M. et al. (Appl. Phys. Lett., submitted).

18. Rose, A., Phys. Rev. **97**, 1537 (1955).
19. Lampert, M. A. and Mark, P., Current Injection in Solids (Academic Press, New York, 1970).
20. Scott, J. F., Galt, D., et al., Integ. Ferroelec. **6**, 189 (1995).
21. Outzourhit, A., et al., Integ. Ferroelec. **8**, 227 (1995).
22. Takahashi, T. and Ishiwara, H., Mat. Res. Bull. **13**, 1447 (1978).
23. Zafar, S., et al., J. Appl. Phys. **82**, 4469 (1997).
24. Amanuma, K. and Kunio, T., Jpn. J. Appl. Phys. **35**, 5229 (1996).
25. Hirantani, M., et al., Jpn. J. Appl. Phys. **36**, 5219 (1997).
26. DeOrnellas, S., et al., Semiconduct. Internat. Sept. 1997, p.103.
27. Budai, J. D., et al., Nature **390**, 384 (1997).

# AUTHOR INDEX

## A

Adler, D. L., 96
Akbas, M., 1

## B

Bellaiche, L., 11
Borstel, G., 87, 207, 217
Boyer, L. L., 227
Burton, B. P., 20
Bussmann-Holder, A., 81, 202

## C

Ceder, G., 20
Christensen, N. E., 207, 265
Cockayne, E., 61, 71
Cohen, R. E., 43
Cross, J. O., 96

## D

Davies, P. K., 1
Dmowski, W., 1

## E

Egami, T., 1
Eglitis, R. I., 87, 207

## F

Fanning, D. M., 96
Frenkel, A. I., 96, 238
Fu, L., 107

## G

Garcia, A., 53
Glazounov, A. E., 118

Grigoriev, D. A., 284

## H

Hafner, J., 20

## J

Jastrabik, L., 184

## K

Kotomin, E. A., 207
Krakauer, H., 32, 43, 139, 165
Kresse, G., 20
Ktitorov, S. A., 184
Kukushkin, S. A., 284

## L

Lanceros-Mendez, S., 129
LaSota, C., 139
Lewis, S. P., 156

## M

Markovin, P. A., 87
Marzari, N., 146
Mazin, I. I., 251
McCormack, R. P., 20
Mehl, M. J., 227
Mele, E. J., 156
Meschia, S., 129
Michel, K.-H., 202
Migoni, R. L., 274

## N

Novikov, D. L., 265

## P

Padilla, J., 11
Postnikov, A. V., 207, 217

## R

Rabe, K. M., 32, 61, 71
Ramer, N. J., 156
Rappe, A. M., 156
Resca, L., 107
Resta, R., 107, 174
Robinson, I. K., 96
Rodriguez, C. O., 265, 274

## S

Saghi-Szabo, G., 43
Schmidt, V. H., 129, 192
Scott, J. F., 294
Selinger, R. L. B., 20
Singh, D. J., 251
Spliarsky, M., 274
Stachiotti, M. G., 265, 274
Stern, E. A., 238

Stokes, H. T., 227

## V

Vanderbilt, D., 11, 53, 146
Vikhnin, V. S., 87

## W

Waghmare, U. V., 32
Wang, C.-Z., 32, 139
Wensell, M., 165

## Y

Yacoby, Y., 238
Yaschenko, E., 107
Yu, R., 32, 139

## Z

Zhang, Q. M., 118
Zhao, J., 118